Methods in Enzymology

Volume 220
MEMBRANE FUSION TECHNIQUES
Part A

METHODS IN ENZYMOLOGY

EDITORS-IN-CHIEF

John N. Abelson   Melvin I. Simon

DIVISION OF BIOLOGY
CALIFORNIA INSTITUTE OF TECHNOLOGY
PASADENA, CALIFORNIA

FOUNDING EDITORS

Sidney P. Colowick and Nathan O. Kaplan

Methods in Enzymology

Volume 220

Membrane Fusion Techniques

Part A

EDITED BY

Nejat Düzgüneş

DEPARTMENT OF MICROBIOLOGY
UNIVERSITY OF THE PACIFIC SCHOOL OF DENTISTRY
SAN FRANCISCO, CALIFORNIA

ACADEMIC PRESS, INC.

A Division of Harcourt Brace & Company
San Diego   New York   Boston   London   Sydney   Tokyo   Toronto

Academic Press, Inc.
1250 Sixth Avenue, San Diego, California 92101-4311

United Kingdom Edition published by
Academic Press Limited
24–28 Oval Road, London NW1 7DX

International Standard Serial Number: 0076-6879

International Standard Book Number: 0-12-182121-8

PRINTED IN THE UNITED STATES OF AMERICA
93 94 95 96 97 98   EB   9 8 7 6 5 4 3 2 1

Table of Contents

Section I. Fusion of Liposomes and Other Artificial Membranes

Section II. Induction of Cell–Cell Fusion

Section III. Fusion of Viruses with Target Membranes

Contributors to Volume 220

Article numbers are in parentheses following the names of contributors.
Affiliations listed are current.

KLAUS ARNOLD (12), *Institute of Biophysics, Faculty of Medicine, University of Leipzig, D-7010 Leipzig, Germany*

JOCELYN M. BALDWIN (13), *Department of Biochemistry and Chemistry, Royal Free Hospital School of Medicine, University of London, London NW3 2PF, England*

YECHEZKEL BARENHOLZ (22), *Department of Membrane Biochemistry and Neurochemistry, The Hebrew University–Hadassah Medical School, Jerusalem 91-010, Israel*

ROBERT BLUMENTHAL (21), *Section on Membrane Structure and Function, National Cancer Institute, National Institutes of Health, Bethesda, Maryland 20892*

ROMKE BRON (23), *Department of Physiological Chemistry, University of Groningen, 9712 KZ Groningen, The Netherlands*

KOERT N. J. BURGER (27), *Department of Cell Biology, Medical School AZU, University of Utrecht, 3584 CX Utrecht, The Netherlands*

STEPHEN W. BURGESS (4), *Department of Biochemistry and Nutrition, University of North Carolina at Chapel Hill, Chapel Hill, North Carolina 27599*

LESLEY J. CALDER (27), *Division of Virology, National Institute of Medical Research, London NW7 1AA, England*

LEONID V. CHERNOMORDIK (9), *Laboratory of Theoretical and Physical Biology, National Institute of Child Health and Development, National Institutes of Health, Bethesda, Maryland 20892*

MICHAEL J. CLAGUE (21), *Section on Membrane Structure and Function, National Cancer Institute, National Institutes of Health, Bethesda, Maryland 20892*

FREDRIC S. COHEN (5), *Department of Physiology, Rush Medical College, Chicago, Illinois, 61612*

JAN DIJKSTRA (23), *Department of Physiological Chemistry, University of Groningen, 9712 KZ Groningen, The Netherlands*

NEJAT DÜZGÜNEŞ (1, 2), *Department of Microbiology, University of the Pacific School of Dentistry, San Francisco, California 94115, and Department of Pharmaceutical Chemistry, University of California, San Francisco, San Francisco, California 94143*

PETER M. FREDERIK (27), *EM Unit, Department of Pathology, BMC, University of Limburg, 6200 MD Maastricht, The Netherlands*

KLAUS GAWRISCH (12), *Department of Physics, University of Leipzig, D-7010 Leipzig, Germany*

CHRISTIANE A. HELM (11), *Department of Chemical and Nuclear Engineering, University of California, Santa Barbara, Santa Barbara, California 93106*

YOAV I. HENIS (26), *Department of Biochemistry, Faculty of Life Sciences, Tel Aviv University, Tel Aviv 69978, Israel*

DICK HOEKSTRA (2, 20), *Laboratory of Physiological Chemistry, University of Groningen, 9712 KZ Groningen, The Netherlands*

SEK WEN HUI (16), *Membrane Biophysics Laboratory, Department of Biophysics, Roswell Park Cancer Institute, Buffalo, New York 14263*

JACOB N. ISRAELACHVILI (11), *Department of Chemical and Nuclear Engineering, University of California, Santa Barbara, Santa Barbara, California 93106*

UWE KARSTEN (17), *Max-Delbrück Centrum für Molekulare Medizin, D-13125 Berlin-Buch, Germany*

KARIN KLAPPE (20), *Laboratory of Physiological Chemistry, University of Groningen, 9712 KZ Groningen, The Netherlands*

KAZUMICHI KURODA (24), *Department of Microbiology, Osaka University for Pharmaceutical Science, Matsubara, Osaka 580, Japan*

JOHN LENARD (25), *Department of Physiology and Biophysics, University of Medicine and Dentistry of New Jersey, Robert Wood Johnson Medical School, Piscataway, New Jersey 08854*

BARRY R. LENTZ (4), *Department of Biochemistry and Biophysics, University of North Carolina at Chapel Hill, Chapel Hill, North Carolina 27599*

RANIA LEVENTIS (3), *Department of Biochemistry, McGill University, Montreal, Quebec, Canada H3G 1Y6*

JACK A. LUCY (13), *Department of Biochemistry and Chemistry, Royal Free Hospital School of Medicine, University of London, London NW3 2PF, England*

MARK MARSH (19), *MRC Laboratory for Molecular Cell Biology, University College London, London WC1E 6BT, England*

DAVID NEEDHAM (10), *Department of Mechanical Engineering and Materials Science, Duke University, Durham, North Carolina 27708*

GARRY A. NEIL (14), *Department of Internal Medicine, The University of Iowa College of Medicine, Iowa City, Iowa 52242*

WALTER D. NILES (5), *Department of Physiology, Rush Medical College, Chicago, Illinois, 61612*

SHLOMO NIR (28), *The Seagram Centre for Soil and Water Sciences, Faculty of Agriculture, The Hebrew University of Jerusalem, Rehovot 76100, Israel*

SHINPEI OHKI (7, 8), *Department of Biophysical Sciences, School of Medicine, State University of New York at Buffalo, Buffalo, New York 14214*

SHUN-ICHI OHNISHI (24), *Department of Biophysics, Faculty of Science, Kyoto University, Kyoto 606, Japan*

ANTONIO ORTIZ (23), *Department of Biochemistry, Veterinary Faculty, University of Murcia, E-30071 Murcia, Spain*

RANAJIT PAL (22), *Department of Cell Biology, Bionetics Research, Inc., Rockville, Maryland 20850*

ANU PURI (21), *Section on Membrane Structure and Function, National Cancer Institute, National Institutes of Health, Bethesda, Maryland 20892*

CHRISTIAN SCHOCH (21), *Section on Membrane Structure and Function, National Cancer Institute, National Institutes of Health, Bethesda, Maryland 20892*

BERTOLT SEIDEL (17), *Institute of Neurobiology and Brain Research, Academy of Sciences, D-3090 Magdeburg, Germany*

JOHN R. SILVIUS (3), *Department of Biochemistry, McGill University, Montreal, Quebec, Canada H3G 1Y6*

ARTHUR E. SOWERS (15), *Department of Biophysics, University of Maryland School of Medicine, Baltimore, Maryland 21201*

TOON STEGMANN (23), *Department of Biophysical Chemistry, Biocenter of the University of Basel, 4056 Basel, Switzerland*

DAVID A. STENGER (16), *Naval Research Laboratory, Center for Bio-Molecular Science and Engineering, Washington D.C. 20375*

PETER STOLLEY (17), *Max-Delbrück Centrum für Molekulare Medizin, D-13125 Berlin-Buch, Germany*

MASAHIRO TOMITA (18), *Department of Chemistry for Materials, Faculty of Engineering, National Mie University, Mie 514, Japan*

TIAN YOW TSONG (18), *Department of Biochemistry, College of Biological Sciences, University of Minnesota, St. Paul, Minnesota 55108, and Department of Biochemistry, Hong Kong University of Science and Technology, Clear Water Bay, Kowloon, Hong Kong*

ARIE J. VERKLEIJ (27), *Department of Molecular Cell Biology, University of Utrecht, 3584 CH Utrecht, The Netherlands*

ROBERT R. WAGNER (22), *Department of Microbiology, University of Virginia School of Medicine, Charlottesville, Virginia 22908*

JAN WILSCHUT (1, 23), *Department of Physiological Chemistry, University of Groningen, 9712 KZ Groningen, The Netherlands*

PHILIP L. YEAGLE (6), *Department of Biochemistry, School of Medicine and Biomedical Sciences, State University of New York at Buffalo, Buffalo, New York 14214*

ULRICH ZIMMERMANN (14), *Lehrstuhl für Biotechnologie, Universität Würzburg, D-8700 Würzburg, Germany*

Preface

To commemorate the twenty-first anniversary of the publication of J. D. Watson and F. H. C. Crick's famous article on the structure of DNA, the April 26, 1974, issue of *Nature* featured a special section entitled "Molecular biology comes of age." While the origin of the field of membrane fusion research cannot be traced to a single article, two comprehensive reviews on virus-induced cell fusion and on membrane fusion appeared in 1972 and 1973, respectively (G. Poste, *Int. Rev. Cytol.* **33,** 157–252; G. Poste and A. C. Allison, *Biochim. Biophys. Acta* **300,** 421–465). In the two decades since, there has been a rapid growth in the number of studies on the molecular mechanisms of membrane fusion, culminating in several books on the subject (A. E. Sowers, ed., "Cell Fusion," Plenum Press, 1987; S. Ohki, D. Doyle, T. D. Flanagan, S. W. Hui, and E. Mayhew, eds., "Molecular Mechanisms of Membrane Fusion," Plenum Press, 1988; N. Düzgüneş, ed., "Membrane Fusion in Fertilization, Cellular Transport, and Viral Infection," Academic Press, 1988; J. Wilschut and D. Hoekstra, eds., "Membrane Fusion," Marcel Dekker, 1991). With the publication of Volumes 220 and 221 of *Methods in Enzymology* dedicated to this subject, it is not entirely inappropriate to declare the field of membrane fusion as having come of age.

The chapters in this and the accompanying Volume 221 present not only the details of methods used in membrane fusion research, but also a critical analysis of the methods, their advantages and shortcomings, and possible artifacts. While several sections focus on the elucidation of the mechanisms of fusion in various experimental systems (Fusion of Liposomes and Other Artificial Membranes; Fusion of Viruses with Target Membranes; Cell–Cell Fusion Mediated by Viruses and Viral Proteins; Conformational Changes of Proteins during Membrane Fusion; Membrane Fusion during Exocytosis; Intracellular Membrane Fusion; Membrane Fusion in Fertilization), several others describe applications of membrane fusion technology (Induction of Cell–Cell Fusion; Introduction of Macromolecules into Cells by Membrane Fusion; Protoplast Fusion). The methodology presented should be of value not only to newcomers to membrane fusion research who wish to employ some of the techniques described in these books, but also to researchers in the field who need to adopt an alternative technique.

I would like to thank the contributors to this volume, without whose willing and able collaboration this work would not even have begun. I would also like to express my appreciation for their patience with me and with their fellow authors, not all of whom were able to submit their manuscripts at the same time. I thank Shirley Light of Academic Press for her patience, understanding, encouragement, and persistence in producing this volume, and Cynthia Vincent for her invaluable editorial assistance. I also thank my wife Diana Flasher for her constant support and enthusiasm for this project, despite countless weekends I spent editing manuscripts. Finally, I wish to dedicate this volume to my dear father, Professor Orhan Düzgüneş, and to the memory of my dear mother, the late Professor Zeliha Düzgüneş, both at the University of Ankara, in deep appreciation of their ceaseless encouragement, understanding, and love.

NEJAT DÜZGÜNEŞ

METHODS IN ENZYMOLOGY

VOLUME XXXVI. Hormone Action (Part A: Steroid Hormones)
Edited by BERT W. O'MALLEY AND JOEL G. HARDMAN

VOLUME XXXVII. Hormone Action (Part B: Peptide Hormones)
Edited by BERT W. O'MALLEY AND JOEL G. HARDMAN

VOLUME XXXVIII. Hormone Action (Part C: Cyclic Nucleotides)
Edited by JOEL G. HARDMAN AND BERT W. O'MALLEY

VOLUME XXXIX. Hormone Action (Part D: Isolated Cells, Tissues, and Organ Systems)
Edited by JOEL G. HARDMAN AND BERT W. O'MALLEY

VOLUME XL. Hormone Action (Part E: Nuclear Structure and Function)
Edited by BERT W. O'MALLEY AND JOEL G. HARDMAN

VOLUME XLI. Carbohydrate Metabolism (Part B)
Edited by W. A. WOOD

VOLUME XLII. Carbohydrate Metabolism (Part C)
Edited by W. A. WOOD

VOLUME XLIII. Antibiotics
Edited by JOHN H. HASH

VOLUME XLIV. Immobilized Enzymes
Edited by KLAUS MOSBACH

VOLUME XLV. Proteolytic Enzymes (Part B)
Edited by LASZLO LORAND

VOLUME XLVI. Affinity Labeling
Edited by WILLIAM B. JAKOBY AND MEIR WILCHEK

VOLUME XLVII. Enzyme Structure (Part E)
Edited by C. H. W. HIRS AND SERGE N. TIMASHEFF

VOLUME XLVIII. Enzyme Structure (Part F)
Edited by C. H. W. HIRS AND SERGE N. TIMASHEFF

VOLUME XLIX. Enzyme Structure (Part G)
Edited by C. H. W. HIRS AND SERGE N. TIMASHEFF

VOLUME L. Complex Carbohydrates (Part C)
Edited by VICTOR GINSBURG

VOLUME LI. Purine and Pyrimidine Nucleotide Metabolism
Edited by PATRICIA A. HOFFEE AND MARY ELLEN JONES

VOLUME LII. Biomembranes (Part C: Biological Oxidations)
Edited by SIDNEY FLEISCHER AND LESTER PACKER

VOLUME LIII. Biomembranes (Part D: Biological Oxidations)
Edited by SIDNEY FLEISCHER AND LESTER PACKER

VOLUME LIV. Biomembranes (Part E: Biological Oxidations)
Edited by SIDNEY FLEISCHER AND LESTER PACKER

VOLUME 213. Carotenoids (Part A: Chemistry, Separation, Quantitation, and Antioxidation)
Edited by LESTER PACKER

VOLUME 214. Carotenoids (Part B: Metabolism, Genetics, and Biosynthesis)
Edited by LESTER PACKER

VOLUME 215. Platelets: Receptors, Adhesion, Secretion (Part B)
Edited by JACEK J. HAWIGER

VOLUME 216. Recombinant DNA (Part G)
Edited by RAY WU

VOLUME 217. Recombinant DNA (Part H)
Edited by RAY WU

VOLUME 218. Recombinant DNA (Part I)
Edited by RAY WU

VOLUME 219. Reconstitution of Intracellular Transport
Edited by JAMES E. ROTHMAN

VOLUME 220. Membrane Fusion Techniques (Part A)
Edited by NEJAT DÜZGÜNEŞ

VOLUME 221. Membrane Fusion Techniques (Part B) (in preparation)
Edited by NEJAT DÜZGÜNES

VOLUME 222. Proteolytic Enzymes in Coagulation, Fibrinolysis, and Complement Activation (Part A: Mammalian Blood Coagulation Factors and Inhibitors) (in preparation)
Edited by LASZLO LORAND AND KENNETH G. MANN

VOLUME 223. Proteolytic Enzymes in Coagulation, Fibrinolysis, and Complement Activation (Part B: Complement Activation, Fibrinolysis, and Nonmammalian Blood Coagulation Factors) (in preparation)
Edited by LASZLO LORAND AND KENNETH G. MANN

VOLUME 224. Molecular Evolution: Producing the Biochemical Data (in preparation)
Edited by ELIZABETH ANNE ZIMMER, THOMAS J. WHITE, REBECCA L. CANN, AND ALLAN C. WILSON

VOLUME 225. Guide to Techniques in Mouse Development (in preparation)
Edited by PAUL M. WASSARMAN AND MELVIN L. DEPAMPHILIS

VOLUME 226. Metallobiochemistry (Part C: Spectroscopic and Physical Methods for Probing Metal Ion Environments in Metalloenzymes and Metalloproteins) (in preparation)
Edited by JAMES F. RIORDAN AND BERT L. VALLEE

VOLUME 227. Metallobiochemistry (Part D: Physical and Spectroscopic Methods for Probing Metal Ion Environments in Metalloproteins) (in preparation)
Edited by JAMES F. RIORDAN AND BERT L. VALLEE

Section I

Fusion of Liposomes and Other Artificial Membranes

[1] Fusion Assays Monitoring Intermixing of Aqueous Contents

By NEJAT DÜZGÜNEŞ and JAN WILSCHUT

Introduction

The fusion of phospholipid vesicles has been used as a model system for understanding the molecular mechanisms of membrane fusion in biological systems.[1-6] Fusion has been monitored by a variety of techniques, including differential scanning calorimetry,[7] NMR,[8] electron microscopy,[7,9,10] dynamic light scattering,[11] and gel filtration.[7,12] These techniques essentially measure changes in the system under study before and after the induction of fusion. Methods to study the kinetics of the fusion process include measuring the dilution of spin-labeled phospholipids from labeled to unlabeled vesicles as an indicator of lipid mixing,[13] monitoring the intermixing or dilution of fluorescent phospholipids,[14-19] and following the intermixing of aqueous contents of vesicles.[20-24]

[1] D. Papahadjopoulos, G. Poste, and W. J. Vail, *Methods Membr. Biol.* **10**, 1 (1979).

[2] S. Nir, J. Bentz, J. Wilschut, and N. Düzgüneş, *Prog. Surf. Sci.* **13**, 1 (1983).

[3] N. Düzgüneş, *in* "Subcellular Biochemistry" (D. B. Roodyn, ed.), Vol. 11, p. 195. Plenum, New York, 1985.

[4] N. Düzgüneş, J. Wilschut, and D. Papahadjopoulos, *in* "Physical Methods on Biological Membranes and Their Model Systems" (F. Conti, W. E. Blumberg, J. DeGier, and F. Pocchiari, eds.), p. 193. Plenum, New York, 1985.

[5] J. Wilschut and D. Hoekstra, *Chem. Phys. Lipids* **40**, 145 (1986).

[6] J. H. Prestegard and M. P. O'Brien, *Annu. Rev. Phys. Chem.* **38**, 383 (1987).

[7] D. Papahadjopoulos, G. Poste, B. E. Schaeffer, and W. J. Vail, *Biochim. Biophys. Acta* **352**, 10 (1974).

[8] J. H. Prestegard and B. Fellmeth, *Biochemistry* **13**, 1122 (1974).

[9] A. J. Verkleij, C. Mombers, W. J. Gerritsen, L. Leunissen-Bijvelt, and P. R. Cullis, *Biochim. Biophys. Acta* **555**, 358 (1979).

[10] E. L. Bearer, N. Düzgüneş, D. S. Friend, and D. Papahadjopoulos, *Biochim. Biophys. Acta* **693**, 93 (1982).

[11] S. Sun, E. P. Day, and J. T. Ho, *Proc. Natl. Acad. Sci. U.S.A.* **75**, 4325 (1978).

[12] D. Lichtenberg, E. Freire, C. F. Schmidt, Y. Barenholz, P. L. Felgner, and T. E. Thompson, *Biochemistry* **20**, 3462 (1981).

[13] T. Maeda and S. I. Ohnishi, *Biochem. Biophys. Res. Commun.* **60**, 1509 (1974).

[14] G. A. Gibson and L. M. Loew, *Biochem. Biophys. Res. Commun.* **88**, 135 (1979).

[15] P. Vanderwerf and E. F. Ullman, *Biochim. Biophys. Acta* **596**, 302 (1980).

[16] P. S. Uster and D. W. Deamer, *Arch. Biochem. Biophys.* **209**, 385 (1981).

[17] D. K. Struck, D. Hoekstra, and R. E. Pagano, *Biochemistry* **20**, 4093 (1981).

[18] D. Hoekstra, *Biochim. Biophys. Acta* **692**, 171 (1982).

[19] J. Rosenberg, N. Düzgüneş, and C. Kayalar, *Biochim. Biophys. Acta* **735**, 173 (1983).

We envision the following processes taking place during membrane fusion[2,25]: (1) adhesion or aggregation of two membranes or vesicles, (2) establishment of molecular contact between the membranes, (3) membrane destabilization at the area of intermembrane contact, (4) merging of membrane components, and (5) coalescence of the internal aqueous contents of the vesicles or compartments bounded by the membranes. The two assays that we describe in this chapter monitor the intermixing of aqueous contents of phospholipid vesicles. Other assays have been used to detect this step[25]; however, the terbium/dipicolinic acid (Tb/DPA) and aminonaphthalenetrisulfonic acid/p-xylylene bis(pyridinium) bromide (ANTS/DPX) assays have been thoroughly examined for their reliability.

Terbium/Dipicolinic Acid Assay

Principles of Assay

In the Tb/DPA assay, Tb and DPA are initially encapsulated in different populations of phospholipid vesicles.[21,22] Fusion results in the formation of the fluorescent $[Tb(DPA)_3]^{3-}$ chelation complex that is excited at a wavelength near the absorption maximum of DPA.[26,27] Fluorescence is generated through internal energy transfer from DPA to Tb; fluorescence intensity then increases by four orders of magnitude. For a fusion assay monitoring the intermixing of aqueous contents within the confines of the fusing vesicles, it is important that any contents that are released into the medium do not interact or give rise to a fluorescence signal. In the Tb/DPA assay, the external medium contains 0.1 mM ethylenediaminetetraacetic acid (EDTA), which inhibits the chelation complex of Tb and DPA. In addition, the presence of divalent cations used to induce fusion further inhibits this complex. The main elements of the Tb/DPA assay are shown in Fig. 1.[28]

[20] D. Hoekstra, A. Yaron, A. Carmel, and G. Scherphof, *FEBS Lett.* **106**, 176 (1979).

[21] J. Wilschut and D. Papahadjopoulos, *Nature (London)* **281**, 690 (1979).

[22] J. Wilschut, N. Düzgüneş, R. Fraley, and D. Papahadjopoulos, *Biochemistry* **19**, 6011 (1980).

[23] H. Ellens, J. Bentz, and F. C. Szoka, *Biochemistry* **24**, 3099 (1985).

[24] N. Düzgüneş, R. M. Straubinger, P. A. Baldwin, D. S. Friend, and D. Papahadjopoulos, *Biochemistry* **24**, 3091 (1985).

[25] N. Düzgüneş, and J. Bentz, in "Spectroscopic Membrane Probes" (L. M. Loew, ed.), Vol. 1, p. 117. CRC Press, Boca Raton, Florida, 1988.

[26] I. Grenthe, *J. Am. Chem. Soc.* **83**, 360 (1961).

[27] T. D. Barela and A. D. Sherry, *Anal. Biochem.* **71**, 351 (1976).

[28] N. Düzgüneş, K. Hong, P. A. Baldwin, J. Bentz, S. Nir, and D. Papahadjopoulos, in "Cell Fusion" (A. E. Sowers, ed.), p. 241. Plenum, New York, 1987.

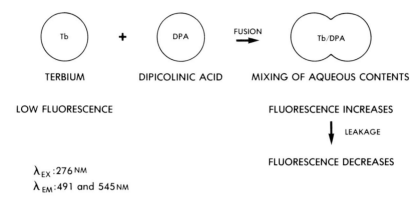

FIG. 1. The Tb/DPA fluorescence assay for membrane fusion. (Reproduced from Düzgüneş et al.,[28] with permission.)

Reagents for Assay

We describe two protocols, one for large unilamellar vesicles and the other for small unilamellar vesicles. The following reagents are encapsulated in large unilamellar vesicles:

Tb vesicles: 2.5 mM TbCl$_3$, 50 mM sodium citrate, 10 mM 2-[tris(hydroxymethyl)methylamine]-1-ethanesulfonic acid (TES), pH 7.4
DPA vesicles: 50 mM sodium dipicolinate, 20 mM NaCl, 10 mM TES, pH 7.4
Tb/DPA vesicles: A 1:1 mixture of the above reagents

Higher concentrations of terbium citrate and NaDPA can be encapsulated in small unilamellar vesicles of approximately 30 nm diameter, since these vesicles are not osmotically active. Because the encapsulated volume of these vesicles is small, it is also necessary to use higher concentrations of reagents to be able to obtain a good signal/noise ratio in the fluorescence measurements. The following reagents are encapsulated in small unilamellar vesicles:

Tb vesicles: 15 mM TbCl$_3$, 150 mM sodium citrate, 10 mM TES, pH 7.4
DPA vesicles: 150 mM sodium-dipicolinate, 10 mM TES, pH 7.4
Tb/DPA vesicles: A 1:1 mixture of the above reagents

Terbium chloride can be obtained from Alfa (Danvers, MA), DPA from Sigma (St. Louis, MO), and TES from Sigma or Calbiochem (San Diego, CA). It should be noted that preparing a sodium-dipicolinate solution adjusted to neutral pH may be time-consuming and should be approached patiently, since solvation is lengthy and it is easy to overshoot the pH while titrating. The concentrations of the encapsulated reagents have an osmolality similar to that of 100 mM NaCl to maintain osmotic balance across the vesicle membrane. For experiments to be carried out in higher salt concentrations, the osmolality of the encapsulation solutions should be increased by adding more NaCl. The osmolality of the solutions is checked with a Wescor vapor pressure osmometer (Logan, UT). Citrate is encapsulated in the Tb vesicles in order to chelate the Tb^{3+} which would otherwise interact with negatively charged phospholipids.

Preparation of Phospholipid Vesicles

Large unilamellar liposomes can be prepared by any number of techniques. Our laboratories have mostly utilized vesicles prepared by reversed-phase evaporation followed by extrusion through polycarbonate membranes to achieve a uniform size distribution.[22,29-31]

A phospholipid mixture in chloroform is placed in a screw-cap high-quality glass tube, then dried under vacuum in a rotary evaporator (Büchi, Brinkmann, Westbury, NY). Preparations involving 10 μmol of lipid for each vesicle type are convenient. A few milliliters of diethyl ether are washed with water (in a glass tube closed with a screw cap with a Teflon lining) by gentle shaking and allowed to settle. One milliliter of the ether (top) layer is placed on the dried phospholipid film, using a glass pipette or syringe. Care should be taken at this point to assure that the lipid is completely dissolved in the ether. The buffer to be encapsulated (0.34 ml) is added to the phospholipid solution, the mixture is purged with argon, and the tube is sealed with Teflon tape and the screw cap. The mixture is then sonicated in a bath-type sonicator (Laboratory Supply Co., Hicksville, NY) for 2–5 min. A stable emulsion should result at this point.

The tube is again sealed with Teflon tape and placed inside a second tube that fits onto a rotary evaporator. About 1 ml of water is placed in the the outer tube, both to maintain thermal contact and to minimize the evaporation of the aqueous solution in the inner tube. The outer tube is placed in a water bath at 30°. The ether is carefully evaporated under controlled vacuum (~ 350 mm Hg), occasionally purging with argon gas attached to the evaporator in order to maintain the level of vacuum and prevent excessive effervescence. When a gel is formed, the vacuum is allowed to build up. Then, the inner glass tube is removed and briefly (5 to 10 sec), but vigorously, vortexed to break up the gel. The tube is again

[29] F. Szoka, F. Olson, T. Heath, W. Vail, E. Mayhew, and D. Papahadjopoulos, *Biochim. Biophys. Acta* **601**, 559 (1980).
[30] N. Düzgüneş, J. Wilschut, R. Fraley, and D. Papahadjopoulos, *Biochim. Biophys. Acta* **642**, 182 (1981).
[31] N. Düzgüneş, J. Wilschut, K. Hong, R. Fraley, C. Perry, D. S. Friend, T. L. James, and D. Papahadjopoulos, *Biochim. Biophys. Acta* **732**, 289 (1983).

placed in the rotary evaporator, and evaporation is resumed. This process is repeated once or twice more. An aqueous opalescent suspension results. An additional aliquot (0.66 ml) of the encapsulation buffer is added to the vesicles, and the suspension is evaporated for 20 min to remove any residual ether. The vesicles are extruded several times through polycarbonate membranes of 0.1 μm pore diameter (Poretics, Plesanton, CA) to achieve a uniform size distribution. A high-pressure extrusion device (Lipex Biomembranes, Vancouver, BC) may also be utilized at this step.

Small unilamellar vesicles are prepared by sonication of a multilamellar liposome suspension.[22,32-34] A phospholipid mixture (10 μmol) in chloroform is placed in a screw-capped high-quality glass tube, then dried under vacuum in a rotary evaporator followed by evaporation with a vacuum pump. The dried film is hydrated in the appropriate buffer to be encapsulated and vortexed under an argon atmosphere, achieved by purging the tube with high-purity argon gas. The tube is purged with argon again and sealed with Teflon tape followed by the screw cap. The multilamellar suspension is sonicated in a bath-type sonicator for 0.5 to 1 hr. The water in the bath is maintained at room temperature. The resulting opalescent suspension is centrifuged at 100,000 g for 1 hr to pellet any large liposomes.

Unencapsulated material is removed by gel filtration of the liposomes on Sephadex G-75 or G-50, or other appropriate filtration media. The chromatography column is equilibrated with 100 mM NaCl, 1 mM EDTA, 10 mM TES, pH 7.4, which is also used for elution. EDTA is added to the buffer to prevent Tb^{3+} binding to the vesicle membrane as the citrate is diluted during chromatography. After removal of the unencapsulated material, the lipid concentration of the liposomes is determined by inorganic phosphate analysis[35]. The Tb vesicles and DPA vesicles are then mixed 1:1 at a 10-fold higher concentration (usually 0.5 mM) than that required for the actual assay (usually 50 μM). For the assay, the vesicles are diluted into a solution of 100 mM NaCl, 10 mM TES, pH 7.4, whereby the EDTA concentration is diluted to 0.1 mM.

Calibration of Assay

To calibrate the fluorescence scale, two methods can be used. In the first method, an aliquot of the chromatographed Tb vesicles (about 1/3) is rechromatographed on a Sephadex G-75 or G-50 column, but equilibrated with 100 mM NaCl, 10 mM TES, pH 7.4, to eliminate the EDTA in the

[32] D. Papahadjopoulos and J. C. Watkins, *Biochim. Biophys. Acta* **135**, 630 (1967).
[33] C. C. Huang, *Biochemistry* **8**, 344 (1969).
[34] N. Düzgüneş and S. Ohki, *Biochim. Biophys. Acta* **467**, 301 (1977).
[35] G. R. Bartlett, *J. Biol. Chem.* **234**, 466 (1959).

first buffer. An aliquot (usually 25 μM) equivalent to the amount of Tb vesicles used in the actual assay is lysed in the presence of 20 μM free DPA, using 0.5% (w/v) sodium-cholate (Calbiochem) or 0.8 mM $C_{12}E_8$ (octaethylene glycol dodecyl ether; Calbiochem). EDTA is removed from this vesicle preparation, since it would interfere with the formation of the Tb/DPA complex when the vesicles are lysed. The free DPA is sufficient to chelate all the Tb^{3+}; higher concentrations of DPA should be avoided because of the absorbance of the compound at the excitation wavelength. The fluorescence of this preparation is set to 100% (F_{max}). The low fluorescence of the mixture of Tb vesicles and DPA vesicles is then set to 0% F_{max}, and any necessary adjustments to the 100% level are made.

For fluorometers in which data are acquired with a computer, the calibration (i.e., normalizing the scale) can be done by data manipulation. For example, to normalize the scale to 100%, the initial level of fluorescence [$I(0)$] is subtracted from the data set, and the resulting data are divided by the numerical (fluorescence intensity) difference between the new 0% level and the fluorescence intensity of the calibration vesicles [$I(\infty)$], with the result being multiplied by 100 to obtain percentage values. Otherwise, normalization to 1 is also possible. Thus, if the fluorescence intensity at time t is $I(t)$, the extent of fusion $F(t)$ may be written as

$$F(t) = 100 \ [I(t) - I(0)]/[I(\infty) - I(0)]$$

The second calibration method is based on the principle that the Tb/DPA vesicles represent the fusion product of all the Tb vesicles and DPA vesicles in the fusion assay when they have the same size distribution and are used at the same lipid concentration as the combination of the two vesicle populations. Thus, 50 μM Tb/DPA vesicles should, ideally, have the same amount of Tb and DPA as the individual vesicle populations, but in a completely complexed form.[36,37] We have found that when such vesicles are transferred from 0°, at which temperature they are stored to minimize leakage, to 25°, the fluorescence takes a rather long time (0.5 hr) to equilibrate. This aspect may be an inconvenience, but the calibration is performed only a few times, once at the beginning of the experiment and again after a series of measurements to ensure that the light intensity of the fluorometer is stable.

Release of Aqueous Contents

In many phospholipid vesicle systems, the aqueous contents may be released into the medium as a result of fusion. This process may be very slow compared to fusion, or it may be very rapid and extensive.[2,3,25] The

[36] J. Bentz, N. Düzgüneş, and S. Nir, *Biochemistry* **22**, 3320 (1983).
[37] S. Nir, N. Düzgüneş, and J. Bentz, *Biochim. Biophys. Acta* **735**, 160 (1983).

release of aqueous contents may be monitored by encapsulating carboxy-fluorescein (Eastman/Kodak, Rochester, NY, or Molecular Probes, Eugene, OR) at self-quenching concentrations.[38] In the case of large unilamellar vesicles, 50 mM carboxyfluorescein (Na salt) with 10 mM TES, pH 7.4, is encapsulated, as described above.[22,30] For small unilamellar vesicles, higher concentrations of carboxyfluorescein can be used (100 mM). The maximal fluorescence (100%) is established by lysing the vesicles with 0.1% (w/v) Triton X-100 or 0.8 mM $C_{12}E_8$. Impurities in some commercial preparations of carboxyfluorescein may be eliminated by chromatographing a concentrated solution in water on Sephadex LH-20.[39]

The Tb/DPA vesicles can also be used to measure the increase in permeability of the liposome membrane, that is, leakage of aqueous contents and the entry of the external medium into the vesicles (if the medium contains quenchers of the Tb/DPA reaction, such as Ca^{2+} and EDTA).[36,40,41] The decrease of Tb/DPA fluorescence is more rapid than the increase of carboxyfluorescein fluorescence during Ca^{2+}-induced fusion of large unilamellar phosphatidylserine vesicles. This observation indicates that the dissociation of Tb/DPA is not only caused by the release of the complex into the medium and its interaction with Ca^{2+} and EDTA, but also by the entry of Ca^{2+} and EDTA into the vesicles.[25]

The use of ions such as Co^{2+} and Mn^{2+} may cause interference with the formation of the Tb/DPA complex if they enter the liposome interior during fusion, and may produce erroneous results. Likewise, investigating the effects of low pH on membrane fusion is hampered by the protonation of DPA at about pH 5.

The results of the Tb/DPA assay may be corrected for the leakage of aqueous contents and dissociation of the Tb/DPA complex. A mathematical expression describing this correction has been developed.[42,43] A simpler expression for this correction for the early stages in fusion in a 1 : 1 population of Tb vesicles and DPA vesicles is given by[36,37,43]

$$I(t) = F(t) + 0.5Q(t)$$

where $F(t)$ is the measured fluorescence intensity (expressed as the percent-

[38] J. N. Weinstein, S. Yoshikami, P. Henkart, R. Blumenthal, and W. A. Hagins, *Science* **195**, 489 (1977).

[39] E. Ralston, L. M. Hjelmeland, R. D. Klausner, J. N. Weinstein, and R. Blumenthal, *Biochim. Biophys. Acta* **649**, 133 (1981).

[40] J. Bentz and N. Düzgüneş, *Biochemistry* **24**, 5436 (1985).

[41] D. Hoekstra, N. Düzgüneş, and J. Wilschut, *Biochemistry* **24**, 565 (1985).

[42] S. Nir, J. Bentz, and J. Wilschut, *Biochemistry* **19**, 6030 (1980).

[43] S. Nir, *in* "Membrane Fusion" (J. Wilschut and D. Hoekstra, eds.), p. 127. Dekker, New York, 1991.

age of maximal fluorescence), $Q(t)$ is the percentage of fluorescence quenching obtained with the Tb/DPA vesicles, and the factor 0.5 reflects the fact that during the fusion experiment the Tb/DPA complex is produced in half of the fused doublets. At later stages of fusion when higher order fusion products are formed, the correction term should be increased.

Fluorescence Measurements

The excitation wavelength is set at 276–278 nm. The Tb/DPA complex has two major emission maxima, at 491 and 545 nm. Except for the initial report on this assay,[21] we have found it more useful to detect the emission at 545 nm, with intermediate slit widths, and a high-pass cutoff filter (e.g., Corning 3–68; Corning, NY), transmitting light above 530 nm, to minimize any contributions from light scattering. Crossed polarizers can also be used to minimize light scattering contributions.[44] We have used SLM 4000, SLM 8000 (SLM-Aminco, Urbana, IL), Spex Fluorolog (Edison, NJ), and Perkin-Elmer MPF43 (Mountain View, CA) fluorometers for our experiments. However, other instruments with similar light intensity and sensitivity can be used to perform the assays.

Figure 2 shows the time course of fusion of large unilamellar phosphatidylserine liposomes, induced by Ca^{2+} and monitored by the Tb/DPA assay.[44] With 3 mM Ca^{2+}, the initial rapid rise in fluorescence intensity indicating membrane fusion is followed by a decrease in fluorescence, owing to the release of aqueous contents into the medium. The release is measured independently by the increase in carboxyfluorescein fluorescence. With 2 mM Ca^{2+}, no appreciable leakage of contents is observed for several minutes, although considerable fusion occurs. Figure 2 also shows that the release of contents measured by carboxyfluorescein fluorescence is identical to that measured by the chelation of Tb leaking from the Tb vesicles with free DPA in the medium.[44]

It is conceivable that the released contents of aggregated vesicles can intermix outside the vesicles and contribute to a spurious fusion signal. Experiments were designed to address this issue. Under conditions where phosphatidylserine vesicles are aggregated (in the presence of Mg^{2+} and Ca^{2+}) and release their contents after a lag phase, there is no increase in fluorescence due to the released Tb and DPA.[44] Similarly, when large unilamellar phosphatidylserine/dipalmitoylphosphatidylcholine (1:1) liposomes, which aggregate in the presence of Ca^{2+} but do not undergo significant fusion within a particular temperature range, are lysed with

[44] J. Wilschut, N. Düzgüneş, K. Hong, D. Hoekstra, and D. Papahadjopoulos, *Biochim. Biophys. Acta* **734**, 309 (1983).

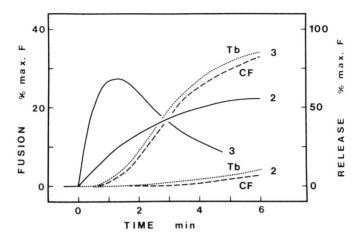

FIG. 2. Fusion of large unilamellar liposomes composed of pure phosphatidylserine monitored by the Tb/DPA assay, and the release of carboxyfluorescein (CF) or Tb into the
medium. Ca^{2+} was added at $t = 0$ at the indicated concentrations (mM). Solid lines denote
mixing of aqueous contents; dashed lines, release of CF; dotted lines, release of Tb. (Reproduced from Düzgüneş et al.,[28] with permission.)

detergent in the aggregated state, no significant increase in fluorescence is
detected.[45]

The Tb/DPA assay has also been utilized in monitoring the fusion of
biological membranes. For example, Hoekstra et al.[46] have encapsulted Tb
and DPA in different populations of erythrocyte ghosts and monitored the
induction of fusion by calcium phosphate. No interference in the assay due
to Tb binding to membrane proteins was noted. The effect of proteins on
liposome fusion has also been monitored successfully with the Tb/DPA
assay.[47]

Aminonaphthalenetrisulfonic Acid/p-Xylylene Bis(pyridinium)
    Bromide Assay

Principles of Assay

Studies on the kinetics of liposome fusion induced by mildly acidic pH
necessitated the development of an assay that was not affected appreciably
by low pH, since the protonation of DPA around pH 5 interferes with the

[45] N. Düzgüneş, unpublished data (1982).
[46] D. Hoekstra, J. Wilschut, and G. Scherphof, *Eur. J. Biochem.* **146**, 131 (1985).
[47] K. Hong, N. Düzgüneş, and D. Papahadjopoulos, *J. Biol. Chem.* **256**, 3541 (1981).

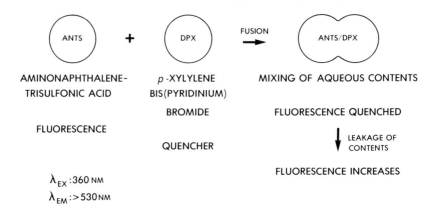

FIG. 3. The ANTS/DPX fluorescence assay for membrane fusion. (Reproduced from Wilschut and co-workers,[41] with permission.)

formation of the Tb/DPA complex.[23,24] The ANTS/DPX assay is based on the collisional quenching of ANTS fluorescence by DPX.[48,49] ANTS and DPX (Molecular Probes) are encapsulated in different populations of vesicles. Fusion results in the interaction of ANTS and DPX within the confines of the two vesicles and in the quenching of ANTS fluorescence.[23] Because high concentrations of DPX are required for quenching, contents of vesicles that may be released into the medium are sufficiently diluted not to quench the fluorescence of any released ANTS. The ANTS/DPX assay is outlined in Fig. 3.

Reagents for Assay

Three large unilamellar vesicle populations are prepared for the assay, encapsulating the following solutions:

ANTS vesicles: 25 mM ANTS, 40 mM NaCl, 10 mM TES, pH 7.4
DPX vesicles: 90 mM DPX, 10 mM TES, pH 7.4
ANTS/DPX vesicles: A 1 : 1 mixture of the above solutions

These solutions are appropriate for use with 100 mM NaCl in the external medium; that is, their osmolalities are similar to that of 100 mM NaCl. Because the counterions in different reagents can differ, the osmolality of the solutions should be checked. For assays involving physiological levels of saline (i.e., 150 mM NaCl), the NaCl concentrations in the ANTS solution should be increased to 90 mM and that in the DPX solution to

[48] M. Smolarsky, D. Teitelbaum, M. Sela, and C. A. Gitler, *J. Immunol. Methods* **15**, 255 (1977).
[49] H. Ellens, J. Bentz, and F. C. Szoka, *Biochemistry* **23**, 1532 (1984).

50 mM NaCl. Because adjusting the pH of the solutions can change their osmolality, it may be worthwhile to adjust the pH first, measure the osmolality, and then add the appropriate amount of NaCl to adjust the osmolality to the desired level. Solutions containing ANTS should be kept in the dark, as with all fluorophores. Solutions should be stored at 4°.

Attempts to use the ANTS/DPX assay for the fusion of small unilamellar vesicles have not produced reliable results, presumably because ANTS binds excessively to such vesicles, although it does not bind appreciably to large unilamellar vesicles.[25]

Calibration of Assay

The fluorescence level of the ANTS vesicles (for a typical assay, 25 nmol lipid/ml) is set to 100%. The ANTS/DPX vesicles (50 nmol lipid/ml) are used to set the fluorescence to 0%, since they represent the fusion product of all the ANTS vesicles and all the DPX vesicles (also 25 nmol lipid/ml) used in the assay.

Release of Aqueous Contents

The ANTS/DPX vesicles can also be used to measure the release of aqueous contents of vesicles during fusion,[23,24] or for investigating the effect of proteins or peptides on membrane permeability.[50-53] In this case 100% fluorescence is set by lysing the vesicles with 0.1% (w/v) Triton X-100 or 0.8 mM $C_{12}E_8$.

Fluorescence Measurements

Fluorescence is excited at 360 nm, and the emission is monitored at or above 530 nm with a high-pass filter (e.g., Corning 3-68) to eliminate any effects of light scattering.

The kinetics of fusion of large unilamellar liposomes composed of cardiolipin/dioleoylphosphatidylcholine (1:1), monitored by either the ANTS/DPX or Tb/DPA assay, are shown in Fig. 4. The Tb/DPA assay reveals a faster fusion process than the ANTS/DPX assay. In the case of phosphatidylserine vesicles, however, the ANTS/DPX assay showed a faster kinetics than the Tb/DPA assay.[54] The reasons for this discrepancy

[50] K. Shiffer, J. Goerke, N. Düzgüneş, J. Fedor, and S. B. Shohet, Biochim. Biophys. Acta 937, 269 (1988).
[51] K. Shiffer, S. Hawgood, N. Düzgüneş, and J. Goerke, Biochemistry 27, 2689 (1988).
[52] N. Düzgüneş and S. Shavnin, J. Membr. Biol. 128, 71 (1992).
[53] R. A. Parente, S. Nir, and F. C. Szoka, Jr., Biochemistry 29, 8720 (1990).
[54] N. Düzgüneş, T. M. Allen, J. Fedor, and D. Papahadjopoulos, Biochemistry 26, 8435 (1987).

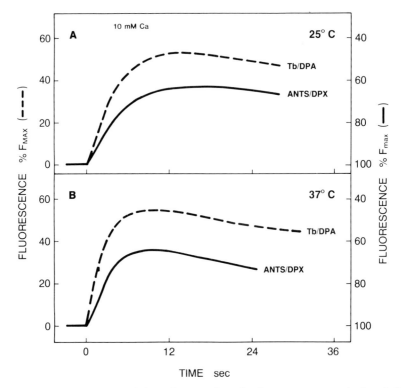

Fig. 4. Calcium-induced fusion of large unilamellar liposomes composed of cardiolipin/dioleoylphosphatidylcholine (1:1), monitored by either the Tb/DPA or the ANTS/DPX assay, at (A) 25° or (B) 37°. Fusion is shown as the increase in fluorescence for the Tb/DPA assay (left ordinate) and as the decrease in fluorescence for the ANTS/DPX assay (right ordinate) [N. Düzgüneş, unpublished data (1986)].

are not understood well. Experiments utilizing liposomes encapsulating one of the components of both the Tb/DPA and the ANTS/DPX assays indicated that differences in the vesicle size distribution in the case of the two assays does not account for the differential kinetics.[55]

Acknowledgments

This work was supported by National Institutes of Health Grant AI-25534, and NATO Collaborative Research Grant CRG 900333. Part of this paper was written during the visit of N. Düzgüneş to the University of Coimbra, Portugal; the hospitality of Dr. M. C. Pedroso de Lima during that visit is gratefully acknowledged. We thank Dr. S. Nir (University of the Pacific and Hebrew University of Jerusalem) for comments on the manuscript.

[55] D. Alford, J. Bentz, and N. Düzgüneş, unpublished data (1987).

[2] Lipid Mixing Assays to Determine Fusion in Liposome Systems

By DICK HOEKSTRA and NEJAT DÜZGÜNEŞ

Introduction

Lipid vesicles or liposomes constitute an important and valuable model to study fundamental elements involved in the mechanism of membrane fusion. Their amenability to variations in composition, including the possibility of reconstituting proteins in their membranes, their access to investigating at the molecular level interactions with fusogens such as certain ions, polypeptides, proteins, and polymers, and their ability to recognize intra- and intermembrane forces in and between liposomal membranes during close approach and aggregation have made liposomes a popular experimental system to unravel basic features of membrane fusion. These features, including the technology developed to reveal these features, may then be exploited and serve as a guide to facilitate the understanding of physiological fusion processes as they occur, for example, during endocytosis and during biosynthetic transport processes in mammalian cells, but also in pathogenic processes like viral infectivity.

Fusion of liposomes or vesicles prepared from synthetic amphiphiles is an uncontrolled event, that is, in contrast to biological fusion it is not a transient phenomenon. Rather, once initiated, fusion of liposomes proceeds until an equilibrium is reached, and under these conditions the original vesicle structure may have been entirely lost. This is observed by electron microscopic techniques in the case of, for example, Ca^{2+}-induced fusion of vesicles consisting of phosphatidylserine (PS), which revert to so-called cochleate cylinders,[1] and it is even more extreme in the case of divalent cation-induced fusion of didodecyl phosphate vesicles, where long tubular hexagonal structures are formed.[2] In both cases, only after the addition of ethylenediaminetetraacetic acid (EDTA) can the formation of large, fused vesicles be revealed. These examples illustrate and emphasize the point, however, that in order to gain insight into the mechanism of fusion per se, it is pertinent to focus on early events during the fusion reaction. Furthermore, an accurate analysis of the initial fusion kinetics

[1] D. Papahadjopoulos, W. J. Vail, K. Jacobson, and G. Poste, *Biochim. Biophys. Acta* **394**, 483 (1975).
[2] L. A. M. Rupert, J. F. C. van Breemen, E. F. J. van Bruggen, J. B. F. N. Engberts, and D. Hoekstra, *J. Membr. Biol.* **95**, 255 (1987).

METHODS IN ENZYMOLOGY, VOL. 220

also allows for a proper evaluation of the significance of distinct structural and physical (i.e., thermotropic or ionotropic) membrane changes frequently accompanying membrane fusion. By correlating such changes with the *rate* of fusion, valuable mechanistic insight can be provided as to whether such changes actually cause fusion or, rather, merely result from it.

Obviously, when lipid vesicles fuse, their membranes merge and their encapsulated volumes mix. Consequently, fusion assays have been developed that monitor the mixing either of liposomal membranes ("lipid mixing") or of their aqueous contents. Ideally, fusion should be followed by measuring both these events because this may provide a more detailed understanding of the molecular aspects involved in fusion, and may also exclude or point to artifacts, particularly in the case of lipid mixing assays.[3,4] On the other hand, aqueous contents mixing assays are not readily or not at all applicable in semiartificial systems, involving fusion of liposomes with biological membranes. Furthermore, some vesicle systems are highly leaky, precluding efficient entrapment of aqueous markers, as is often the case for vesicles prepared from synthetic amphiphiles. In those cases one has to rely, therefore, on the lipid mixing assays to monitor fusion. In the following sections we briefly describe some aspects of these assays (for contents mixing, see Article 1 in this volume).

Lipid Mixing during Membrane Fusion

In principle, two approaches exist to monitor membrane lipid mixing during fusion of vesicles. One approach, frequently applied in early work on liposome fusion, relies on the use of differential scanning calorimetry (DSC).[5] The procedure involves the use of two differently composed vesicle preparations, each consisting of a lipid with a distinctly different main endothermic phase transition temperature (T_c), for example, dimyristoyl-phosphatidylglycerol (DMPG; T_c 24.5°) and dipalmitoyl-PG (T_c 42°). Premixing of equimolar proportions of the two lipids, as would be formed on fusion of the separate vesicles, displays a single phase transition at 31.5°. When the separate vesicles are incubated with 10 mM Ca^{2+}, followed by addition of EDTA (necessary to reveal an intermediate transition temperature, because the T_c shifts to higher temperatures in the presence

[3] N. Düzgüneş and J. Bentz, *in* "Spectroscopic Membrane Probes" (L. M. Loew, ed.), p. 117. CRC Press, Boca Raton, Florida, 1988.

[4] D. Hoekstra, *in* "Membrane Fusion" (J. Wilschut and D. Hoekstra, eds.), p. 289. Dekker, New York, 1991.

[5] D. Papahadjopoulos, W. J. Vail, W. A. Pangborn, and G. Poste, *Biochim. Biophys. Acta* **448**, 265 (1976).

of the cation), only a single peak is found at 31.5°, while the two original peaks disappear, indicating that complete mixing of the lipids had occurred. The occurrence of fusion was confirmed by freeze–fracture electron microscopy. The advantage of this approach is that the membrane is unperturbed, as it relies on mixing of the "natural" lipids. Disadvantages are that the method is quite laborious and insensitive (it requires relatively large amounts of lipids), and it provides insufficient kinetic insight. Furthermore, the application is limited to pure lipid systems with well-defined phase transition temperatures.

Subsequent developments to detect membrane mixing have therefore relied on the use of specific lipid probes. Probes of interest are nitroxide spin labels or fluorophores, substituted onto a lipophilic molecule. One aspect to take into account is that by necessity, such probes may perturb the bilayer packing properties, the extent of which will depend on the localization of the probe (in case of a lipid, head group versus acyl chain attachment) and the nature of the probe, nitroxide labels being less perturbing than fluorescent probes (which are usually aromatic compounds, often comprising more than one ring).

The principle of the application of spin-labeled lipids relies on their incorporation at a relatively high concentration (10–20 mol %, with respect to total lipid) in the liposomal bilayer. At this concentration, the electron spin resonance spectrum shows an exchange broadened signal, arising from strong, concentration-dependent spin–spin exchange interactions. When such vesicles fuse with unlabeled membranes, the probe will be diluted, and the broadened signal is converted to a sharp signal with a concomitant increase in peak height.[6] By comparing the signal to that of a "mock" fusion product, the relative increase in peak height can be correlated with the degree of dilution, which is taken as a measure of fusion. Obviously, this approach has a much wider application than DSC techniques, but, as for DSC, it also suffers from insensitivity since fairly high amounts of lipids are usually needed to record the spectra, while interpretation of the data can be difficult and quite cumbersome. The use of high probe concentrations is also less desirable.

Many of the disadvantages referred to above were overcome by the introduction of assays based on fluorescence spectroscopic techniques. Over other biochemical and biophysical techniques, they offer the advantages that they (1) show a versatile use, (2) are sensitive, that is, require relatively low probe concentrations, (3) have the ability to monitor the kinetics of fusion continuously, (4) allow for rapid acquisition of quantitative data, and, in several instances, (5) allow the monitoring of fusion by fluorescence microscopy (e.g., in the case of liposomes with biological

[6] T. Maeda, A. Asano, K. Ohki, Y. Okada, and S. Ohnishi, *Biochemistry* **14**, 3736 (1975).

membranes).[7,8] In spite of several shortcomings, application of these fluorimetric assays has considerably advanced research in the area of membrane fusion.

Use of Fluorescence Assays

A series of fluorescent lipids, suitable for the purpose of measuring membrane fusion, can be synthesized by conventional procedures, but many of them are now commercially available from Avanti (Pelham, AL), Molecular Probes (Eugene, OR), and KSV Chemicals (Helsinki, Finland). Obviously, a number of criteria have to be met for proper use of a lipid analog in a lipid mixing assay. Apart from sensitivity, these include that (1) its fluorescence quantum yield should be relatively insensitive to the microenvironment of incorporation (also when this environment changes during fusion), (2) it display an optimal miscibility in the bilayer with a minimal degree of perturbation of surrounding lipids, and (3) its rate of spontaneous or collision-mediated transfer between membranes be negligible. Also, when applied in liposome–biological membrane systems, the probe should be nontoxic.

With respect to phospholipid analogs and irrespective of the probe to be used, the fluorophore can be either attached to the polar head group or linked via a spacer as a fluorescent fatty acid analog in the acyl portion, thus substituting for one of the fatty acids of the lipid. It should be noted, however, that certain acyl chain analogs have a strong tendency to transfer spontaneously between labeled and unlabeled membranes because lipid analogs with relatively short acyl chains are quite soluble in water. Notable examples are nitrobenzoxadiazole (NBD) derivatives attached by C_6 and C_{12} spacers to the glycerol backbone of the lipid.[9] Hence, these properties make these probes useless as a tool to monitor membrane fusion.[10] Acyl chain-labeled probes become fairly nonexchangeable when attached to a chain of at least 16 carbons in length, as has been reported, for example, for pyrene and (dialkylamino)coumarin acyl chain-labeled phospholipid analogs.[11,12] Head group-labeled analogs, such as N-(7-nitrobenz-2-oxa-1,3-diazol-4-yl)phosphatidylethanolamine (N-NBD-PE) and N-(lissamine rhodamine B sulfonyl)PE (N-Rh-PE), also display an essentially nonex-

---

[7] D. K. Struck, D. Hoekstra, and R. E. Pagano, *Biochemistry* **20**, 4093 (1981).

[8] T. Kobayashi and R. E. Pagano, *Cell (Cambridge, Mass.)* **55**, 797 (1988).

[9] J. W. Nichols and R. E. Pagano, *Biochemistry* **20**, 2783 (1981).

[10] D. Hoekstra, *Biochemistry* **21**, 1055 (1982).

[11] H. J. Pownall and L. C. Smith, *Chem. Phys. Lipids* **50**, 191 (1989)

[12] J. R. Silvius, R. Leventis, P. M. Brown, and M. Zuckermann, *Biochemistry* **26**, 4279 (1987).

changeable behavior.[7,13,14] In this case the fluorophores protrude into the interface, thus causing no perturbation of the bilayer as such.

Profitable use of these fluorophores in a lipid mixing membrane fusion assay can rely on one of three principles: (1) excimer/monomer formation, (2) fluorescence self-quenching, or (3) resonance energy transfer. The common characteristic of all three approaches is that the signals obtained depend on the concentration of fluorophore(s). Excimer/monomer (E/M) formation is usually carried out with pyrene-labeled lipid analogs. A host of probes have been used in the other procedures. Measuring lipid mixing by changes in the excimer/monomer ratio or changes in self-quenching requires a relatively high concentration of lipid probe, namely, at least 5 mol % with respect to total lipid. On the other hand, resonance energy transfer assays can be used at lipid probe concentrations of less than 1 mol %. The application of an assay based on self-quenching of fluorescence is discussed elsewhere in this volume.[15] Here, we limit our discussion to the use of pyrene derivatives to monitor excimer/monomer formation and the use of fluorescence energy transfer couples as a means to measure fusion by lipid mixing.

Lipid Mixing Monitored by Changes of Excimer/Monomer Ratio

Pyrene is a fluorophore that has been widely used to study many aspects of membrane dynamics,[11] including membrane fusion.[16-20] The probe is usually attached to the acyl chain (of at least 16 carbons in length, to avoid spontaneous transfer) of the phospholipid of choice [phosphatidylcholine (PC), PE, etc]. Incorporation in a liposomal bilayer is accomplished by drying down the probe, mixed with other lipids, from a chloroform phase. Vesicles are then prepared by classic procedures.[21] The fluorescence emission spectrum of pyrene is sensitive to the concentration of the probe (Fig. 1). When pyrene is excited ($\lambda_{ex} \sim 330$ nm), a complex series of reactions occur subsequently. Because of the long lifetime of pyrene in the excited state (P*, up to ~ 300 nsec), P* can not only decay by

[13] D. Hoekstra, *Biochemistry* **21**, 2833 (1982).
[14] J. W. Nichols and R. E. Pagano, *J. Biol. Chem.* **258**, 5368 (1983).
[15] D. Hoekstra and K. Klappe, this volume [20].
[16] D. P. Via, Y. J. Kao, and L. C. Smith, *Biophys. J.* **25**, 2620 (1979).
[17] C. S. Owen, *J. Membr. Biol.* **54**, 13 (1980).
[18] S. Schenkman, P. S. Araujo, R. Dijkman, F. H. Quina, and H. Chaimovich, *Biochim. Biophys. Acta* **649**, 633 (1981).
[19] R. Pal, Y. Barenholz, and R. R. Wagner, *Biochemistry* **27**, 30 (1988).
[20] S. Amselem, Y. Barenholz, A. Loyter, S. Nir, and D. Lichtenberg, *Biochim. Biophys. Acta* **860**, 301 (1986).
[21] F. C. Szoka and D. Papahadjopoulos, *Annu. Rev. Biophys. Bioeng.* **9**, 467 (1980).

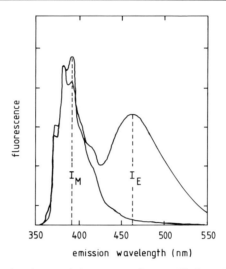

FIG. 1. Monomer and excimer emission spectra of pyrene. Excimers (I_E) are formed when two pyrenes can interact during the excited state lifetime of one pyrene (PP* interaction, see text). This occurs when the probe is incorporated into bilayers at a certain minimal concentration. The degree of excimer formation reduces when probe dilution occurs during fusion. The extent of fusion can be derived from calculating the excimer/monomer ratio (I_E/I_M) as function of time (see text). The excitation wavelength is usually set around 320 nm.

the usual single-molecule radiative route and fluoresce, but it can also react with ground state (i.e., unexcited) pyrene P to form excimers (excited dimers, PP*) as a result of a collision. The radiative decay of the excimeric state occurs at a longer and broad wavelength range (~470 nm) quite distinct from the sharp monomer emission peak (Fig. 1). The number of excimers formed is a function only of the concentration of P, provided that the membrane viscosity and P* lifetime remain constant.

For lipid mixing to be monitored by this approach, vesicles are prepared containing approximately 5 mol % of pyrene-labeled lipid probe. The I_E/I_M ratio is measured at the appropriate wavelengths (cf. Fig. 1) and monitored during the course of fusion with unlabeled vesicles. From the decrease in I_E/I_M ratio due to dilution, the extent of fusion after appropriate time intervals is then determined. Up to approximately 7 mol %, the I_E/I_M ratio is a linear function of the concentration of pyrene-labeled lipid analog.[18] At such conditions, therefore, the decrease in the I_E/I_M ratio is directly proportional to the extent of fusion. It should be noted that the method is somewhat laborious, given the fact that both excimer and monomer emission have to be determined at separate wavelengths. This also implies that fusion is measured following inhibiting the reaction after a certain time interval, precluding continuous monitoring of the process. These shortcomings do not arise with assays based on resonance energy transfer.

Lipid Mixing Monitored by Resonance Energy Transfer

Lipid mixing assays based on the principle of resonance energy transfer (RET) are the most popular assays currently applied in monitoring mixing of membrane phases during fusion. This approach offers a variety of advantages over other techniques. First, the technique is very sensitive, requiring low concentrations of probes (less than 1 mol %, with respect to total lipid), thus minimizing the degree of membrane perturbation. Second, lipid mixing can be monitored continuously, allowing accurate quantitation of the initial fusion kinetics, which is particularly relevant in the case of fast-fusing membrane systems. Third, the procedure is relatively simple and applicable not only to vesicle–vesicle fusion studies, but also to studies involving the fusion of lipid vesicles with biological membranes. Apart from fluorometric data, fluorescence microscopy offers an addi-

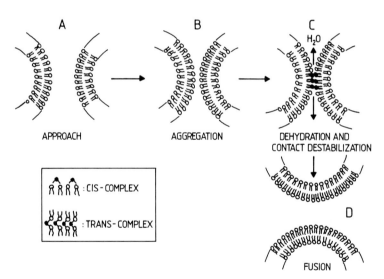

FIG. 2. Schematic view of sequential steps in cation-induced fusion of lipid vesicles containing acidic phospholipids. By virtue of binding in a cis manner, cations can bring about the close approach and aggregation of the vesicles (A and B). A trans cation/acidic phospholipid complex (C) is thought to form subsequently, which is accompanied by dehydration and destabilization of interbilayer contact sites (D). As a result, membrane fusion takes place. The assays discussed in this chapter monitor the kinetics of the overall process (A through D).

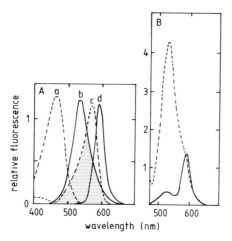

FIG. 3. (A) Excitation (---) and emission (——) spectra of N-NBD-PE (a and b, respectively) and N-Rh-PE (c and d, respectively), incorporated separately in PS vesicles at a concentration of 1 mol %. The shaded area represents the overlap between the emission spectrum of N-NBD-PE (donor) and the excitation spectrum of N-Rh-PE (acceptor), allowing energy transfer to occur when the probes are in close proximity. Under such conditions (B), that is, when both probes are incorporated (at 0.8–1 mol % each) into PS vesicles, the fluorescence of N-NBD-PE is almost negligible at 530 nm when excited at 470 nm. Rather, a peak characteristic for N-Rh-PE (at 585 nm) is seen owing to energy transfer between donor and acceptor (solid line). When detergent is added an infinite dilution of the probes is obtained, eliminating the occurrence of energy transfer and resulting in "unquenched" NBD fluorescence (---).

tional tool for analyzing fusion in the latter systems.[7] Finally, in this context is should also be noted that the data obtained can be readily analyzed in the framework of a mass action kinetic model.[22] This model views the overall fusion reaction, which is actually monitored with the assay, as consisting of two distinct steps: the second-order aggregation reaction and the fusion reaction itself, which is first-order (Fig. 2). Thus, under a variety of experimental conditions the rate constants for aggregation and fusion can be established, providing insight into molecular factors that affect aggregation, fusion itself, or both.

The approach of RET relies on the interactions which occur between two different fluorophores if the emission band of one, the energy donor, overlaps with the excitation band of the second, the energy acceptor, and the two probes exist in close physical proximity[23,24] (Fig. 3A). Fluorescence energy transfer involves the transfer of the excited state energy, derived

[22] S. Nir, this volume [28].
[23] T. Förster, Z. Naturforsch. A 4A, 321 (1949).
[24] L. Stryer, Annu. Rev. Biochem. 47, 819 (1978).

from a photon absorbed by the energy donor on excitation, from a donor to an acceptor. The acceptor will then fluoresce as though it had been excited directly. Thus, energy transfer does *not* involve the emission and reabsorption of photons, but occurs in a nonradiative way. The rate and efficiency of energy transfer is dependent on the extent of overlap of the emission spectrum of the donor with the absorption spectrum of the acceptor and on their spatial separation. The rate of energy transfer from a specific donor to a specific acceptor is given by

$$V_T = \frac{1}{\tau_d} \left(\frac{R_0}{r} \right)^6$$

where τ_d is the lifetime of the donor in the absence of acceptor, r is the distance between donor and acceptor, and R_0 is a characteristic distance, called the Förster distance, at which the efficiency of energy transfer is 50%. The transfer efficiency, E, can be derived from the equation

$$E = 1 - F/F_0$$

in which F is the fluorescence of the donor measured at its emission wavelength in the presence of the acceptor and F_0 is the fluorescence in the absence of the acceptor (i.e., at infinite dilution of donor and acceptor; see Fig. 3B and below).

A large variety of lipid probes exist that satisfy the criteria for efficient energy transfer to occur and which have been applied in lipid mixing assays to monitor fusion. A number of these probes ("energy transfer couples") are listed in Table I. For the sake of illustration, we limit our discussion to presenting mainly a more detailed description of the practical application and limitations of a frequently applied energy-transfer couple, namely, that consisting of N-NBD-PE and N-Rh-PE.

The practical application of energy transfer in monitoring lipid mixing offers several possibilities, depending on the experimental system. In studies involving fusion of lipid vesicles, the probes can be incorporated into separate vesicles. Fusion is then reported by an increase in energy transfer efficiency, occurring when the probes mix within the same ("fused") membrane.[13,25] The disadvantage of this approach is that RET which might occur during aggregation has to be excluded (for the NBD/Rh couple R_0 is ~ 50 Å). To establish its contribution, control experiments are necessary, for example, by adding EDTA to systems involving Ca^{2+}-induced fusion of acidic phospholipid vesicles.[13] A decrease in RET efficiency will reflect the occurrence of energy transfer caused by vesicle aggregation.

[25] P. S. Uster and D. W. Deamer, *Arch. Biochem. Biophys.* **209**, 385 (1981).

TABLE I

FLUORESCENT ANALOGS USED AS PROBES IN
LIPID MIXING ASSAYS[a]

Analogs[b] (donor/acceptor)	Ref.
N-NBD-PE/N-Rh-PE	c
Dansyl-PE/Rh-PE	d
Chlorophyll b/chlorophyll a	e
Dansyl-PE/ASPPS	f
DH-NBD/CA9C	g
(12-CPS)-18-PC/(12-DABS)-18-PC	h

[a] Assays are based on RET to monitor lipid vesicle fusion.

[b] Dansyl, 5-Dimethylaminonaphthalene-1-sulfonyl; ASPPS, N-(3-sulfopropyl)-4-(p-didecylaminostyryl)pyridinium; DH, dihexadecyl; CA9C, cholesterylanthracene 9-carboxylate; CPS, [(N-{4-[7-(diethylamino) - 4 - methylcoumarin - 3 - yl]phenyl}carbamoyl)methyl]thio; DABS, [(4-([4-(dimethylamino)phenyl]azo)-phenyl)-sulfonyl]methylamino.

[c] D. K. Struck, D. Hoekstra, and R. E. Pagano, *Biochemistry* **20**, 4093 (1981).

[d] P. Vanderwerf and E. F. Ullman, *Biochim. Biophys. Acta* **596**, 302 (1980).

[e] A. E. Gad and G. D. Eytan, *Biochim. Biophys. Acta* **727**, 170 (1983).

[f] G. A. Gibson and L. M. Loew, *Biochem. Biophys. Res. Commun.* **88**, 135 (1979).

[g] P. S. Uster and D. W. Deamer, *Arch. Biochem. Biophys.* **209**, 385 (1981).

[h] J. R. Silvius, R. Leventis, P. M. Brown, and M. Zuckermann, *Biochemistry* **26**, 4279 (1987).

A more convenient approach, therefore, is to incorporate both probes in the same vesicle population (Fig. 4). An initially high energy transfer efficiency will thus be established. When such vesicles fuse with a nonlabeled vesicle population the probes will dilute; that is, their mutual distance increases and consequently the efficiency of energy transfer decreases. This approach is particularly versatile: instead of fusion with unlabeled vesicles, fusion of the labeled vesicles with any target membrane, including biological membranes,[7,8,26–29] can be investigated.

[26] M. C. Harmsen, J. Wilschut, G. Scherphof, C. Hulstaert, and D. Hoekstra, *Eur. J. Biochem.* **149**, 591 (1985).

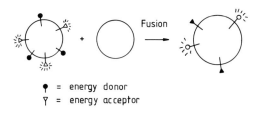

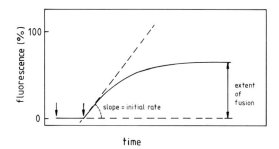

FIG. 4. Schematic representation of the principle of the resonance energy transfer assay. When exciting the donor, the acceptor will fluoresce, provided that the probes are in close proximity. On fusion with unlabeled vesicles, the energy transfer efficiency will decrease when the probes dilute. In this case, donor fluorescence at its emission wavelength will increase, which can be monitored continuously (bottom graph). Prior to starting a fusion experiment, the fluorescence scale is calibrated (first arrow, see text). Fusion is triggered on addition of an appropriate fusogen (e.g., Ca^{2+}, second arrow). Wavelength settings are those of the fluorescent donor.

At low surface densities of N-Rh-PE (1 mol % with respect to total lipid), the efficiency of energy transfer is nearly linearly related to the ratio of acceptor lipid to total lipid in the vesicle membrane, and it is essentially independent of the surface density of the energy donor, N-NBD-PE. This is of obvious convenience because any decrease in surface density of the energy acceptor will thus result in a proportional decrease in energy transfer efficiency.

In practice, a typical fusion experiment is carried out as follows. The desired lipids and fluorescent lipid analogs, from stock solutions in chloroform/methanol, are mixed in a tube, such that the amounts of the analogs

[27] A. J. M. Driessen, D. Hoekstra, G. Scherphof, R. D. Kalicharan, and J. Wilschut, *J. Biol. Chem.* **260**, 10880 (1985).

[28] H. J. P. Marvin, M. B. A. ter Beest, D. Hoekstra, and B. Witholt, *J. Bacteriol.* **171**, 5268 (1989).

[29] T. Stegmann, D. Hoekstra, G. Scherphof, and J. Wilschut, *Biochemistry* **24**, 3107 (1985).

are 0.8–1.0 mol % each with respect to total lipid. Lipid vesicles are then prepared by a variety of procedures.[21] Similarly, unlabeled vesicles are prepared. Subsequently, the labeled and unlabeled vesicles, at concentrations as used under fusion conditions, are taken for calibration of the fluorescence scale. To this end the vesicles (premixed in the desired ratio of labeled and unlabeled vesicles) are suspended in a buffer-containing cuvette, placed in the fluorimeter. The incubation volume is usually 2 ml; the final lipid concentration of the order of 50 μM. For practical purposes, it has been shown to be most convenient to monitor energy transfer changes for the NBD/Rh couple by following the changes; that is, the *increase*, in fluorescence intensity of the donor (N-NBD-PE, λ_{ex} 460,λ_{em} 530 nm) as a function of time. It is equally possible to follow the *decrease* in fluorescence emission of the acceptor (N-Rh-PE). However, the increase of donor fluorescence is of greater magnitude than the decrease of acceptor fluorescence, as seen for other couples as well,[25] whereas the Rh peak often appears as a small shoulder on the NBD emission spectrum, particularly when the extent of dilution increases.[7] Having thus set the wavelengths at the excitation and emission maximum of NBD, the fluorescence is read. The residual fluorescence seen, usually less than 5% (at 1 mol % probe) of the fluorescence seen at infinite dilution, is then taken as the zero level and adjusted as such on the chart recorder.

The extent of dilution to be obtained on fusion (in terms of the percentage of fluorescence increase) will depend on the concentration of unlabeled vesicles, that is, it will correlate with the final surface density of the probes in the fused membrane. Two procedures have been used in this respect. One approach is to lyse the vesicles by adding detergents (at final concentrations of 1%) such as Triton X-100[7] or cetyltrimethylammonium bromide.[30] This treatment will eliminate energy transfer (cf. Fig. 3B), and the signal yields the NBD fluorescence obtained at infinite dilution. It should be noted that the signal obtained needs to be corrected for sample dilution and, in the case of Triton X-100, for the effect of this detergent on the quantum yield of N-NBD-PE. This factor (1.4–1.5) can be obtained by incorporating the lipid analog in the vesicles at a concentration of 0.05–0.1 mol % (i.e., a concentration where essentially no self-quenching of the probe occurs) and measuring fluorescence before and after addition of Triton X-100.

The fluorescence level thus obtained is set at 100%, representing the extent of fluorescence that would be reached if the labeled vesicles were to fuse with a large excess of unlabeled vesicles, resulting in complete dilution of the probes. Obviously, when the ratio of labeled to unlabeled vesicles is

---

[30] L. A. M. Rupert, J. B. F. N. Engberts, and D. Hoekstra, *Biochemistry* **27**, 8232 (1988).

1, dilution over the entire membrane area available would lead to a fluorescence level of 50% (on the scale of 100% as obtained by the detergent procedure). It is therefore also possible to prepare a "mock" vesicle fusion product containing a probe density as would be attained had complete fusion between labeled and unlabeled vesicles taken place. These vesicles are then placed in the cuvette at the same lipid concentration as the labeled and unlabeled vesicles to be used for the assay, and the fluorescence level is determined. Having thus calibrated the fluorescence scale, the fusogen is added to a mixture of labeled and unlabeled vesicles, and the fluorescence

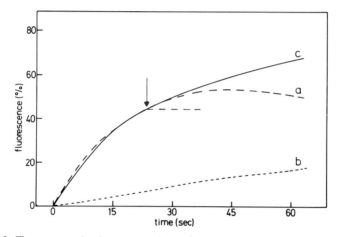

Fig. 5. Fluorescence development during Ca^{2+}-induced fusion of CL/DOPC vesicles, monitored by lipid mixing and coalescence of aqueous contents. Large unilamellar vesicles consisting of CL/DOPC (1:1) were labeled with N-NBD-PE and N-Rh-PE, mixed with an equimolar amount of unlabeled CL/DOPC vesicles (final lipid concentration was 50 μM), and injected into the fusion medium that contained 11 mM $CaCl_2$. The increase of N-NBD-PE fluorescence (cf. Fig. 4) was monitored as a function of time (curve c). Alternatively, fusion was also monitored with a contents mixing assay in which a 1:1 mixture of vesicles containing terbium (Tb) and dipicolinic acid (DPA) were injected into the medium, under otherwise identical conditions as for curve c. In this case, fusion results in formation of a fluorescent Tb/DPA complex, which can be monitored continuously (curve a). Note the similarity of initial rates obtained with both assays for this particular vesicle composition. After approximately 30 sec, significant destabilization of the vesicle bilayer becomes apparent, causing leakage of vesicle contents as reflected by a decrease in the Tb/DPA signal. The decrease is due to dissociation of the complex by EDTA, which is present in the external medium to prevent complex formation outside the vesicle interior compartment. In a separate experiment, leakage can also be determined independently by encapsulating the preformed Tb/DPA complex and monitoring its dissociation (reflected by a *decrease* in fluorescence) in the EDTA-containing fusion medium (curve b). For details, see Ref. 31.

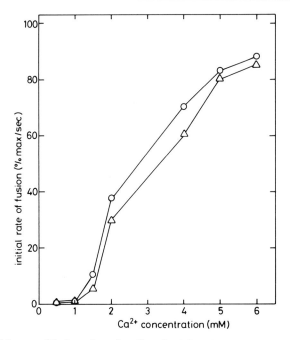

FIG. 6. Initial rates of fusion of small unilamellar PS vesicles as a function of the Ca^{2+} concentration. Fusion was monitored by the N-NBD-PE/N-Rh-PE lipid mixing assay (O), using equimolar amounts of labeled and unlabeled vesicles, or by the Tb/DPA assay (△; see legend to Fig. 5). Initial rates were determined as indicated in Fig. 4 and plotted as a function of the Ca^{2+} concentration.

increase, as a measure of fusion, is monitored directly and continuously (Fig. 4). The relative change in fluorescence is given by

$$I(t) = \frac{[F(t) - F_0]}{(F_\infty - F_0)}$$

where $F(t)$ represents the fluorescence intensity after a given time t, F_0 is the initial fluorescence of the labeled vesicles, and F_∞ is the maximal level of fluoroscence attainable. $I(t) \times 100\%$ then represents the percentage of fusion obtained.

Figure 5 shows the results of a typical fusion experiment, obtained on Ca^{2+}-induced fusion of lipid vesicles consisting of cardiolipin (CL) and dioleoylphosphatidylcholine (DOPC).[31] Curve c in Fig. 5 represents the time-dependent fluorescence development during lipid mixing. For comparison, fusion of the same vesicles as monitored by aqueous contents

[31] J. Wilschut, S. Nir, J. Scholma, and D. Hoekstra, *Biochemistry* **24,** 4630 (1985).

mixing is also shown (curve a, Fig. 5). Note that Ca^{2+}-induced leakage of vesicles occurs, a phenomenon that often accompanies vesicle fusion, and which may underestimate the extent of fusion when registered by contents mixing. Obviously, leakage does not interfere with assays based on lipid mixing. However, initial fusion events of many vesicle systems are essentially nonleaky, which in principle should give rise to very similar initial rates when measured by lipid and contents mixing assays (Fig. 6; however, see also below). Continuous monitoring of the fusion process also offers the advantage of correlating the occurrence of structural alterations in the bilayers with the mechanism of fusion. When appropriate assays are available, a direct kinetic comparison can thus be made. The results of such a study[13] show that the kinetics of Ca^{2+}-induced lipid-phase separation, occurring in PS-containing lipid vesicles, are much slower than those of fusion (Fig. 7). From these studies it was concluded[13] that macroscopic lipid phase separation is secondary to fusion, rather than acting as a trigger that initiates the process.

Finally, in contrast to contents mixing assays, assays based on lipid

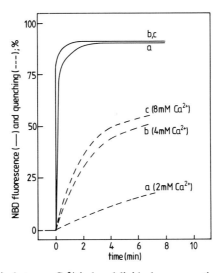

FIG. 7. Relationship between Ca^{2+}-induced lipid phase separation and fusion of small unilamellar PS-containing vesicles. At various Ca^{2+} concentrations, the kinetics of PS vesicle fusion were monitored continuously using the N-NBD-PE/N-Rh-PE RET assay. In separate experiments, the kinetics of lipid phase separation were monitored by following the increase of self-quenching of N-NBD-PE, which occurs when the local concentration of the probe increases on Ca^{2+}-induced phase separation. Note that in this vesicle system fusion (solid curves) proceeds much more rapidly than the process of phase separation (dashed curves) (for details, see Ref. 13).

mixing may provide a convenient tool to characterize and analyze the fusion of systems involving biological membranes.[7,8,15] Figure 8 shows an elegant example of such an application, involving fusion between liposomes and Golgi membranes in permeabilized cells.[8]

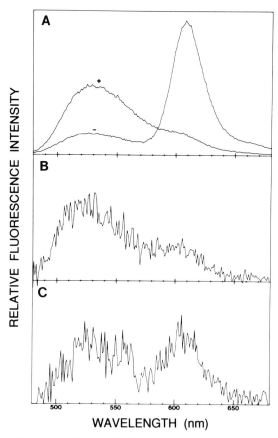

FIG. 8. Fusion of liposomes with the Golgi apparatus in perforated cells. Small unilamellar vesicles consisting of DOPC and labeled with 1 mol % each of N-NBD-PE and N-Rh-PE were incubated with perforated human skin fibroblasts. The fluorescence emission spectra (λ_{ex} 470 nm) of the vesicles per se, prior to (−) and after (+) addition of Triton X-100 are shown (A). After incubation with the cells for 10 min at 30° in the presence of an ATP-generating system, a spectrum (B) was obtained, indicating that substantial dilution of the probes had occurred, presumably owing to fusion of the vesicles with intracellular membranes (compare with + spectrum in A). Spectrum (C) is obtained when the experiment as in (B) is carried out with an ATP-depleting system. In conjunction with fluorescence microscopy and studies performed with calcein as an aqueous contents marker, it was demonstrated that fusion occurred with Golgi membranes (for details, see Ref. 8).

Hazards of Lipid Mixing Assays

Lipid mixing assays do not require painstaking corrections of the leakage that usually occurs during vesicle fusion and that interferes with assays based on contents mixing. Also, as noted above, lipid mixing assays have a broad range of applications. On the other hand, in each system in which lipid mixing assays are used, it is imperative to include control experiments that exclude monomeric transfer of the probe between membranes which may occur at either step illustrated in Fig. 2. Thus, in the absence of a fusogen no change in energy transfer should occur, nor should it take place when vesicles are aggregated.[32,33] It is therefore highly recommended to embed the probes in the same rather than in separate bilayers as the latter conditions, except in fast-fusing systems (cf. Fig. 6), may give rise to energy transfer during aggregation. It also has to be taken into account that during Ca^{2+}-induced fusion of mixed or pure acidic phospholipid vesicles, ionotropic phase separations may occur. This could in principle lead to an enhanced local concentration of the fluorophore, giving rise to fluorescence quenching.[10] In the case of N-NBD-PE and N-Rh-PE, these effects will be negligible at the concentrations of probe applied, since such effects, at Ca^{2+} concentrations up to 10 mM, can be revealed only at lipid analog concentrations above about 1.5 mol %.

It has been shown recently[34,35] that lipid polymorphic transitions can be accompanied by a small but significant increase in fluorescence quantum yield of N-NBD-PE when incorporated in bilayers undergoing such a change. This increase is observed even when both N-NBD-PE and N-Rh-PE are incorporated in bilayers consisting of cardiolipin, which readily adopts the hexagonal H_{II} phase in the presence of Ca^{2+} (T. Fonteijn and D. Hoekstra, unpublished observation). Although the effect is completely reversible on addition of EDTA, alternative fluorophores, less sensitive to such alterations, are preferred in these systems.[12]

Finally, during fusion of certain vesicle systems, it has also been observed that lipid mixing can proceed considerably faster than contents mixing, which, moreover, may also depend on the type of contents mixing assay used (see Ref. 3). Some of these differences have been explained in terms of the occurrence of semifusion, that is, fusion of outer monolayers

---

[32] L. A. M. Rupert, J. B. F. N. Engberts, and D. Hoekstra, *J. Am. Chem. Soc.* **108**, 3920 (1986).

[33] N. Düzgüneş, T. M. Allen, J. Fedor, and D. Papahadjopoulos, *Biochemistry* **26**, 8435 (1987).

[34] K. Hong, P. A. Baldwin, T. M. Allen, and D. Papahadjopoulos, *Biochemistry* **27**, 3947 (1988).

[35] C. D. Stubbs, B. W. Williams, L. T. Boni, J. B. Hoek, T. F. Taraschi, and E. Rubin, *Biochim. Biophys. Acta* **986**, 89 (1989).

(see Fig. 2) preceding inner monolayer destabilization that leads to intermingling of contents. These notions thus emphasize the point that, whenever possible, fusion should be measured by both lipid mixing and aqueous contents mixing assays, and both assays should be seen as complementary rather than as alternatives.

Acknowledgments

We thank Dr. Richard E. Pagano (Carnegie Institution, Baltimore, MD) for kindly providing us with Fig. 8, and Mrs. Rinske Kuperus for expert secretarial assistance with the preparation of the manuscript. This work was supported by National Institutes of Health Grant AI 25534.

[3] Intermembrane Lipid-Mixing Assays Using Acyl Chain-Labeled Coumarinyl Phospholipids

By RANIA LEVENTIS and JOHN R. SILVIUS

Introduction

Fluorescence-based assays of intermembrane lipid mixing are useful to monitor the coalescence of lipid bilayers that accompanies fusion, as well as certain other interactions of artificial and biological membranes.[1-4] Several sensitive assays based on resonance energy transfer (RET) measurements have been developed for this purpose,[1-3] most of which utilize phospholipid-derived probes with fluorescent-labeled polar headgroups. Although such assays are convenient and versatile, the fluorescence signal for various head group-labeled probes can be sensitive to changes in interfacial potential or hydration,[5,6] and possibly to resonance energy transfer

[1] P. Vanderwerf and E. F. Ullman, *Biochim. Biophys. Acta* **596**, 302 (1980).
[2] P. S. Uster and D. W. Deamer, *Arch. Biochem. Biophys.* **209**, 385 (1981).
[3] D. K. Struck, D. Hoekstra, and R. E. Pagano, *Biochemistry* **20**, 4093 (1981).
[4] N. Düzgüneş and J. Bentz, in "Spectroscopic Membrane Probes" (L. M. Loew, ed.), p. 117. CRC Press, Boca Raton, Florida, 1988.
[5] K. Hong, P. A. Baldwin, T. M. Allen, and D. Papahadjopoulos, *Biochemistry* **27**, 3947 (1988).
[6] C. D. Stubbs, B. W. Williams, L. T. Boni, J. B. Hoek, T. F. Taraschi, and E. Rubin, *Biochim. Biophys. Acta* **986**, 89 (1989).

between probes in apposed but unfused bilayers,[7] as well as to interbilayer lipid mixing.

We describe here the preparation and applications of an alternative class of fluorescent lipid probes, carrying a dialkylaminocoumarinyl moiety on one acyl chain, which can be combined with a variety of energy-transfer acceptors for sensitive and versatile assays of intermembrane lipid mixing. These probes provide a strong fluorescence signal which shows good stability under a variety of perturbations, including drastic changes in the ionic environment of the membrane and in the composition of the immediate lipid (or amphiphile) environment. These properties, and the ability to utilize phospholipid probes with head groups that match those found in natural membranes, can offer significant advantages in many applications to monitor lipid mixing between natural and artificial membranes.

Preparation of Donor and Acceptor Probes

12-(N-tert-*Butoxycarbonyl-methylamino*)*octadecanoic Acid*

The methanesulfonate derivative of 12-hydroxyoctadecanoic acid is prepared essentially as described by Kimura and Regen.[8] Briefly, a solution of methanesulfonyl chloride in dry pyridine (0.7 ml in 3 ml) is added over 15 min at 0° to a solution of 12-hydroxyoctadecanoic acid (1 g) in pyridine and dry tetrahydrofuran (THF) (7 ml each). The mixture is stirred at 25° for 3 hr, the products are partitioned between chloroform and 5% aqueous HCOOH (100 ml each), and the chloroform layer is washed twice with 1% aqueous NaCl, then concentrated *in vacuo*. The residue is stirred for 30 min at 0° with 40 ml THF plus 1.6 ml 10 N NaOH, then worked up as above.

The crude product from the above reaction is incubated for 48 hr at 60° with 6 ml each of ethanol and 40% aqueous methylamine in a sealed tube. The contents of the tube are concentrated under nitrogen with warming, then partitioned between chloroform and 1:1 (v/v) methanol/ 1% aqueous HCOOH. The chloroform layer is washed once with the same methanol/aqueous mixture, then concentrated *in vacuo*. The residue is dissolved in 25 ml of 2:1:0.15 (v/v/v) acetone/water/triethylamine, then 920 mg of Boc-ON (2-*tert*-butoxycarbonyloxyimino-2-phenylacetonitrile) is added, and the mixture is stirred for 5 hr at 25°. Water (25 ml) and

[7] N. Düzgüneş, T. M. Allen, J. Fedor, and D. Papahadjopoulos, *Biochemistry* **26**, 8435 (1987).
[8] Y. Kimura and S. L. Regen, *J. Org. Chem.* **48**, 1533 (1983).

HCOOH (to pH 4) are added, the mixture is extracted with chloroform (50 ml), and the chloroform layer is washed with 1% aqueous NaCl, then concentrated *in vacuo*. The residue is applied in chloroform (100 ml) to a 2.2 × 10 cm column of BioSil A (200–400 mesh, Bio-Rad, Richmond, CA; packed in 50:50:0.5 hexane/chloroform/triethylamine), which is eluted successively with chloroform (200 ml, discarded), 1% (v/v) methanol in chloroform (200 ml, discarded), and 2% methanol in chloroform (400 ml, collected). The yield of purified 12-(*N-tert*-Boc-methylamino)octadecanoic acid is 480 mg (35%).

Preparation of Labeled Phosphatidylcholines

12-(*N-tert*-Boc-methylamino)octadecanoic acid (165 mg) is converted to the anhydride by stirring with dicyclohexylcarbodiimide (49.5 mg) in dry CCl_4 overnight at 25°, then filtering the mixture and concentrating the filtrate under nitrogen.[9] The anhydride is reacted with 1-palmitoyllysophosphatidylcholine (50 mg) and 4-pyrrolidinopyridine (15 mg) in 1 ml dry (P_2O_5-redistilled) chloroform for 3 hr at 37°.[10] The reaction products are partitioned between chloroform (50 ml) and 1:1 methanol/aqueous HCOONa, pH 3.0, at 0°, and the recovered chloroform layer is washed twice with the same methanol/aqueous mixture and once with 1:1 methanol/1% aqueous NaCl, then concentrated *in vacuo*. The residue is applied in chloroform to a 2.2 × 7 cm column of BioSil A (packed in chloroform), which is eluted successively with 20% methanol in chloroform (150 ml, discarded), 35% methanol in chloroform (100 ml, discarded), 50% methanol in chloroform (50 ml), and 60% methanol in chloroform (3 × 50 ml). The 50 and 60% methanol fractions are examined by thin-layer chromatography (on silica gel G, developing with 50:20:10:10:5 chloroform/acetone/methanol/acetic acid/water), and the fractions containing pure phosphatidylcholine (PC) are pooled.

The purified 1-palmitoyl-2-(12′-*N-tert*-Boc-methylamino)octadecanoylphosphatidylcholine is dissolved in 25 ml dry $CHCl_3$ (redistilled from P_2O_5) in a vessel protected from moisture by a $CaCl_2$ guard tube. The solution is bubbled at 0° with dry N_2 for 15 min followed by dry HCl for 30 min. The sealed vessel is then stirred for 2 hr at 0°. After bubbling again with dry N_2 for 15 min the mixture is combined with 1% aqueous $NaHCO_3$ and methanol (25 ml each) and shaken well. The chloroform layer is recovered, washed once with 1:1 methanol/100 mM aqueous ethylenediaminetetraacetic acid (EDTA) (50 ml), then concentrated *in vacuo*. Deprotection is quantitative.

---

[9] Z. Selinger and Y. Lapidot, *J. Lipid Res.* **7**, 174 (1966).
[10] J. T. Mason, A. V. Broccoli, and C.-H. Huang, *Anal. Biochem.* **113**, 96 (1981).

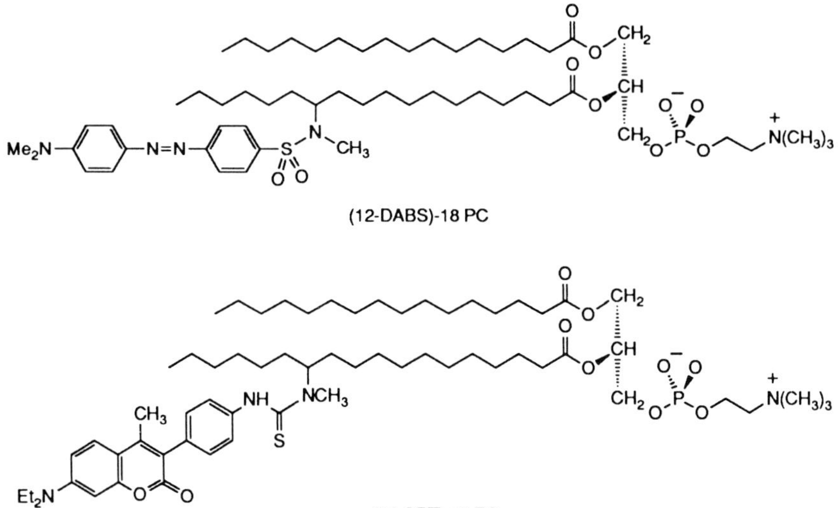

FIG. 1. Structures of (12-CPT)-18 PC and (12-DABS)-18 PC.

Labeling of Phosphatidylcholines

The deprotected phosphatidycholine is labeled overnight at 25° in the dark with a 50% molar excess of either 4-(dimethylamino)azobenzene-4'-sulfonyl (DABSyl) chloride or 7-diethylamino-3-(4'-isothiocyanato-phenyl)-4-methylcoumarin in 2:1:0.5 methanol/chloroform/100 mM aqueous EDTA, pH 8.0. The labeled phosphatidylcholines are purified by preparative thin-layer chromatography (on silica gel G, in 65:30:2.5:2.5 chloroform/methanol/concentrated NH_4OH/water) or, for larger quantities (> 20 mg), by silicic acid column chromatography as described above for the protected phosphatidylcholine.

Basic Properties of Probes

The structures of the fluorescent phosphatidylcholine (12-CPT)-18 PC and the quencher (12-DABS)-18 PC are shown in Fig. 1. The fluorescence properties of (12-CPT)-18 PC and its related analog (12-CPS)-18 PC in various lipid systems are described in detail elsewhere.[11] Most importantly, (12-CPT)-18 PC exhibits very strong fluorescence [roughly five times more intense than that of N-(7-nitrobenz-2-oxa-1,3-diazol-4-yl)phosphatidyleth-anolamine (NBD-PE) at optimal instrumental settings for both probes], the intensity of which is virtually constant in a number of different lipid and amphiphile environments.[11] These labeled phosphatidycholines are

---

[11] J. R. Silvius, R. Leventis, P. M. Brown, and M. J. Zuckerman, *Biochemistry* **26**, 8435 (1986).

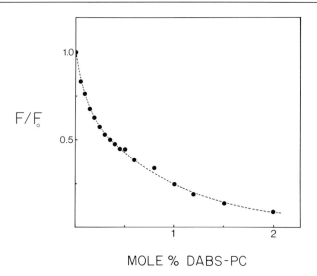

MOLE % DABS-PC

FIG. 2. Efficiency of quenching of fluorescence of (12-CPT)-18 PC by (12-DABS)-18 PC in POPC vesicles. Lipid samples containing 1 mol % (12-CPT)-18 PC and the indicated molar percentages of (12-DABS)-18 PC were thoroughly dried down from chloroform, dispersed by bath sonication in 150 mM KC1, 10 mM 2-[tris(hydroxymethyl)methylamino]-1-ethanesulfonic acid (TES), pH 7.4, and freeze–thawed three times before fluorescence reading.

substrates for cabbage and *Streptomyces* phospholipase D and *Bacillus cereus* phospholipase C, so that a variety of other fluorescent (phospho)glycerolipids can be prepared from them (Silvius *et al.*,[11] and J. Silvius, unpublished results, 1990).

The efficiency of quenching of fluorescence of (12-CPT)-18 PC by various concentrations of (12-DABS)-18 PC in egg phosphatidylcholine vesicles is shown in Fig. 2. A bilayer concentration of 0.36 mol % (12-DABS)-18 PC gives half-maximal quenching of the coumarinyl PC fluorescence in these vesicles. As shown in Table I, various other energy-transfer acceptors also quench the fluorescence of (12-CPT)-18 PC in PC vesicles (although generally with somewhat lower efficiencies). Similar quenching efficiencies are observed using vesicles with other lipid compositions, such as bovine brain phosphatidylserine or 1:3 phosphatidylserine/egg phosphatidylethanolamine.

Lipid Mixing Assay

A "standard" assay of lipid mixing using (12-CPT)-18 PC and (12-DABS)-18 PC is carried out as follows.

TABLE I
EFFICIENCY OF QUENCHING OF (12-CPT)-18 PC
FLUORESCENCE BY DIFFERENT ENERGY-
TRANSFER ACCEPTORS IN
PHOSPHATIDYLCHOLINE BILAYERS

Acceptor probe	Amount of acceptor (mol %) giving 50% quenching
(12-DABS)-18 PC	0.36
1-Palmitoyl-2-(16'-trinitro-phenylaminopalmitoyl)-PC	0.68
N-NBD-PE	0.42
N-Fluoresceinyl PE	0.80

Sample Preparation

A population of labeled lipid or lipid/protein vesicles is prepared containing 0.1–1.0 mol % (12-CPT)-18 PC together with 0.35 mol % (12-DABS)-18 PC (giving ~50% quenching of the donor fluorescence). Labeled vesicles may be prepared by any standard method (e.g., reversed-phase evaporation,[12] high-pressure extrusion,[13] detergent dialysis,[14] or sonication of lipid mixtures dried down from solvent). The labeled lipids should initially be mixed with the other lipids used in the vesicle preparation (e.g., by drying or colyophilization from a solvent) to ensure a homogeneous distribution of probes in the final vesicle preparation. A parallel sample of vesicles is also prepared to calibrate the fluorescence signal expected for complete lipid mixing (see below).

Sample Incubation

A portion of labeled vesicles, typically comprising 2–20 nmol of total lipid, is mixed with a 9-fold excess of unlabeled vesicles. The sample fluorescence can be monitored continuously during the vesicle incubation (see below), or the mixed samples may be preincubated outside the fluorimeter, preferably with exclusion of light and/or under N_2, before measuring their fluorescence.

[12] F. Szoka and D. Papahadjopoulos, *Proc. Natl. Acad. Sci. U.S.A.* **75,** 4194 (1978).
[13] L. D. Mayer, M. J. Hope, and P. R. Cullis, *Biochim. Biophys. Acta* **858,** 161 (1986).
[14] L. T. Mimms, G. Zampighi, Y. Nozaki, C. Tanford, and J. A. Reynolds, *Biochemistry* **20,** 833 (1981).

Fluorescence Measurements

The sample fluorescence is measured, using excitation and emission wavelengths of 394 and 472 nm, respectively (slitwidths 15/20 nm). The following quantities are determined, as illustrated in Fig. 3A: F_0(sample), the fluorescence recorded before lipid mixing has taken place; F_t(sample), the fluorescence measured at some fixed time (or continuously) after initiation of lipid mixing; and F_{Tx}(sample) the fluorescence measured after

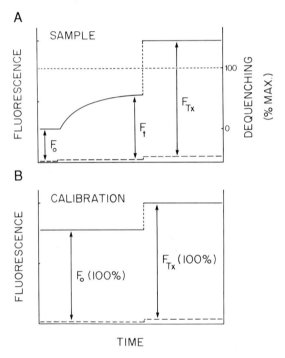

FIG. 3. Representative fluorescence time courses obtained in a typical lipid mixing experiment, indicating the fluorescence values that are measured to quantitate and to calibrate the extent of dequenching of (12-CPT)-18 PC through intermembrane lipid mixing. (A) A sample of vesicles labeled with 1 mol % (12-CPT)-18 PC and 0.35 mol % (12-DABS)-18 PC is incubated with unlabeled vesicles (representing a nine times greater amount of lipid), giving gradual fluorescence dequenching through lipid mixing. Triton X-100 is added subsequently (to 1%, w/v) to determine accurately the amount of donor probe in the sample. (B) To calibrate the fluorescence signal expected on complete intermixing of lipids in the above experiment, the fluorescence of a sample of vesicles labeled with 0.1 mol % CPT-PC and 0.035 mol % (12-DABS)-18 PC is measured before and after the addition of Triton X-100. For simplicity, the total amount of donor probe in the sample is shown as equal in the two experiments. The dashed lines represent the fluorescence of vesicles containing CPT-PC and DABS-PC in the absence of unlabeled vesicles.

the addition of Triton X-100 (to calibrate the amount of probe present). In all cases, fluorescence readings are corrected for the background fluorescence measured for otherwise identical samples omitting the labeled vesicles.

Calibration of Lipid Mixing Signal

In studies of lipid mixing between lipid vesicles or reconstituted lipid/protein vesicles, determination of the fluorescence signal corresponding to complete lipid mixing is straightforward. A control sample of vesicles is prepared with a lipid composition and acceptor probe density that match those expected on complete mixing of the lipids between the labeled and unlabeled vesicle populations in the above experiment.[14a] The fluorescence of this reference sample is recorded before and after the addition of Triton X-100, giving values designated as F (100%) and $F_{Tx}(100\%)$ (see Fig. 3B). The maximum fluorescence expected for any given sample in the above experiment is then

$$F_{max}(\text{sample}) = F(100\%) \, [F_{Tx}(\text{sample})/F_{Tx}(100\%)]$$

Fluorescence readings from lipid mixing assays are then presented on a scale with $F_0(\text{sample})$ set as 0% and F_{max} (sample) as 100%. In many cases, no further calibration of the data is done. However, with certain assumptions it is possible to calibrate the fluorescence signal to account for the variable extents of dequenching expected through successive "rounds" of lipid mixing events between vesicles.[4,15] At a minimum, it is often useful (and straightforward) to calibrate initial rates of intervesicle lipid mixing by scaling to the maximum dequenching expected through a single round of intervesicle fusion, namely, that expected if each vesicle in a sample intermixes lipids completely with one other vesicle.

Studies of lipid mixing between lipid (or lipid/protein) vesicles and natural membranes can be complicated by difficulties in estimating the fluorescence expected in the limit of complete lipid mixing. This difficulty can be readily overcome in cases where the dilution of lipid probes from a lipid vesicle into a (typically larger) membrane vesicle essentially eliminates energy transfer between the donor and the acceptor species. The most direct approach to calibrate the 100% lipid mixing signal in this case is to introduce a low level of the donor probe into a sample of the membrane

[14a] N. Düzgüneş, R. M. Straubinger, P. A. Baldwin, D. S. Friend, and D. Papahadjopoulos, *Biochemistry* **24**, 3091 (1985).
[15] J. Bentz and H. Ellens, *Colloids Surf.* **30**, 65 (1989).

vesicles (e.g., by the use of phospholipid exchange proteins, or by incorporating an exchangeable analog of the donor probe with shortened acyl chains[11,16]). The fluorescence of the labeled vesicles is then measured in the presence and absence of 1% Triton X-100 (the latter to calibrate the amount of probe present).

When necessary, an alternative method may be used to estimate the extent of lipid mixing between labeled lipid (or lipid/protein) vesicles and unlabeled membrane vesicles, assuming that donor–acceptor energy transfer is abolished on transfer of the probes into the membrane vesicles.[17] In this approach, parallel samples of lipid vesicles, one labeled with the donor only and the other labeled with both the donor and the acceptor species, are incubated under identical conditions with replicate samples of unlabeled membrane vesicles. Let F_0(sample), F_t(sample), and F_{Tx}(sample) be as defined above, and let F_0(C), F_t(C), and F_{Tx}(C) represent the analogous fluorescence readings obtained when the labeled vesicle population contains only the donor probe. The extent of lipid mixing between the labeled and the unlabeled vesicles (which is assumed to be the same in the two parallel incubations) can then be calculated by solving the system of equations

$$F_t(C)/F_{Tx}(C) = f_{mix}F_{memb} + (1 - f_{mix}) [F_0(C)/F_{Tx}(C)]$$
$$F_t(sample)/F_{Tx}(sample) = f_{mix}F_{memb} + (1 - f_{mix}) [F_0(sample)/F_{Tx}(sample]$$

where f_{mix} represents the fraction of lipid mixing between the labeled and unlabeled membranes and F_{memb} represents the fluorescence (normalized to the Triton value) for the donor species when it becomes incorporated into the membrane vesicles.

Potential Artifacts and Remedies

Certain artifacts can complicate these and other lipid mixing assays when interactions between vesicles lead not only to lipid mixing but also to extensive structural rearrangements within the system. Extensive vesicle aggregation that greatly increases light scattering may cause a spurious enhancement of fluorescence that mimics dequenching due to lipid mixing. Such aggregation may be followed by gradual vesicle precipitation which causes a secondary decline in the sample fluorescence. Such artifacts are readily detected by incubating a sample of vesicles, labeled with the

[16] M. Gardam, J. Itovich, and J. R. Silvius, *Biochemistry* **28**, 884 (1989).
[17] L. Stamatatos, R. Leventis, M. J. Zuckermann, and J. R. Silvius, *Biochemistry* **27**, 3917 (1988).

donor species, but without the quencher, together with unlabeled vesicles under the standard assay conditions.

A second class of artifacts can arise when the conditions that promote intermixing of lipids between vesicles also lead to large-scale lipid lateral phase separations. Systems exhibiting such behavior include the widely studied phosphatidylserine/calcium system.[4,7,11,18] Two remedies are possible for artifacts arising from this source. First, the (12-DABS)-PC acceptor probe can be replaced by another DABS-labeled phospholipid which co-distributes more uniformly with the majority phospholipid species [e.g., (12-DABS)-18 phosphatidic acid (PA) in the above case[11]]. Second, because the rate of lipid mixing in such strongly interacting systems is often aggregation-limited,[19,20] lipid mixing rates (i.e., aggregation rates) in these systems can often be reliably monitored using a lipid mixing/quenching assay,[7,21] in which the donor and acceptor probes are initially incorporated into separate vesicle populations.

Variations

While the above protocols allow a variety of modifications for particular experimental systems and designs, two types of variations are especially noteworthy. First, a given lipid mixing process can often give very different time courses of fluorescence depending on the initial level of the acceptor probe in the labeled vesicle population, and depending on the ratio of labeled to unlabeled vesicles in the population. Adjustment of these variables may permit one to weight the fluorescence signal differently to examine different aspects of the overall process of lipid mixing between vesicles. For example, for the many lipid model systems that exhibit successive rounds of intervesicle fusion,[4,15] an experiment using vesicles initially labeled with a high mole fraction of the acceptor probe will give greater weight to later, higher-order fusion events than to initial fusion events.[14a] In contrast, in the standard protocol given, the later, higher-order fusion events contribute less to the overall fluorescence time course than does the first round of intervesicle fusion.

Another noteworthy variation of the above protocols involves the use of energy-transfer acceptors other than (12-DABS)-18 PC. The alternative quencher 1-palmitoyl-2-(16-trinitrophenyl)palmitoyl-PC is a less efficient

[18] D. Papahadjopoulos, W. J. Vail, K. Jacobson, and G. Poste, *Biochim. Biophys. Acta* **394**, 483 (1975).

[19] D. Hoekstra, *Biochim. Biophys. Acta* **692**, 171 (1982).

[20] R. Leventis, J. Gagné, N. Fuller, R. P. Rand, and J. R. Silvius, *Biochemistry* **25**, 6978 (1986).

[21] D. Hoekstra, *Biochemistry* **21**, 2833 (1982).

energy-transfer acceptor than is (12-DABS)-18 PC but has proved useful to minimize fluorescence artifacts in some systems that exhibit massive vesicle–vesicle aggregation.[11] Alternatively, the commercially available acceptor fluoresceinylphosphatidylethanolamine can be used to monitor energy transfer through acceptor emission rather than donor quenching. However, the fluorescence of this head group-labeled probe is highly sensitive to surface characteristics such as interfacial pH. Other potential acceptors (e.g., NBD-labeled phospholipids) typically do not provide equally sensitive emission when paired as acceptors with the highly fluorescent coumarinyl probes.

[4] Fluorescence Lifetime Measurements to Monitor Membrane Lipid Mixing

By STEPHEN W. BURGESS and BARRY R. LENTZ

Introduction

Several fluorescent probes have been developed to monitor the intermembrane mixing of lipids during membrane fusion. These assays are listed elsewhere by Düzgüneş and Bentz.[1] Most of the assays described are based on Förster energy transfer between two fluorophores and therefore rely on monitoring an extensive probe property, namely, the fluorescence intensity. The only lipid mixing assay utilizing an intensive probe property is the DPHpPC fluorescence lifetime assay introduced by Parente and Lentz.[2,3] The probe was shown to conform to all criteria for usefulness as a membrane probe for lipid exchange.[3]

In the following sections, we describe how fluorescence lifetime measurements have been used to monitor lipid exchange in model membrane systems induced to fuse by polyethylene glycol (PEG). We describe also the features of the probe which make it useful not only for this fusion assay, but also as a probe for fusion-independent lipid exchange, phase separation, and formation of membrane domains. Finally, we describe aspects of the technique that limit its effectiveness as a tool for studying membrane fusion or the other membrane processes mentioned.

[1] N. Düzgüneş and J. Bentz, in "Spectroscopic Membrane Probes" (L. M. Loew, ed.), p. 117. CRC Press, Boca Raton, Florida, 1988.
[2] R. A. Parente and B. R. Lentz, Biochemistry 24, 6178 (1985).
[3] R. A. Parente and B. R. Lentz, Biochemistry 25, 1021 (1986).

Sample Preparation

Phospholipids, including the lipid probe 1-palmitoyl-2-[(2-[4-(6-phenyl-*trans*-1,3,5-hexatrienyl)phenyl]ethyl}oxy)carbonyl]-3-*sn*-phosphatidylcholine (DPHpPC), are obtained from Avanti Polar Lipids (Birmingham, AL)[4] and are stocked in nitrogen-saturated chloroform and stored under argon in light-protected vials at −20°. Phospholipids are aliquotted into glass sample vials and the chloroform evaporated using argon. Lipids are redissolved in either benzene or cyclohexane, frozen, and lyophilized overnight. During all phases of sample manipulation, the probe is protected from light by using amber glassware, painted glassware, or by wrapping glassware in aluminum foil.

Probe vesicles containing 4, 1.33, and 0.4 mol % (25:1, 75:1, 250:1 lipid/probe) of DPHpPC are prepared. In some cases, different probe concentrations (10, 5, and 1 mol %) are used in order to obtain greater sensitivity, as explained below. In addition, a vesicle population without probe (blank) is prepared. Dried samples are hydrated, for approximately 30 min at a temperature above the phase transition of the lipid system under study, with 2 mM N-[tris(hydroxymethyl)methyl]-2-aminoethanesulfonic acid (TES) buffer prepared from doubly distilled water and containing 100 mM NaCl and 1 mM ethylenediaminetetracetic acid (EDTA), pH 7.4. Large, unilamellar vesicles (LUVET) are prepared using the extrusion technique described elsewhere.[5,6] The hydrated lipid is forced repeatedly (7 times) through a 0.1 μm polycarbonate filter (Nuclepore Corp., Pleasanton, CA) above the phase transition under a pressure of approximately 200 psi of argon. This technique for vesicle preparation is preferred to older techniques (reversed-phase evaporation, octylglucoside dialysis, ethanol injection, etc.) because there are no residual organic solvents or detergents to perturb the bilayer structure. In addition, these vesicles are stable for at least 1 week if stored at 4°. The concentrations of all vesicle samples are determined by phosphate analysis using a modification of the procedure of Chen *et al.*[7]

---

[4] We should note that the DPHpPC probe is available from other suppliers. Although we have never used DPHpPC obtained from other suppliers, carboxyethyl-DPH fatty acid so obtained and converted to DPHpPC degraded within 1 month (as evidenced by a decrease in the extinction coefficient from 80,000 to 10,000). Probe obtained from Avanti Polar Lipids (Birmingham, AL) incorporates carboxyethyl-DPH obtained from Emrys Thomas (Salford, UK) and was stable for up to 4 years.

[5] M. J. Hope, M. B. Bally, G. Webb, and P. R. Cullis, *Biochim. Biophys. Acta* **812,** 55 (1985).

[6] L. D. Mayer, M. J. Hope, and P. R. Cullis, *Biochim. Biophys. Acta* **858,** 161 (1986).

[7] P. S. Chen, Jr., T. Y. Toribara, and H. Warner, *Anal. Chem.* **28,** 1756 (1956).

Fluorescence

All fluorescence measurements are made on an SLM 48000 spectro-fluorometer (Urbana, IL) equipped with a modified, three-position, multi-temperature cuvette holder[8] and 200-W mercury–xenon lamp mounted horizontally in a Photon Technology International (Princeton, NJ) lamp housing. The 366-nm mercury line is used to excite DPHpPC for lifetime measurements, while emission is monitored through a 3-mm high-pass KV-450 filter (50% transmittance at 450 nm; Schott Optical Glass, Duryea, PA). Vertically polarized and modulated light from the Pockel cell is rotated to 35° from vertical by placing a Soliel-Babinet compensator (Karl Lambrecht, Chicago, IL) rotated 17.5° from vertical and set for half-wave at 366 nm in the excitation path. This allows excitation with linearly polarized light at the magic angle[9] so as to avoid errors in lifetime calculations without the loss of intensity inherent in the use of polarizers. Exciting light is modulated at 30 MHz for all lifetime measurements.

Calibration Curves

To monitor mixing of membrane phospholipid compartments by using the DPHpPC fluorescent lifetime lipid mixing assay,[10] calibration curves must be generated by measuring the lifetime of the probe at different lipid/probe ratios. Because the lifetime of a fluorophore is dependent on the environment in which it resides,[11] separate calibration curves must be generated for each lipid system studied (see Fig. 1). In addition, any alterations in the system which affect the membrane structure (e.g., addition of high concentrations of PEG) may affect the fluorescence lifetime of the probe. Under these circumstances, separate calibration curves for each experimental condition must be generated. For example, Fig. 2 shows the effect of increasing PEG concentration on the fluorescence lifetime of DPHpPC in dipalmitoylphosphatidylcholine (DPPC) vesicles. The presence of the dehydrating polymer has a pronounced effect on the probe lifetime as compared to the lifetime observed for the probe in DPPC vesicles in an aqueous environment (dashed line, Fig. 2).

We have found, using extensive calibration data sets,[2] that the functional form of the calibration curve is well represented by Eq. (1):

$$\tau(R) = \tau_\infty - (\tau_\infty - \tau_o) \exp[-k(R - R_i)] \tag{1}$$

where τ_∞ is the lifetime of the highest lipid/probe ratio measured, τ_o is the

[8] D. A. Barrow and B. R. Lentz, *Biophys. J.* **48** 221 (1985).
[9] M. Badea and L. Brand, This series, Vol. 61, p. 378.
[10] R. A. Parente and B. R. Lentz, *Biochemistry* **25**, 6678 (1986).
[11] J. R. Lakowicz, "Principles of Fluorescence Spectroscopy." Plenum, New York, 1983.

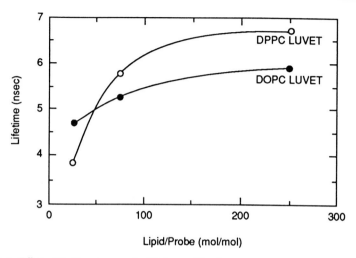

FIG. 1. Effect of lipid system on the lifetime calibration curve. Lifetime calibration curves for the DPHpPC lipid mixing assay are shown for dipalmitoylphosphatidylcholine (DPPC) LUVET at 48° and dioleoylphosphatidylcholine (DOPC) LUVET at 23° in 10% (w/w) PEG.

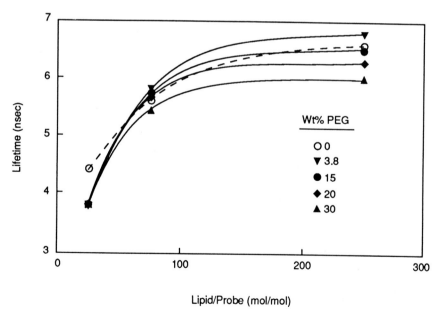

FIG. 2. Effect of PEG on the lifetime calibration curve. Lifetime curves for DPPC LUVET at 48° were generated for PEG concentrations ranging from 0 to 30% (w/w).

lifetime of the lowest lipid/probe ratio measured, R is the lipid/probe ratio of the sample, R_i is the initial or lowest lipid/probe ratio used, and k is an exponential constant. All values except k are known. A Simplex curve-fitting algorithm is used to optimize k for each calibration curve.[12] Once k is determined, the lifetime at each lipid/probe ratio can be calculated to give a complete calibration curve. Because this functional form is known, the calibration curve can be established to reasonable accuracy by determination of the probe lifetime at only three lipid/probe ratios (see Fig. 2).

Lipid Mixing Assay

Once the calibration curves have been established for a particular lipid system, a lipid mixing experiment can be performed by injecting probe vesicles and blank (unlabeled) vesicles (in a ratio of 1:10) into a cuvette containing a fusogen, in our case, a solution of PEG. Following mixing, the phase angle of the sample can be continuously monitored on the SLM 48000 or equivalent phase fluorometer. The phase angle of an isochromal reference fluorophore[13] at 23° (DPH in heptane; τ 6.75 nsec, 2×10^{-7} M) can be measured before and after the scan to determine the average phase angle of the reference (no change in the reference phase angle is observed on the SLM 48000 over a time period of 30 min). The average reference phase angle is used to convert sample phase angles to lifetimes using the method of Spencer and Weber.[14] The average lifetime after leveling of the sample phase angle can be compared to the appropriate calibration curve to obtain a lipid/probe ratio (R) for the sample, which is used to calculate the extent of lipid mixing according to Eq. (2):

$$\% \text{ Lipid mixing} = \frac{T - R_i}{R_f - R_i} \times 100 \tag{2}$$

where R_i is the initial lipid/probe ratio of the probe vesicles used and R_f is the final lipid/probe ratio expected if complete lipid exchange occurs (this depends on the probe vesicle/blank vesicle ratio during the experiment). By this procedure the exchange of probe to unlabeled vesicles under fusogenic and nonfusogenic conditions can be monitored.[15] This procedure assumes that the distribution of probe among the vesicles is random (i.e., probe enters all vesicles with equal probability, and probe-rich and probe-free populations cease to exit), a condition that can obtain only near the

[12] M. S. Caceci and W. P. Cacheris, Byte (May Issue), (1984).
[13] D. A. Barrow and B. R. Lentz, J. Biophys. Methods 7, 217 (1983).
[14] R. D. Spencer and G. Weber, Ann. N.Y. Acad. Sci. 158 361 (1969).
[15] S. W. Burgess, D. Massenburg, J. Yates, and B. R. Lentz, Biochemistry 30, 4193 (1990).

completion of the lipid mixing process. In the initial stages of the lipid mixing process, a more realistic treatment assumes the existence of two compartments (probe-free and probe-rich) and estimates the amounts of each on the basis of lifetime heterogeneity analysis.[15a]

Advantages

Fluorescence lifetime measurements as an assay for lipid exchange between membranes offer several advantages over fluorescence intensity measurements. Fluorescence lifetime is an intrinsic property of the probe and is not dependent on the concentration of membranes in the exciting beam. Instead, the lifetime of DPHpPC is dependent on the surface concentration of the probe in the membrane.[16] However, at a given lipid/probe ratio, the lifetime is invariant with the total lipid concentration. In contrast, fluorescence intensity is an extrinsic property of the probe and is entirely dependent on the concentration. Monitoring the lifetime of a probe makes it possible to compare experiments without normalizing the results with an intensity standard, since the lifetime, or changes in lifetime on lipid exchange, do not depend on day-to-day instrument setup. In addition, experiments in which aliquots of lipid are removed at timed intervals to determine the amount of lipid exchange are possible, since the concentration of lipid does not influence the lifetime of the probe removed from the stock solution.

Although the surface concentration-dependent lifetime behavior of DPHpPC is an obvious advantage, other features make it an attractive membrane probe.[2,17] The fluorescent moiety diphenylhexatriene (DPH) is esterified to the glycerol backbone and is located in the hydrophobic core of the membrane. This makes the probe inaccessible to hydrophilic perturbants in the aqueous medium. In addition, DPH is a linear molecule similar in its geometry to that of fatty acids in phospholpids, resulting in little perturbation of the bilayer. The probe also has a high fluorescence yield (extinction coefficient of 60,000 OD units M^{-1} cm^{-1} at 355 nm in $CHCL_3$),[2] which allows it to be studied at low membrane concentrations. Finally, it has been shown that DPHpPC transfers spontaneously between vesicles only very slowly.[18]

Not only do fluorescence lifetime determinations lend themselves to lipid exchange measurements, but with the increasing availability of multi-

[15a] J. R. Wu and B. R. Lentz, *Biochemistry* **30,** 6780 (1991).
[16] B. R. Lentz and S. W. Burgess, *Biophys. J.* **56,** 723 (1989).
[17] B. R. Lentz, *Chem. Phys. Lipids* **50,** 171 (1989).
[18] J. B. Morgan, E. W. Thomas, T. S. Moras, and Y. P. Yianni, *Biochim. Biophys. Acta* **692,** 196 (1982).

frequency phase and modulation spectrofluorometers, the ability to resolve multiple lifetime species provides the possibility to detect the presence of different domains within a membrane. When DPHpPC has lifetimes in different local environments differing by at least a factor of 2–3, it should be possible, with modern instrumentation, to resolve not only the lifetimes associated with these two environments but also the fractional intensities associated with these two contributions.[19,20] These domains might represent discrete domains within a single membrane (i.e., phase separation) or the existence of separate populations of vesicles. We are currently using this approach in attempts to determine initial rates of lipid exchange by monitoring changes in discrete vesicle populations on a millisecond time scale using the SLM 48000-MHF multi-harmonic Fourier transform spectrofluorometer.[21] Parente and Lentz[3] have already shown that DPHpPC can be used to detect Ca^{2+}-induced phase separation of phosphatidylserine vesicles.

Disadvantages

The main disadvantage of the DPHpPC lipid exchange assay derives from the fact that the probe lifetime is an insensitive and nonlinear function of lipid/probe ratio at high lipid/probe ratios. This causes a large increase in the error associated with the calculated extent of lipid mixing as 100% lipid mixing is approached (see Fig. 3). As the average error of the lifetime measurement is approximately 0.05 nsec (certainly <0.1 nsec), whereas the lifetime change from 50 to 100% exchange may be as small as 0.2 ns in some cases, the error can be quite large. This situation is exacerbated by environmental conditions that affect the slope of the lifetime change as a function of the lipid/probe ratio. In general, as the slope of the lifetime calibration curve approaches zero, the error associated with the calculated extent of lipid mixing goes up. It is possible to compensate for this problem by working in the nearly linear initial region of the standard curve. This can be done in a number of ways, either by increasing the initial probe concentration in the probe vesicle population (lowering the starting lifetime of the probe) or decreasing the ratio of blank vesicles to probe vesicles. It is for this reason that some experiments are performed with probe vesicles containing a lipid/probe ratio of 10:1 while others can be performed with vesicles with a 25:1 ratio.

---

[19] T. Parasassi, F. Conti, M. Glaser, and E. Gratton, *J. Biol. Chem.* **259**, 14011 (1984).
[20] D. A. Barrow and B. R. Lentz, *Biophys. J.* **48**, 221 (1985).
[21] S. W. Burgess, J. R. Wu, K. Swift, and B. R. Lentz, *J. Fluorescence* **1**, 105 (1991).

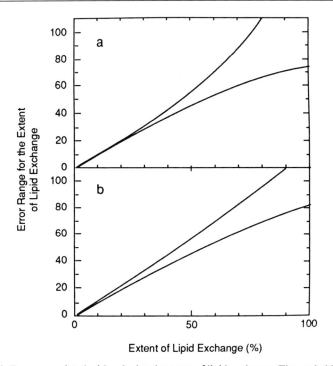

FIG. 3. Errors associated with calculated extents of lipid exchange. The probable range of extent of lipid exchange, based on an average error of 0.05 nsec for lifetime determinations, is bordered by solid lines for (a) DPPC LUVET at 48° and (b) DOPC LUVET at 23°, both in the presence of 10% (w/w) PEG. Extents between boundary lines are indistinguishable based on lifetime measurements.

Another disadvantage of the DPHpPC lipid mixing assay is the requirement that a calibration curve be generated for each new temperature, lipid system, and reactant concentration. On the other hand, the impetus to develop this assay was the observation that PEG quenched the fluorescence of N-7-nitro-2,1,3-benzoxadiazolyl-4-phosphatidylethanolamine (NBD-PE) used in the NBD-PE/N-(lissamine rhodamine B sulfonyl)-PE assay for lipid mixing.[22] The fluorescence quenching was presumably due to the interaction of PEG in the aqueous phase with the probe located at the water–membrane interface. Correction for this quenching would have required that separate calibration curves be prepared for each lipid mixing experiment performed, making data analysis extremely cumbersome if not impossible. The fluorescent moiety of DPHpPC is located in the hydrocarbon core of the bilayer. This protects it from direct quenching by agents in

[22] D. K. Struck, D. Hoekstra, and R. E. Pagano, Biochemistry 20, 4093 (1981).

the aqueous medium. However, the indirect influence of the aqueous environment must still be considered when interpreting the lifetime data. Nonetheless, the fact that fluorescence lifetime is an intrinsic probe property means that a calibration curve need be constructed only once for any number of experiments performed under the same conditions.

Conclusion

The DPHpPc lifetime lipid mixing assay is a reliable indicator of lipid mixing between membranes. The applicability of the technique to many different systems and the variety of detailed information that can be obtained, especially when used with the new generation of harmonic Fourier transform fluorometers, should increase the usefulness of this technique in the years to come.

Acknowledgments

Supported by U.S. Public Health Service Grants GM32707 and HL45916 to B.R.L. and by the University of North Carolina SCOR in Thrombosis and Hemostasis (USPHS HL26309).

[5] Reconstituting Channels into Planar Membranes: A Conceptual Framework and Methods for Fusing Vesicles to Planar Bilayer Phospholipid Membranes

By Fredric S. Cohen and Walter D. Niles

Introduction

In 1962 when Mueller et al.[1] developed a procedure to form planar phospholipid bilayer membranes, it was immediately apparent that if ion channels could be incorporated into such membranes, the properties of the protein channels could be studied under controlled conditions. Planar bilayers can be easily voltage-clamped, the lipid composition selected by the investigator, and both bathing aqueous phases manipulated at will. Therefore, the incorporation of membrane proteins into the bilayer, termed channel "reconstitution," would allow detailed electrophysiological studies of channels. Reports of reconstitution of channels (and even

[1] P. Mueller, D. O. Rudin, H. T. Tien, and W. C. Wescott, *Circulation* **26**, 1167 (1962).

METHODS IN ENZYMOLOGY, VOL. 220

pumps) into planar bilayers appeared sporadically. It was not, however, until methods were developed to fuse vesicles which contain channels to planar membranes that a routine and reliable method for channel reconstitution became available.

The first successful reconstitution via fusion was reported by Miller and Racker.[2] Vesicles derived from fragmented skeletal sarcoplasmic reticulum (SR) membrane were fused to planar bilayers with the hope of reconstituting a Ca^{2+}-release channel. They found that if acidic lipids were included in the bilayer and small amounts of Ca^{2+} were added to the vesicle-containing (cis) solution, electrical increases occurred (later identified as due to incorporation of K^+ and Cl^- channels). Osmotically loading the vesicles with a membrane-impermeable solute (sucrose) was absolutely essential in order to obtain discrete conductance increases. Although not fully appreciated at the time, it is now clear that fusion was induced by the development of an intravesicular pressure that resulted from the rapid dissipation of osmotic gradients between the interiors of the vesicles and the bathing cis medium.

Zimmerberg et al.[3] and Cohen et al.[4] showed that fusion of phospholipid vesicles to planar bilayer membranes occurred if an osmotic gradient was imposed across the planar membrane, the vesicle-containing side (cis) being hyperosmotic with respect to the trans side of the planar membrane. Divalent cations greatly augmented the process. They also presented evidence that an osmotic gradient across the planar membrane induces vesicular swelling.[5-7]

Establishing an osmotic gradient across the planar membrane yields greater fusion than the method of Miller and Racker[2] and does not require that the vesicles be loaded with an agent such as sucrose. Because of its simplicity and efficacy, this method became the most general and reliable procedure for reconstituting channels into planar membranes.[3,4,8-10] Often the asymmetric cis and trans solutions used are given without noting the existence of an osmotic gradient, however, and the conceptual basis for establishing the gradient is not explained. In this chapter, therefore, we refer to this technique as the Standard 1980 Method for inserting channels into bilayers.

[2] C. Miller and E. Racker, J. Membr. Biol. 26, 319 (1976).

[3] J. Zimmerberg, F. S. Cohen, and A. Finkelstein, J. Gen. Physiol. 75, 241 (1980).

[4] F. S. Cohen, J. Zimmerberg, and A. Finkelstein, J. Gen. Physiol. 75, 251 (1980).

[5] F. S. Cohen, M. H. Akabas, and A. Finkelstein, Science 217, 458 (1982).

[6] M. H. Akabas, F. S. Cohen, and A. Finkelstein, J. Cell Biol. 98, 1063 (1984).

[7] W. D. Niles and F. S. Cohen, J. Gen. Physiol. 90, 703 (1987).

[8] P. Labarca, R. Coronado, and C. Miller, J. Gen. Physiol. 76, 397 (1980).

[9] W. Hanke, H. Eibl, and G. Boheim, Biophys. Struct. Mech. 7, 131 (1981).

[10] R. Latorre, C. Vergara, and C. Hidalgo, Proc. Natl. Acad. Sci. U.S.A. 79, 805 (1982).

The majority of studies in this area have used vesicles derived from biological membranes, rather than using phospholipid vesicles with reconstituted channels. One reason for this was a paucity of functional purified channels reconstituted into vesicles. Biologically derived vesicles are also easier to fuse to planar bilayers than are phospholipid vesicles. Because naturally derived vesicles are permeable to many ions, they swell more readily than the relatively impermeable phospholipid vesicles when swelling is induced by the Standard Method.[7,11,12]

Protocols that promote greater swelling (and thus greater intravesicular pressures) than occurs with the Standard Method would greatly aid in the fusion of reconstituted vesicles to planar membranes. Procedures that would induce swelling via water flow across the entire membrane of the vesicle (as opposed to flow only across the region of contact between the vesicular and planar membranes as in the Standard Method) would accomplish this. The first procedures using such water flows showed the viability of the method, but they were not adequate to ensure success with any vesicle preparation[2] or were cumbersome and not practical for daily use.[5,6] The more recently developed methods[7,13] are convenient and reliable. With the increase in the number of purified channels that have been reconstituted into vesicles, these improved swelling protocols could find increased use in the 1990s. Accordingly, we refer to the more recent procedure as the 1990 Method.

Bound State and Induction of Fusion

When acidic lipids are present in membranes, divalent cations promote the tight and virtually irreversible binding of vesicles to planar membranes, with little if any aqueous space between the membranes.[6,7] The divalent cations, in and of themselves, do not produce fusion. However, fusion can be driven by intravesicular pressure. When an intravesicular pressure is generated by swelling, tension is created on a bound vesicle, thereby stretching the membrane. Because of the tight coupling between the membranes of the bound vesicles and planar membrane, in the region of contact the planar bilayer is stretched along with the vesicular membrane. This stretching results in a small increase in area per phospholipid, allowing molecular rearrangements that can culminate in fusion. If the tension producing the increased area exceeds that which can be sustained in the region of contact, fusion occurs. For bilayers in isolation, a tension of

[11] D. J. Woodbury and J. E. Hall, *Biophys. J.* **54,** 1053 (1988).
[12] W. D. Niles, F. S. Cohen, and A. Finkelstein, *J. Gen. Physiol.* **93,** 211 (1989).
[13] F. S. Cohen, W. D. Niles, and M. H. Akabas, *J. Gen. Physiol.* **93,** 201 (1989).

about 3 dyne/cm causes lysis.[14] Because there is little water between the bound membranes, it may be easier to stretch the membranes in the dehydrated region of contact. However, vesicles can break in the region not in contact with the planar membrane.

By loading vesicles with a fluorescent dye (calcein) and adding a quencher (Co^{2+}) to either the *cis* or *trans* compartment, the percentage of fusion events and nonfusion lysis events can be determined.[7,15,16] For asolectin vesicles, in which binding to solvent-free asolectin bilayers has been induced with 20 mM Ca^{2+}, fusion and lysis each occur about 50% of the time. When binding is induced with 100 mM Ca^{2+}, the percentage of fusing vesicles increases to 70%, with only 30% lysing. These large percentages for fusion indicate either that there is an appreciable area of contact between vesicles and planar membrane or that the area of contact is weaker than free isolated bilayers.

Vesicles fuse more readily to solvent-containing (e.g., decane-based) than to solvent-free planar membranes.[17] This is due in part to partial melding in the region of contact of vesicular and solvent-containing planar membranes, a process that is absent with solvent-free planar membranes. When vesicles are bound to decane-based films, decane microlenses (solvent separated out of the bilayer into lens-shaped objects of approximately 1 μm diameter) form and accumulate around the vesicles. When fluorescent lipids [e.g., rhodamine-phosphatidylethanolamine (Rh-PE)] are incorporated into the vesicles as probes, the probes can dissolve into the microlenses.[18] This mixing of probe with microlenses may be due to hemifusion (the merger of the outer monolayer of the vesicle with the *cis*-facing monolayer of the planar membrane). There is a partial breakdown of the individual membranes in the region of contact for solvent-containing membranes. In addition to partial membrane breakdown, reasons for easier fusion of vesicles to solvent-containing planar membranes probably include (1) the greater area of contact of vesicles with the bilayer due to the accumulated microlenses and (2) the partial merger of membranes, which ensures that osmotically driven water flow across the planar membrane is completely into the vesicles.

Although the binding of vesicles to solvent-free membranes is virtually irreversible, each membrane retains its individual integrity. Microlenses do

---

[14] E. A. Evans and R. Skalak, "Mechanics and Thermodynamics of Biomembranes." CRC Press, Boca Raton, Florida, 1983.

[15] D. J. Woodbury and J. E. Hall, *Biophys. J.* **54**, 345 (1988).

[16] M. S. Perin and R. C. MacDonald, *Biophys. J.* **55**, 973 (1989).

[17] F. S. Cohen, M. H. Akabas, J. Zimmerberg, and A. Finkelstein, *J. Cell Biol.* **98**, 1054 (1984).

[18] M. S. Perin and R. C. MacDonald, *J. Membr. Biol.* **109**, 221 (1989).

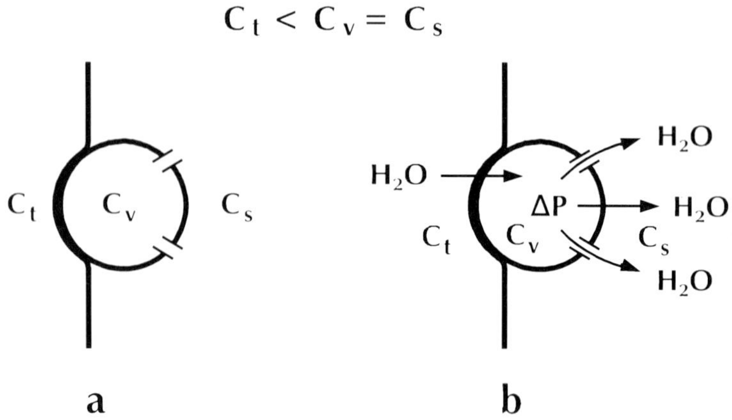

$$C_t < C_v = C_s$$

FIG. 1. An osmotic gradient across the planar membrane, *cis* side hyperosmotic, promotes fusion. (a) The osmolality of the salt in the *cis* compartment, C_s, is taken to be the same as that inside the vesicle, C_v, and greater than that of the *trans* compartment, C_t. (b) Because $C_v > C_t$, the osmotic difference forces water flow into the vesicle from the *trans* compartment. Because $C_v = C_s$, water does not exit the vesicle from osmotic pressure differences; a hydrostatic pressure ΔP develops which opposes the water influx and causes water efflux into the *cis* compartment. As the number of channels in the membrane increases, the water efflux increases for a given pressure. Therefore, a smaller steady-state pressure develops as the number of channels increases in this configuration.

not form around the vesicles, and fluorescent lipids placed in the vesicles remain confined there.[18] Hemifusion does not occur.[7] With solvent-free membranes, there may also be small corridors of water between the vesicular and planar membranes that would shunt some of the water flow across the planar membrane in the region of contact around, rather than into, the vesicles, thereby reducing fusion.

Basis of Standard 1980 Method: Swelling with Gradients across Planar Bilayer

Preexisting Osmotic Gradient

To reconstitute channels into planar membranes, vesicles are added to the *cis* compartment, which is hyperosmotic with respect to the *trans* side. A vesicle, isotonic with the *cis* compartment, is shown bound to the planar membrane in Fig. 1a. The osmotic gradient between the interior of the vesicle and the *trans* compartment produces an influx of water from the *trans* side into the interior of the vesicle. The influx of water causes the vesicle to swell until the membrane becomes taut. Because bilayer membranes are relatively indistensible,[19] once the membrane is taut there is little further volume increase or expansion of area. An intravesicular pressure develops that forces water out of the vesicle through the bilayer and through open channels into the *cis* compartment. Thus, water influx occurs via osmotically induced water flow, and efflux occurs via hydrostatically driven flow through the bilayer and through channels (Fig. 1b). Note that solute (e.g., salt) can also exit the vesicle into the *cis* compartment via solvent–solute flux coupling, thereby lowering the intravesicular solute concentration compared to that shown in Fig. 1b. This lower solute concentration results in a smaller influx of water (there is a smaller osmotic difference than in Fig. 1b) and hence a lower intravesicular pressure. To the extent that a compensating diffusive influx of solute balances this efflux of solute due to solvent drag, the reduction in pressure is minimized. The effect of solvent–solute coupling has been considered in detail[12] and is not discussed further in this chapter.

Quantitatively, let a fraction, f, of the area of the vesicle, A, be in contact with the planar membrane. Let the permeability coefficient of water through the bare bilayer be P_W and the hydraulic permeability through an individual channel be L. The area of contact is modeled as two distinct bilayers in direct apposition; hence, the permeability coefficient of water in this region is $0.5\ P_W$. The osmotic strength of the *trans* compartment is C_t (thus, for 50 mM KCl, $C_t = 100$ mOsm, taking osmotic coefficients to be unitary), the osmotic strength within the interior of the vesicle is C_V, and the osmotic strength of the *cis* compartment is C_S. In Fig. 1, $C_t < C_V = C_S$. The intravesicular pressure is denoted ΔP. For simplicity, we assume that initially the vesicle is fully swollen but has not yet developed an intravesicular pressure, and the solute is impermeant through the bare bilayer (e.g., the solute is a salt). The volume influx of water into the vesicle is given by Eq. (1):

$$J_{in} = 0.5 P_W \overline{V}_W A f [(C_V - C_t) - \Delta P/RT] \qquad (1)$$

The efflux of water out of the vesicle is given by Eq. (2):

$$J_{ef} = [P_W \overline{V}_W A (1 - f)/RT + nL]\Delta P \qquad (2)$$

where n is the number of channels in the vesicle, and all channels are assumed to be identical for notational simplicity. $\overline{V}_W$, R, and T have their usual meanings of partial molar volume of water, universal gas constant,

[19] R. Kwok and E. A. Evans, *Biophys. J.* **35**, 637 (1981).

and temperature in Kelvins, respectively. ΔP increases until at steady state $J_{in} = J_{ef}$, or

$$\Delta P = \frac{0.5 P_w \bar{V}_w A f R T (C_v - C_t)}{P_w \bar{V}_w A (1 - 0.5f) + nLRT} \tag{3}$$

Note that the maximum pressure is attained in the absence of channels. Channels provide an additional pathway for pressure-driven efflux of water, resulting in smaller steady-state pressures. In the absence of channels,

$$\Delta P_{max} = \frac{0.5f}{1 - 0.5f} R T (C_v - C_t) \tag{4}$$

If a vesicle can withstand its steady-state pressure, fusion will not occur. If it cannot, the vesicle will fuse or lyse.

Initially the vesicle is not fully swollen; then in the absence of channels the intravesicular contents will dilute as a consequence of swelling, and the maximal pressure will be less than given by Eq. (4). If, however, the vesicle contains channels through which the solute permeates, then Eq. (3) is valid even for vesicles that are initially shrunken. As the vesicle swells, solute enters the vesicle, and at all times in the swelling process $C_v = C_s$. Thus, for Fig. 1 the open channels are beneficial for fusion in that the initial extent of vesicle inflation is unimportant, but open channels lead to lower pressures than can, in principle, be obtained in their absence. The hydrostatic pressure will also be less than calculated if there is any aqueous space between the vesicle and bilayer. In this case some of the water flowing from the *trans* compartment will move around rather than into the vesicle. This shunting of water has been considered quantitatively.[6]

As an example, if the *trans* compartment contains 50 mM KCl, the vesicle and *cis* compartments contain 250 mM KCl, 10% of the vesicle is in contact with the bilayer, and the vesicle is initially taut, the maximum intravesicular pressure that will develop is about 2.5×10^5 dyne/cm^2. The tension, T, developed by the membrane of a vesicle of radius r and pressure ΔP is given by Laplace's law: $T = \Delta P r / 2$. Thus, in the above example, vesicles of diameter greater than about 2300 Å will fuse (or lyse) whereas smaller vesicles remain intact, assuming a breakdown tension of 3 dyne/cm.

Increasing Osmotic Gradient

If the resulting experimental pressures are not sufficient to cause fusion, additional salt may be added to the *cis* compartment of the configuration shown in Fig. 1a, producing the condition noted in Fig. 2a. If there are

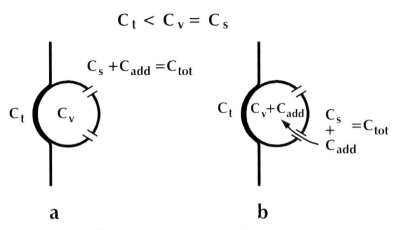

FIG. 2. Addition of salt to the *cis* compartment increases the intravesicular pressure. (a) Conditions are initially as in Fig. 1a. A concentration, C_{add}, of salt is added to the *cis* compartment, resulting in a final concentration of C_{tot}. (b) The additional salt enters the vesicle via the channels. In the steady state the intravesicular concentration is C_{tot}. The hydrostatic pressure is calculated exactly as for Fig. 1.

channels in the membrane through which the salt permeates, the salt concentration inside the vesicle equalizes to that of the surrounding medium as shown in Fig. 2b. The added salt results in a larger osmotic gradient between the vesicle interior and the *trans* compartment, a larger hydrostatic pressure develops, and hence the amount of fusion increases. The initial extent of dilation of the vesicle does not affect the steady-state pressure.

It is important to recognize the role of solute-permeable channels in the vesicles and the permeability of solutes through the bare bilayer. For Fig. 1a, a pressure develops independent of the presence of channels through which solute enters the vesicle; in fact, channels reduce the maximal obtainable pressure. For example, if channels selective to K^+ were in the vesicular membrane, the scheme of Fig. 1a would be suitable to initiate fusion (if the pressures generated were large enough). If however, additional electrolyte (for instance, KCl) were added as in Fig. 2a, only an osmotically insignificant amount of potassium could enter the vesicle (there is no route for Cl^- entry), and an augmented pressure would not develop. In fact, the vesicle would have a net shrinkage because of the osmotic gradient between the vesicle and *cis* compartment. Therefore, in order for an augmented pressure to develop when additional solute is added to the *cis* compartment, channels are required to allow the entry of solute (e.g., the KCl) into the interior of the vesicle.

Now consider the case in which the additional solute is a nonelectrolyte. If the solute (e.g., glucose) enters the vesicle through the channels and is impermeant through the bare bilayer, then the water flow occurs as in Fig. 2b, and a pressure develops as discussed previously. If, however, the nonelectrolyte permeates through bare bilayers, then the case given in Fig. 3 occurs. The solute leaks out of the vesicle through the region of contact into the *trans* compartment. Therefore, the concentration of nonelectrolyte in the vesicle is less than its concentration in the *cis* compartment. The influx of water is less and the efflux to the *cis* chamber is greater than that occurring if the nonelectrolyte is bilayer-impermeant (but channel-permeant). Thus, the permeation of the added nonelectrolyte through the plain bilayer reduces the steady-state pressure. If the solute permeability through the plain bilayer dominates its permeability through the channels, then a hydrostatic pressure does not develop and the additional solute does not augment fusion.

The validity of this statement has been shown and discussed.[11,12] To illustrate the principle, we present a simple case, namely, that in which only the permeant nonelectrolyte is present. The nonelectrolyte attains its steady-state concentration when the influx of water into the vesicle from the *trans* compartment (due to the nonelectrolyte) equals the efflux of water into the *cis* compartment. Similarly, in the steady state there is

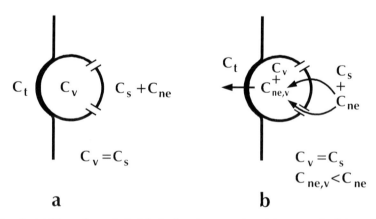

FIG. 3. Addition of nonelectrolyte to the *cis* compartment increases the intravesicular pressure only if it is channel-permeant. (a) Conditions initially as in Fig. 1a. A concentration, C_{ne}, of nonelectrolyte is added to the *cis* compartment. (b) The nonelectrolyte enters the vesicle via the channels and the plain bilayer. The nonelectrolyte exits the vesicle into the *trans* compartment by permeating the plain bilayer in the region of contact. The intravesicular concentration of nonelectrolyte, $C_{ne,v}$, is therefore less than in the *cis* compartment ($C_{ne,v} < C_{ne}$).

equality of the vesicular solute influx and efflux. The mechanism of movement of water and nonelectrolytes through the bare bilayer is the same: via solubility and diffusion. Therefore, if we let the permeability coefficient of the bilayer to solute be P_s, in the region of contact this coefficient is $0.5P_s$. If the intravesicular pressure is denoted by ΔP, in the steady state, the balance of water flux yields

$$0.5fAP_w\overline{V}_w[RT(C_v - C_t) - \Delta P] = (1 - f)AP_w\overline{V}_w[RT(C_s - C_v) + \Delta P] \tag{5}$$

The balance of solute flux yields

$$0.5fAP_s(C_v - C_t) = (1 - f)AP_s(C_s - C_v) \tag{6}$$

Dividing the left- and right-hand sides of Eqs. (5) and (6) yields

$$\frac{R\tilde{T}(C_v - C_t) - \Delta P}{(C_v - C_t)} = \frac{RT(C_s - C_v) + \Delta P}{(C_s - C_v)} \tag{7}$$

Simple inspection shows that the solution is $\Delta P = 0$. Note that the pressure can only increase monotonically. Unlike the configuration in Fig. 1a, pressures do not develop in the solute configurations shown in Figs. 2 and 3 in the absence of channels.

When channels are present, pressures develop in the configuration given in Figs. 2 and 3 as calculated.[11,12] This can be understood intuitively: When channels are present, C_v is larger than in the absence of channels. This larger C_v causes a greater influx of water into the vesicle [$(C_v - C_t)$ is larger], and a reduced efflux of water from the vesicle [$(C_s - C_v)$ is smaller]. The larger influx and smaller efflux of water leads to an intravesicular pressure. This pressure opposes the water influx and augments the efflux, and a steady-state can be reached.

An alternate way to view the process is to start conceptually with a bilayer-impermeant solute and imagine that the permeability of the solute increases. If the pressure is zero, steady-state flux of a solute, permeant through the channels but impermeant through the bilayer, can only be reached if the intravesicular and *cis* concentrations of the solute are equal. But then there would be a greater influx of water from the *trans* compartment into the vesicle than efflux from the vesicle into the *cis* chamber. The vesicle would swell, further solute would enter the vesicle, and the process would repeat indefinitely. However, the vesicle cannot expand to infinite volume. Instead, a pressure develops. This pressure inhibits water influx and augments water efflux, allowing a steady state to be reached. Consider the consequences of allowing the bilayer to become permeant to solute: The concentration of solute within the vesicle decreases because it permeates into the *trans* compartment, the influx of water from the *trans* compartment decreases, the efflux of water into the *cis* compartment increases, and therefore the steady-state pressure is lowered. As the solute becomes sufficiently permeant through the bare bilayer, the relative solute influx via channels becomes insignificant and the pressures decrease. This decrease of pressure as the solute becomes progressively more bilayer-permeant has been experimentally demonstrated[13] and theoretically quantified.[12]

With bilayer-permeant solutes, increasing the number of channels allows greater influx of solute into the vesicle, counteracting solute efflux into the *trans* compartment. On the other hand, increasing the number of channels also causes the pressure-driven water efflux to increase. There is an optimal number of channels that balances solute influx against water efflux to yield the largest pressure obtainable.[11,12]

Use of Nystatin to Aid Fusion

Channels that allow osmotically significant amounts of solute to enter vesicles are essential for swelling, and therefore fusion, when the Standard 1980 Method depicted in Figs. 2 and 3 is used. When vesicles are derived from cells, there are in practice multiple types of channels, allowing movement of both anions and cations, and therefore the swelling schemes described are usually adequate. When ion-selective channels are reconstituted into vesicles, experimental practice has shown that fusion is much more difficult to achieve. This is in part because osmotically significant amounts of solute (e.g., salt) cannot enter the vesicle when added by the stratagem of Fig. 2. Incorporating other channels (in addition to the reconstituted channels) that allow anions and cations to enter the interior of the vesicle but do not contaminate electrical study of the reconstituted channels would alleviate this problem.

The channels formed by the antibiotics nystatin or amphotericin B are suitable choices for these auxiliary channels. These agents form channels that are permeable to both small anions and cations, have reasonable sieving diameters (~ 8 Å), and have small single-channel conductances (5 pS in 2 M KCl[20]). The concentration of antibiotic needed to form channels when added to both sides of the membrane is significantly less than that required when added from one side.[21] Including the antibiotic when preparing the reconstituted vesicle (~ 5 μg nystatin/mg lipid) and adding the antibiotic to the *cis* compartment allow two-sided channels to form in the vesicle, but only the much less likely one-sided channels can occur in the planar membrane. These antibiotics require sterols in the membrane for efficient formation of channels. Including sterol in the vesicle but not the

[20] M. E. Kleinberg and A. Finkelstein, *J. Membr. Biol.* **80,** 257 (1984).
[21] A. Marty and A. Finkelstein, *J. Gen. Physiol.* **65,** 515 (1975).

planar membrane ensures the desired placement of the channels. In the absence of sterols these channels dissociate. After fusion, the antibiotic channels that were in the vesicular membrane will dissociate in the planar membrane and will not contribute to the conductance. Once nystatin or amphotericin is incorporated into the vesicular membrane, a more efficient method to fuse vesicles becomes available.

1990 Method: Optimal Methods to Swell Vesicles Reconstituted with Ion-Selective Channels

When vesicles are reconstituted with an ion-selective channel, the configuration of Fig. 1 promotes swelling and the development of an intravesicular pressure. If the pressures produced by this procedure are not sufficient to produce fusion, the addition of solute to the *cis* compartment, shown in Figs. 2 and 3, requires that solute (e.g., salt) be able to enter the vesicle. As discussed above, use of nystatin facilitates this. However, although swelling of vesicles by establishing an osmotic gradient across the planar membrane has the advantage of convenience, it has the disadvantage that the influx of water occurs only in the region of contact. Thus the pressure that can develop is scaled by the fraction of the vesicle, f, that is in contact with the planar membrane. As f may be small, the pressures may not be large enough to produce fusion. Once nystatin (or amphotericin B) is used, much larger pressures than those produced by the schemes previously presented can be produced by the 1990 Method configuration illustrated in Fig. 4.

For this procedure, vesicles reconstituted with the ion-selective channel are prepared in the presence of nystatin (e.g., nystatin is mixed with lipids or included in the solution used to prepare the vesicles). Sterol, preferably ergosterol as discussed below, is included in the vesicle membrane. The vesicles are loaded with an impermeant solute, such as glucose or sucrose, through nystatin channels (Fig. 4a).[22] The *trans* and *cis* chambers contain the salt of choice (e.g., KCl). All compartments (*cis, trans,* intravesicular) are made isotonic. With nystatin in the *cis* compartment, double-sided nystatin channels form in the vesicular membrane, salt enters the vesicle through the nystatin channels, water follows, and the vesicles develop a pressure (Fig. 4b). If it is not practical to prepare the vesicles in the presence of nystatin, the antibiotic can be added to only the *cis* compartment, thereby forming single-sided nystatin channels. Greater concentrations of nystatin will be required than in the case of two-sided channels.

[22] R. Holz and A. Finkelstein, *J. Gen. Physiol.* **56,** 1835 (1970).

$$C_t = C_I = C_s$$

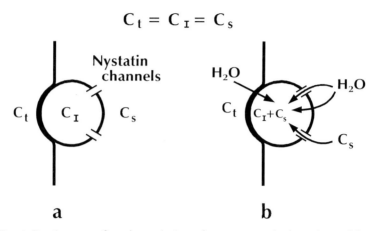

FIG. 4. Development of maximum hydrostatic pressures under isotonic conditions with salt on both sides of the bilayer. (a) The vesicle membrane contains nystatin channels, and the vesicle lumen is loaded with a bilayer- and channel-impermeant molecule. The *cis* and *trans* compartments contain isotonic salt. (b) The salt enters the vesicles and water follows. In the steady state the intravesicular concentrations of the impermeant and salt are C_I and C_s, respectively.

The vesicle is allowed to swell fully as in Fig. 4a, but with an intravesicular pressure of zero. The osmolality of the intravesicular impermeant is C_I with $C_I = C_s = C_t$. The net water flux into the vesicle, J_w, for Fig. 4b is given by Eq. (8):

$$J_w = 0.5fAP_w\overline{V}_w(C_I - \Delta P/RT) + (1 - f)AP_w\overline{V}_w(C_I - \Delta P/RT) + nL(RTC_I - \Delta P) \qquad (8)$$

We have explicitly written the flux terms in the order of flow through the region of contact, through the plain bilayer not in contact, and through the channels. Steady state is reached when $J_w = 0$. Simple inspection of Eq. (8) or Fig. 4 shows that the steady-state pressure is

$$\Delta P = RTC_I \qquad (9)$$

This is the maximum pressure that can develop for any possible configuration. If 250 mM KCl is used on the *cis* side ($C_I = C_s$), a steady-state pressure of 1.2×10^7 dyne/cm² develops. Shunting of water around the vesicle does not occur, and vesicles of all sizes will fuse or lyse. If the vesicle is initially shrunken, the intravesicular impermeant will be diluted (C_I decreases) by the swelling that precedes the development of a pressure. The pressure will be reduced correspondingly. There is, however, a sufficient safety factor: a reduction in the pressure because of swelling can be tolerated.

The practicality of this method has been demonstrated.[7,13] If there is a sufficient number of nystatin channels in the vesicles, fusion will be marked by a stepwise increase in conductance which decays with time as the oligomeric nystatin channels dissociate, leaving only the channel of interest intact in the bilayer. Nystatin has been used to fuse vesicles reconstituted with a Cl^--selective channel, which resisted fusion by traditional methods.[23]

Use of Ergosterol and Nystatin

Nystatin channels form most readily when sterol is present in the membrane. Ergosterol works much better than cholesterol.[24] Unfortunately, commercially available ergosterol is rarely pure; it is hydrated and contains balsams and other aromatic compounds. The nystatin procedure is much more reliable if the ergosterol is recrystallized from ethanol according to the following procedure.

1. Heat approximately 400 ml of absolute ethanol (200 proof) in a covered 600 to 1000-ml beaker in a water bath to about 70°. Do not exceed this temperature as the boiling point of ethanol is 78.5°, but do not let the ethanol cool to much less than 70°, as the solubility of the ergostrol decreases. We use a hot plate with stirring to heat the water bath; *avoid flames* as the flash point of ethanol is very low (~ − 12°).

2. While stirring, add 10–20 g of the crude ergosterol until it no longer dissolves in the hot ethanol.

3. Keep the beaker covered and store at − 20° overnight. The ergosterol precipitates and the impurities remain dissolved.

4. In a cold room, pour the entire mixture through Whatman (Clifton, NJ) No. 50 (hardened) paper in a Büchner funnel connected to a vacuum pump. Prechill all glassware. Otherwise, as the mix becomes warmer, the ergosterol redissolves and passes through the filter paper.

5. Wash the precipitated ergosterol with absolute ethanol that has been chilled to − 20°. Again, keep all materials cold. Ethanol below − 20° can be used, but if too cold it becomes too viscous.

6. Dry the ergosterol in the funnel and then *in vacuo* in a desiccator for 24–36 h or until dry.

7. The ergosterol is stable indefinitely when stored desiccated at − 20°.

[23] D. J. Woodbury, A. G. Goldberg, and C. Miller, *Soc. Neurosci. Abstr.* **15** (Part 1), 79 (1989).

[24] S. M. Hammond, *in* "Progress in Medicinal Chemistry" (G. P. Ellis and G. B. Shaw, eds.), Vol. 14, p. 105. Elsevier/North-Holland, Amsterdam, 1977.

If a mixture of crude asolectin and recrystallized ergosterol (4:1, w/w) is used to prepare vesicles, about 60 μg/ml nystatin on both sides of the membrane produces sufficient channels for the procedure to yield maximum fusion.[7] The exact amount of nystatin required will need to be determined for each mixture of lipids used to make the vesicles. If the procedure does not work, merely increase the concentration of nystatin. The number of channels varies roughly as the eighth power of the nystatin concentration,[25] helping to ensure that nystatin channels can be formed in the vesicles.

Efficacy of Nonelectrolytes with 1990 Method

Pressures similar to the cases presented above can be developed by the configuration illustrated in Fig. 5. Vesicles, reconstituted with ion-selective channels, are prepared with membrane-impermeant contents. In Fig. 5a salt is present in the intravesicular and *trans* compartments, although any bilayer-impermeant solute is suitable. An isotonic concentration of a bilayer permeant (e.g., formamide), is placed in the *cis* compartment (Fig. 5a). Formamide enters the interior of the vesicle and can leak into the *trans* compartment (Fig. 5b). In the steady state the influx of formamide is equal to its efflux. Assuming the *trans* compartment does not fill up with formamide (the compartments have effectively infinite volume), and denoting the concentration of formamide inside the vesicle by C_F, the equality of solute flux yields

$$(1 - f)AP_S(C_S - C_F) = 0.5fAP_SC_F \tag{10}$$

or

$$C_F = \left(\frac{1-f}{1 - 0.5f}\right) C_S \tag{11}$$

As in Eq. (9), the pressure is given by

$$\Delta P = RTC_F = RT \left(\frac{1-f}{1 - 0.5f}\right) C_S \tag{12}$$

For $f = 0.1$, 200 mM formamide, $\Delta P = 4.7 \times 10^6$ dyne/cm^2. All vesicles having diameters greater than 250 Å (essentially all types of vesicles) will fuse or lyse. The permeant nonelectrolyte (e.g., formamide) can be present initially or perfused in for salt after the bilayer is formed, according to experimental procedures. In principle any permeant nonelectrolyte will work, but the higher the permeability, the quicker the vesicle will fill up. Formamide and ethylene glycol with permeability coefficients,

[25] A. Cass, A. Finkelstein, and V. Krespi, *J. Gen. Physiol.* **56**, 100 (1970).

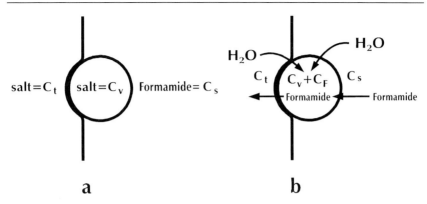

Fig. 5. Development of large pressures in the absence of open channels. (a) Vesicles are loaded with salt. Isotonic salt is in the *trans* compartment, and an isotonic solute (e.g., formamide) permeable through the plain bilayer, is in the *cis* compartment. (b) The formamide enters the vesicle from the *cis* compartment and exits into the *trans* compartment. The resulting intravesicular concentration of formamide is denoted by C_F.

P_s, on the order of 10^{-4} cm/sec for phosphatidylcholine[26] are excellent in this regard. For a 1000 Å diameter vesicle, steady-state concentrations of formamide or ethylene glycol are attained (theoretically) on a time scale of milliseconds. Even after removal by perfusion, if these agents are pharmacologically deleterious to the channel of interest, glycerol or urea with P_s values of 5×10^{-6} and 4×10^{-6} cm/sec, respectively, can be tried.

This last procedure has the merits of producing near-maximal pressures, as well as ease of experimental design. The ion-selective channels can be reconstituted into vesicles of arbitrary lipid composition, and additional vesicular components such as nystatin are not needed. This procedure has the disadvantage of requiring an osmotically significant amount of nonelectrolyte at some point in the experiment. In the 1990 Method shown in Figs. 4 and 5, all vesicles swell and burst, whereas in the Standard 1980 Method schemes given in Figs. 1–3 only bound vesicles swell. It is therefore useful to have the vesicles attach to the membrane as soon as possible for the 1990 Method. For the 1980 scheme, optimization of binding is clearly beneficial. This is accomplished as described in the next section.

Delivery of Vesicles

Vesicles (or any other material added to an aqueous phase bathing a planar membrane) are added directly above a small magnetic stirring bar to ensure dispersal of material throughout the *cis* compartment. However, as

[26] A. Finkelstein, *J. Gen. Physiol.* **68,** 127 (1976).

most of the vesicles never encounter the bilayer with this procedure, it is better to deliver the vesicles directly at the planar membrane.[7,11] As a practical matter, this greatly increases the amount of fusion. Directing vesicles at the bilayer has allowed practical reconstitution of purified Cl⁻ channels from kidney and trachea[27] and channels from pancreatic endoplasmic reticulum.[28] It is also useful to minimize the diffusion times of vesicles through the unstirred layers for efficient reconstitution of channels via the 1990 Method. Fluorescently labeled material can be delivered to the bilayer by injection near the bilayer. To minimize the background fluorescence, the optimal delivery process should be used. Less elaborate procedures are adequate for reconstitution purposes.

Vesicles are ejected from a pipette directed to a vertical planar membrane. It is preferable to fabricate the pipette with a right-angle bend rather than aim a straight pipette obliquely at the membrane. The ease of crafting the pipette is determined by the ability to form the bend without constricting the lumen in the region of the bend. While most glasses are probably adequate, narrow-walled capillaries made of hard glass with good shearing at high temperatures aids fabrication. The requirements for formation of these pipettes are similar to those used by Andersen[29] to patch planar membranes.

A piece of Pyrex capillary tubing (7740, Corning Glass Works, Corning, NY), 15 cm long with 0.8 mm O.D. and 0.5 mm I.D., is pulled by hand over a Bunsen burner with a low to medium flame to yield a long (~ 5 cm), gradually tapered tip. The tubing is cut in the narrowed region with a diamond knife, yielding a right circular break with an outer diameter of approximately 0.2 mm. The bend is made with a homemade forge using a platinum wire having a diameter of approximately 0.4 mm and a manipulator to position the pipette. The bend is made 4–5 mm back from the tip. The pipette is slowly advanced past the wire as the glass melts, producing a smoothly rounded bend. The tip is fire-polished to the desired inner diameter, typically 10–50 μm.

Vesicles are loaded into the pipette by back-filling with a 3-inch-long 30-gauge needle. The pipette is placed in an electrode holder with its top end protruding (EH-2R, E. W. Wright, Guilford, CT) and maneuvered with a hydraulic micromanipulator (MO-103, Narishige Scientific Instruments). The holder is mounted in a Lucite block with a Teflon connector

[27] D. W. Landry, M. H. Akabas, C. Redhead, A. Edelman, E. J. Cragoe, Jr., and Q. Al-Awqati, *Science* **244**, 1469 (1989).
[28] S. M. Simon, G. Blobel, and J. Zimmerberg, *Proc. Natl. Acad. Sci. U.S.A.* **86**, 6176 (1989).
[29] O. S. Andersen, *Biophys. J.* **41**, 119 (1983).

(016-2016, Sealectro Corp., Mamaroneck, NY) and connected to the manipulator. The top end of the pipette is connected with polyethylene tubing to an electric valve (Picospritzer, General Valve Corp., East Hanover, NJ), which is connected to a pressurized nitrogen tank. The pipette is lowered into the *cis* chamber and the tip brought close to the membrane (5–10 μm) with the micromanipulator. Vesicles are ejected at the membrane with 20 psi pressure by opening the valve for 3–4 msec. Varying the diameter of the pipette tip and the number of squirts allows from submicroliter to 50 μl of vesicles to be delivered directly to the planar membrane.

Before use, the glass capillaries are cleaned of residues resulting from the manufacturing process. Without cleansing, the pipettes tend to clog. The cleaning procedure is as follows.

1. Loosely pack 15 cm lengths of tubing in a 25 × 200 mm culture tube (~7/8 full with tubing), fill with 6 N HCl, and cap loosely.
2. In a hood, heat the HCl for 30 min with the tube held in a boiling water bath.
3. Pour off the HCl, replace, and boil in acid again for 15 min.
4. Pour off the acid and wash with distilled water 3 times or until the wash water has a pH of 4–5 as determined with pH strips.
5. To remove the acid that remains within the lumen of the tubing, fill the tube with fresh water and boil for 30 min. Repeat 2 more times or until the pH of the water after boiling is around 5, then pour off the water.
6. Remove the residual water by washing the tube 3 times with absolute ethanol.
7. Wash the tubes once with chloroform/methanol (2 : 1, v/v) and then boil (carefully) once in chloroform/methanol for 15 min.
8. Dry the capillaries in the culture tube (with the cap loose and a wadded Kimwipe in the neck) at 110° for 3 days or until dry.
9. Tighten the cap and store.

Summary

Protocols to reconstitute channels into planar bilayers via fusion methods have now been developed. The greater the intravesicular pressures generated, the greater is the fusion. These pressures can be calculated exactly for any experimental configuration. For some of the configurations, adding nystatin channels to the vesicle membrane will greatly aid fusion. The configurations of the 1990 Method (Figs. 4 and 5) are optimal for fusing vesicles that are reconstituted with ion-selective channels to planar membranes. Greater binding, and ultimately greater fusion, is achieved by ejecting vesicles directly at the membrane rather than by simply adding material to the *cis* compartment.

Acknowledgments

Fredric S. Cohen and Walter D. Niles were supported by National Institutes of Health Grant GM-27367.

[6] Phosphorus-31 Nuclear Magnetic Resonance in Membrane Fusion Studies

By PHILIP L. YEAGLE

Introduction

Progress in the study of membrane fusion has resulted, in part, from the development of technologies for studying various aspects of the membrane fusion events and the development of models for fusion mechanisms that could be tested by using the new technologies. Two of the most significant fusion mechanisms that have been proposed are based on model membrane fusion studies of simple lipid vesicles. One utilizes a mechanism involving calcium and phosphatidylserine (or cardiolipin) which may involve a gel (dehydrated) phase formation by the calcium–phosphatidylserine complex.[1-6] The second pathway has been suggested to involve "isotropic" structures, I_s, identified in ^{31}P nuclear magnetic resonance (^{31}P NMR) spectra of lipid dispersions as intermediates in membrane fusion.[7-10]

The "isotropic" structures may satisfy one of the fundamental demands of fusion mechanisms. It has been suggested that fusion requires the following steps.[11-13] (1) aggregation or adhesion of the membranes that will

[1] J. R. Silvius and J. Gagne, *Biochemistry* 23, 3232 (1984).

[2] D. Papahadjopoulos, W. J. Vail, C. Newton, S. Nir, K. Jacobson, G. Poste, and R. Lazo, *Biochim Biophys. Acta* 465, 579 (1977).

[3] R. Leventis, J. Gagne, N. Fuller, R. P. Rand, and J. R. Silvius, *Biochemistry* 25, 6978 (1986).

[4] D. Hoekstra, *Biochemistry* 21, 1055 (1982).

[5] N. Düzgüneş, J. Paiement, K. B. Freeman, N. G. Lopez, J. Wilschut, and D. Papahadjopoulos, *Biochemistry* 23, 3486 (1984).

[6] D. Papahadjopoulos, S. Nir, and N. Düzgüneş, *J. Bioenerg. Biomembr.* 22, 157 (1990).

fuse; (2) close approach of the lipid bilayers of the membranes, leading to removal of some of the water separating the membranes (partial dehydration); (3) destabilization of the bilayer at the point of fusion (two bilayers closely opposed will not spontaneously fuse by themselves); and (4) mixing of the bilayers and ultimate separation from the point of fusion into the new membrane structure(s). Step (3) requires the localized, and transient, disruption of the stable bilayer structure so that components of the two original membranes can mix. Nonbilayer structures are candidates for this localized defect in the bilayer structure that facilitates membrane fusion. An important technical question is how to observe nonbilayer structures that might be important to the fusion event. This chapter describes one of the few available techniques for detecting nonbilayer structures.

^{31}P NMR and Morphology of Phospholipid Aggregation

^{31}P is a 100% naturally abundant nucleus moderately sensitive to NMR experiments. Because phospholipids have at least one phosphorus in their head group, ^{31}P NMR has proved to be a powerful nonperturbing approach to the study of phospholipid head group behavior. Information can be obtained from both simple phospholipid systems and from biological membranes without the need to introduce probes or to alter the sample.

In addition, because of the anisotropic nature of the phospholipid bilayer, ^{31}P powder patterns have proved to be a sensitive means for analyzing the morphology of phospholipid dispersions. I give a brief background on the origin of the powder patterns and thus demonstrate how these powder patterns can reflect the morphology of the system. More complete reviews on this topic are available.[14,15]

In phospholipid bilayers, particularly in the liquid crystalline state, the phospholipid molecules are capable of axial diffusion which leads to mo-

[7] J. Gagne, L. Stamatatos, T. Diacovo, S. W. Hui, P. L. Yeagle, and J. Silvius, *Biochemistry* **24,** 4400 (1985).

[8] H. Ellens, J. Bentz, and F. C. Szoka, *Biochemistry* **25,** 4141 (1986).

[9] H. Ellens, D. P. Siegel, L. Lis, P. J. Quinn, P. L. Yeagle, and J. Bentz, *Biochemistry* **28,** 3692 (1989).

[10] D. P. Siegel, J. Banschbach, D. Alford, H. Ellens, L. Lis, P. J. Quinn, P. L. Yeagle, and J. Bentz, *Biochemistry* **28,** 3703 (1989).

[11] J. Bentz and H. Ellens, *Colloids Surf.* **30,** (1988).

[12] N. Düzgüneş, *Subcell. Biochem.* **11,** 195 (1985).

[13] S. Nir, J. Bentz, J. Wilschut, and N. Düzgüneş, *Prog. Surf. Sci.* **13,** 1 (1983).

[14] P. L. Yeagle, *in* "Phosphorus NMR in Biology" (C. T. Burt, ed.), p. 95, CRC Press, Boca Raton, Florida, 1987.

[15] J. Seelig, *Biochim. Biophys. Acta* **515,** 105 (1978).

tional averaging of some of the elements of the chemical shift tensor. The result is a pseudoaxially symmetric powder pattern, resulting from the axial diffusion of the phospholipid head group. Figure 1 shows the powder pattern arising from phospholipids in a bilayer. The width of this powder pattern can be influenced by head group dynamics that are beyond the scope of the chapter.

The shape of the ^{31}P NMR powder pattern is indicative of the morphology of the phospholipid dispersion, because the kind of motion possible in the system directly influences the motional averaging of the observed chemical shift anisotropy. Thus, the ^{31}P powder patterns reflect the motions that the phospholipids undergo within the lipid assemblies.

To illustrate this point more clearly, consider the effect of a lamellar-to-hexagonal II (H_{II}) phase transition. The powder pattern from the lamellar phase already discussed arises primarily from the axial diffusion of the phospholipids in the bilayer. In the hexagonal II phase, another important mode of motion is added: diffusion around the cylinders of the hexagonal II phase. The effect of this additional motional averaging is dramatic; the powder pattern is reduced to one-half of its width in the lamellar phase, and the sense of the anisotropy in the powder pattern is reversed from that in the lamellar state (Fig. 1).

If extensive modes of motional averaging are present in the sample, the ^{31}P powder pattern narrows even further. In the extreme of isotropic motional averaging, no powder pattern is seen. Instead, a narrow resonance at the isotropic chemical shift is observed. Isotropic spectra are

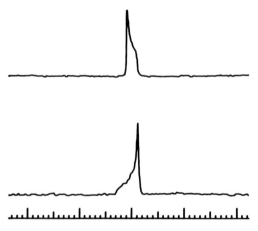

FIG. 1. ^{31}P NMR spectra (109 MHz) of soy phosphatidylethanolamine in the lamellar phase (bottom) and in the hexagonal II phase (top spectrum).

observed from ^{31}P NMR of phospholipids in small vesicles and micelles, where rapid tumbling of the vesicles and micelles motionally averages the chemical shift anisotropy. Isotropic ^{31}P NMR spectra also arise from phospholipids in a cubic phase and from lipidic particles (as seen in freeze–fracture electron microscopy), the latter being referred to as I_s. Figure 2 shows the development of isotropic resonances in LUV of N-monomethyldioleoylphosphatidylethanolamine with increase in temperature. The isotropic ^{31}P NMR spectra alone are not sufficient to distinguish between the various possibilities for the phospholipid morphology of micelles, small vesicles, and cubic structures. Other techniques must be employed in conjunction with ^{31}P NMR.

As shown in Figs. 1 and 2, ^{31}P NMR of phospholipid dispersions is a powerful tool for analyzing the distribution of phospholipids among various morphologies. ^{31}P NMR, therefore, can be used to study phase changes. ^{31}P NMR can concurrently determine quantitatively the percentage of phospholipids in each of the phases. Furthermore, ^{31}P NMR is, in principle, sensitive to all the phospholipids in the sample. It does not

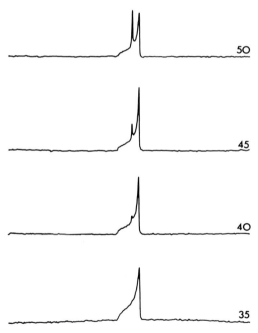

FIG. 2. ^{31}P NMR spectra (109 MHz) of LUV of N-monomethyldioleoylphosphatidylethanolamine at pH 4.5 and at the temperatures indicated. The development of the isotropic resonance, characteristic of I_s, is observed superimposed on the bilayer powder pattern.

depend on ordered arrays, as does X-ray diffraction; thus, some isotropic structures, which are not readily apparent in X-ray experiments, can be seen with ^{31}P NMR. ^{31}P NMR also does not depend on representative sampling, as does electron microscopy, to be quantitative. With these advantages, it is not surprising that ^{31}P NMR has been widely used for the study of the morphology of phospholipid dispersions and biological membranes.

The shape of the ^{31}P NMR powder pattern is sensitive to the morphology of the phospholipid-containing portion of the sample. Therefore, one must obtain undistorted spectra to interpret the results correctly. Unfortunately, this may not be easy to do as most commercial spectrometers are designed specifically for high-resolution NMR and cannot obtain undistorted ^{31}P NMR powder patterns using normal single-pulse methods. As a consequence, many published spectra suffer from some distortion in the shape of the ^{31}P powder patterns. Because interpretations of phospholipid morphology depend on details of the shape of the powder patterns, improperly obtained spectra can lead to incorrect conclusions.[16]

By using a simple Hahn echo pulse sequence, along with extensive phase cycling, it has been shown that these artifacts can be avoided.[17] Because many commercial spectrometers are capable of effectively utilizing this approach, this method is recommended. Otherwise, the results obtained are not reliable. Details of this technique have been discussed elsewhere.[17]

Use of ^{31}P NMR to Identify Phospholipid Morphologies Important to Membrane Fusion

Many investigators have suggested that some sort of nonlamellar structure in a membrane may be an intermediate in the process of membrane fusion. ^{31}P NMR has been a powerful tool to explore the question of putative intermediates, along with freeze–fracture electron microscopy and cryoelectron microscopy.

Some membrane lipids, both synthetic and biological, are unstable in the L_α phase (liquid crystalline bilayer). A notable example is phosphatidylethanolamine. At defined temperatures and buffer conditions, phosphatidylethanolamine undergoes a transition from the L_α phase to the H_{II} phase. The temperature at which this phase transition occurs is highly

[16] P. L. Yeagle, in "Methods for Studying Membrane Fluidity" (R. C. Aloia, C. C. Curtain, and L. M. Gordon, eds.), p. 267. Alan R. Liss, New York, 1988.

[17] M. Rance and R. A. Byrd, J. Magn. Reson. **52,** 221 (1983).

dependent on the hydrocarbon chain composition of the phospholipid and the environment of the bilayer. The H_{II} phase was initially characterized by X-ray diffraction.[18] Subsequently it was shown that ³¹P NMR powder patterns were exquisitely sensitive to the L_α-to-H_{II} phase transition.[19] These powder patterns have been used by many investigators to monitor the phase behavior of phospholipid dispersions.

In studying phase transitions from L_α to H_{II} for various lipid systems, evidence of the I_s structures was obtained from ³¹P NMR. ³¹P NMR powder patterns provide distinctly different spectral shapes for the L_α phase, the H_{II} phase and the I_s structures.[13] As an example, consider the behavior of the mixed soy phosphatidylethanolamine/egg phosphatidyl-choline system.[20] This system undergoes the L_α-to-H_{II} phase transition as a function of the lipid content of the membrane and as a function of temperature. A high phosphatidylethanolamine content in the membrane and elevated temperature favor the formation of the H_{II} phase. In addition, in this study a region of temperature and composition was found where another structure appeared, characterized by an isotropic ³¹P NMR resonance. In freeze–fracture electron micrographs of the system, "lipidic particles" were found in the same temperature range and composition where the isotropic ³¹P NMR resonance was observed. These lipidic particles are possibly the source of the isotropic ³¹P NMR resonance[20,21] and are referred to herein as I_s.

A relationship between the presence of the isotropic ³¹P NMR resonance in the ³¹P NMR spectrum of phospholipid vesicles and membrane fusion was reported[8] for the N-monomethyldioleoylphosphatidylethanola-mine system. Membrane fusion in this system began with the first appearance of I_s[8] and increased with the increase in intensity of I_s when the system was below the L_α-to-H_{II} phase transition temperature.

Subsequently the relationship between the appearance of I_s and membrane fusion was investigated more thoroughly in a concerted study involving fusion assays, ³¹P NMR, differential scanning calorimetry, freeze–fracture electron microscopy and time-resolved X-ray diffraction.[9] The fusion of LUV made of lipids which form I_s was studied. These studies confirmed the correlation between the appearance of I_s and membrane

[18] F. Reiss-Husson, *J. Mol. Biol.* **5**, 363 (1967).
[19] P. R. Cullis and B. deKruijff, *Biochim. Biophys. Acta* **507**, 207 (1978).
[20] S. W. Hui, T. P. Stewart, P. L. Yeagle, and A. D. Albert, *Arch. Biochem. Biophys.* **207**, 227 (1981).
[21] A. J. Verkleij, C. Mombers, W. J. Gerritsen, L. Leunissen-Bijvelt, and P. R. Cullis, *Biochim. Biophys Acta* **555**, 358 (1979).

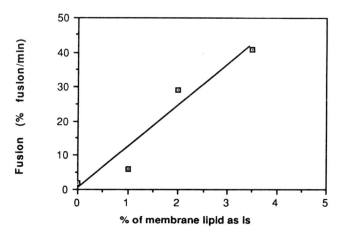

FIG. 3. Relationship between the amount of lipid in I_s and the rate of fusion for LUV of *N*-monomethyldioleoylphosphatidylethanolamine.

fusion, both in the monomethylphosphatidylethanolamine system and in other vesicle systems having different lipid compositions. Figure 3 shows the relationship between the amount of I_s present in the vesicles (measured from spectra of the kind represented in Fig. 2) and the rate of fusion in a representative experiment.[22] The percentage of phospholipid in I_s is simply calculated by a deconvolution of the resonance intensities of the bilayer powder pattern and the isotropic resonance in the spectra (see Fig. 2). These data suggest that the rate of vesicle membrane fusion, measured by fluorescence assay for aqueous contents mixing,[23] is directly proportional to the percentage of the total membrane lipid found in I_s. On the basis of such studies, the hypothesis was advanced that I_s was involved in the mechanism of membrane fusion for these lipid systems as an intermediate in the fusion pathway.[9]

This hypothesis has been tested further. Small amounts of diacylglycerol (1–2 mol %) in the lipid bilayers cause dramatic reductions in the stability of the phospholipids in the bilayer structure. This effect shows most clearly in a reduction in the temperature at which the I_s appear. Fusion in this vesicle system follows the appearance of I_s. When I_s appear at a lower temperature due to the presence of diacylglycerol, fusion occurs at a lower temperature, again in direct proportion to the incidence of I_s.[10]

[22] P. L. Yeagle, J. Young, S. W. Hui, and R. M. Epand, *Biochemistry* **31**, 3177 (1992).
[23] N. Düzgüneş and J. Wilschut, this volume [1].

Practical Use of ^{31}P NMR in Studying Phospholipid Morphology
Related to Membrane Fusion

Sample Preparation

Membrane samples are prepared in the same manner as for fusion experiments, if the morphology observed in the ^{31}P NMR spectrum is to be related to the fusion studies. In many cases, this involves extrusion of vesicles as described elsewhere in this volume.[23] Significantly larger quantities of NMR samples than are normally used in the fusion assays must be prepared. To gain adequate sample material for the NMR experiment, these larger volumes of extruded vesicles are centrifuged into a soft pellet in an ultracentrifuge (e.g., in a Beckman 50 rotor at 45,000 rpm for 20 min). The pellet is then transferred to an NMR sample tube. Often a large sample tube, such as a 10-mm diameter tube, is used to provide sufficient volume to hold the membrane material necessary for the experiment. For many commercial NMR spectrometers, 20 to 40 mg of phospholipid in the sample will produce an adequate signal-to-noise ratio in 1 or 2 hr of accumulation time. Note that the necessary accumulation time is related to the square root of the amount of sample. If the amount of available sample is reduced by one-half, then the necessary accumulation time will be 4 times as long. Therefore, although in principle one can simply accumulate spectra for longer periods of time when using smaller amounts of material, there is a practical limit to this exercise. Note also that there is a difference in the volume occupied by the phospholipid dispersions and by biological membranes. In the former case, one can readily work with phospholipid concentrations as high as 200 mg/ml. In the latter case, one will encounter great difficulty working with concentrations higher than 30–40 mg/ml of phospholipid (even if such large amounts of material are available). This is due to the inability to pack the biological membranes more densely.

Because of the extended time periods required for the ^{31}P NMR experiments, care must be taken in protecting the sample from degradation. Two precautions can be employed: (1) purge the buffers and the gas above the sample of oxygen, by flushing with nitrogen or argon; (2) include small amounts of ethylenediaminetetraacetic acid (EDTA, 0.1–1.0 mM) to chelate metal ions that might catalyze oxidation. Furthermore, the behavior of the temperature controller on the NMR spectrometer should be carefully monitored. When commercial temperature controllers adjust the temperature, serious overshoot can sometimes result. This can lead to sample degradation and in some cases causes the material to enter a metastable state which cannot be readily reversed.

When working with lipids capable of nonlamellar phase behavior, one must be especially alert to the problem just mentioned. The careful temperature-dependent work reported in the literature has all been done as monotonic heating experiments. The samples are prepared and maintained at or below the lowest temperature of interest prior to measurement. Then measurements are made with consecutive increases in temperature. If one allows temperature variations or deliberately lowers the temperature after heating the sample, many phospholipids capable of forming nonlamellar phases can become trapped in a cubic phase. This phase can persist even after lowering the temperature well below that at which the cubic phase appears on sample heating.

^{31}P NMR Spectroscopy

Under typical conditions ^{31}P NMR spectra are obtained in 10-mm tubes. A fully phased cycle (32 pulse) Hahn echo is used. The echo sequence eliminates baseline artifacts, removing the need for first-order phase corrections.[17] Gated proton decoupling (only during acquisition) at a decoupling field of 9 kHz is employed to eliminate sample heating. A 50-kHz spectral width is used. A delay time of 1 sec is used between pulses, and the delay time should be carefully chosen. If there are resonances with different T_1 (spin lattice relaxation) in the spectrum, a delay that is too short can selectively attenuate some of the resonances relative to the others. Fortunately, for phospholipids the ^{31}P T_1 is similar, irrespective of the head group structure or whether the phosphilipids are in small vesicles, large membrane vesicles, or I_s. Given the ^{31}P T_1 of the phospholipid head groups, a 1-sec delay is probably the most efficient manner in which to obtain ^{31}P NMR spectra of phospholipid dispersions.

The observed pulse is located in the center of the powder pattern. Problems can arise if any pulses used exceed 20 μsec. Wide powder patterns will not be adequately excited because of the length of the pulse (a longer observe pulse excites a narrower frequency window in the spectrum). This can also lead to distortions of the observed powder patterns.

In the case of gel-phase phospholipids and poorly hydrated phospholipids, it may be necessary to use cross-polarization ^{31}P NMR to readily obtain the ^{31}P powder patterns. This approach exploits the strong dipolar interactions that develop in the solid state. The application of this approach to phospholipids can be found elsewhere.[24] Finally, the only ^{31}P nuclei in the preparation should be in the phospholipid component of the membranes for the most simple interpretation of the observed spectra.

[24] J. Frye, A. D. Albert, B. S. Selinsky, and P. L. Yeagle, *Biophys. J.* **48**, 547 (1985).

Interpretation of Results

Adhering to the principles enumerated above makes the interpretation of the observed spectra relatively simple. Qualitatively, the L_α, H_{II}, and I_s structures can be readily identified by their characteristic powder patterns (or lack thereof in the case of I_s). Even more simply, the presence of these morphologies can be inferred by the position of the maxima of the powder pattern. The maximum for L_α occurs about 12–15 ppm on the shielded side (upfield or on the right side of the spectrum as usually presented) of I_s depending on T_2 and the resulting shape of the powder pattern. The maximum for H_{II} occurs about 6 ppm deshielded (downfield or on the left of the spectrum as usually presented) from I_s.

However, to be certain of the behavior of the phospholipid dispersions it is necessary to obtain undistorted powder patterns. In the gel phase the spectral shapes vary considerably from the L_α phase ^{31}P NMR powder pattern, because of the dominance of slower motions. Also, the relative amounts of several phases in a sample cannot be determined accurately unless the spectra are undistorted.

^{31}P NMR can reveal several properties of the gel phase of phospholipids. For one, the expression of the chemical shift anisotropy (for the most part, the linewidth) is increased in the gel state. Figure 4 shows an example. This may reflect the ordering of the director of head group axial diffusion. The unusual lineshape is a result of cross-polarization in the liquid crystal state. For details, refer to published data.[24]

Thus, ^{31}P NMR can provide quantitative information about the relative population and properties of phospholipid phase structures, quantitatively, with no perturbations of the sample. ^{31}P NMR is in principle capable of observing all the phospholipids in the sample; this is not necessarily true of other techniques. In particular, ^{31}P NMR is sensitive to I_s which may be important to fusion, whereas other techniques, such as X-ray and differential scanning calorimetry, commonly used to observe phospholipid phase behavior, are not.

The most familiar parameter from NMR is the isotropic chemical shift. For all nuclei, including ^{31}P, the isotropic chemical shift is extremely sensitive to the chemical environment of the nuclei under investigation. However, the isotropic chemical shift is seen only in solution, that is, in systems undergoing rapid, isotropic motion or when using magic angle spinning.[25] In the absence of motion or in the case of anisotropic motion, the observed NMR spectrum does not normally consist of the narrow,

[25] A. D. Albert, J. S. Frye, and P. L. Yeagle, *Biophys. Chem.* **36,** 27 (1990).

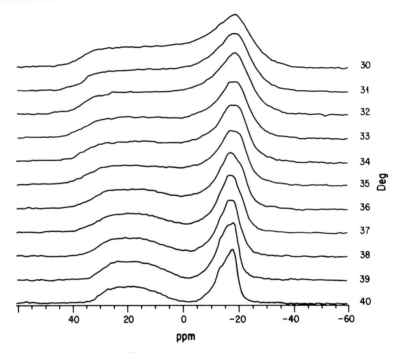

FIG. 4. Cross-polarization ^{31}P NMR spectra (60 MHz) of dipalmitoylphosphatidylcholine through the gel-to-liquid crystalline phase transition. The loss of intensity in the middle of the powder pattern has been explained elsewhere.[20]

isotropic chemical shifts of the observed nuclei, but rather broad, anisotropic lineshapes are recorded. Under the appropriate conditions, these lineshapes reflect aspects of the fundamental chemical shift tensor.

In general, the chemical shift of ^{31}P nuclei is not identical for all orientations of the phosphorus-containing compound with respect to the external magnetic field of the measuring instrument. The chemical shift is a measure of the shielding of the nucleus from the external magnetic field by the surrounding electrons. Because the distribution of electrons around the phosphorus nucleus in the head group of phospholipids is not symmetrical, the shielding is not isotropic. Therefore, if a single crystal of a phosphate-containing compound were examined, the chemical shift would be a function of the orientation of the crystal in the magnetic field. This three-dimensional character of the chemical shift is represented by the chemical shift tensor.

In the case of a dry powder containing phospholipids, all possible orientations of the phospholipid phosphate, with respect to the external magnetic field, are expressed. Therefore, all possible chemical shifts are simultaneously observed. The result is a powder pattern stretching over the full range of chemical shieldings possible from all the orientations of the phospholipid head groups.

Problems

In any fusion system, the true fusion intermediates may be transient structures that do not have a sufficient lifetime to be detected in the ^{31}P NMR experiment. Although the species we have called I_s have been suggested as an intermediate in the fusion event,[10] in fact one cannot rule out short-lived intermediates in the fusion process, which cannot be detected on the time scale of the ^{31}P NMR measurements.[26] In particular, there might be important structural intermediates with lifetimes of seconds or minutes. Because the ^{31}P NMR measurements can take 1 hr or more, any species with lifetimes significantly shorter than that may not be observed, unless the rate of formation is such as to guarantee that a significant population persists through the time of measurement.

[26] E. L. Bearer, N. Düzgüneş, D. S. Friend, and D. Papahadjopoulos, *Biochim. Biophys. Acta* **693**, 93 (1982).

[7] Fusion of Spherical Membranes

By SHINPEI OHKI

Introduction

The events of membrane fusion have been observed or suggested in many cellular processes such as exocytosis, endocytosis, membrane assembly, and fertilization. Because it is difficult to study the molecular mechanism of membrane fusion in biological systems, many researchers have worked on fusion using model membrane systems, especially those composed of lipid bilayers. In the early 1970s, the following lipid membrane systems were explored in studies of membrane fusion. (1) Fusion between lipid vesicles or liposomes as developed by Bangham *et al.*,[1] was studied by

[1] A. D. Bangham, M. W. Hill, and N. G. A. Miller, *Methods Membr. Biol.* **1**, 1 (1974).

METHODS IN ENZYMOLOGY, VOL. 220

Papahadjopoulos and others.[2,3] The methods used to detect fusion were monitoring the mixing of membrane components with the use of differential scanning calorimetry (DSC) and observing the morphological changes in lipid vesicles with the electron microscope. (2) Fusion between bilayers using two semispherical membranes was studied by Liberman and Nenashev[4] and later Neher.[5] These fusion events were monitored by measuring the impedance change of two interacting bilayers. (3) In the mid-1970s, fusion of vesicles with lipid bilayers was also studied by measuring the conductance changes of the lipid bilayers (Moore[6] and others[7,8]). (4) The fusion of spherical bilayer membranes was studied by Breisblatt and Ohki[9] as an alternative system for the semispherical bilayer membranes. The method of monitoring fusion of spherical bilayers was visual observation through an optical microscope. (5) In the late 1970s, one other membrane fusion system, the vesicle–lipid monolayer system, was devised and studied by Ohki and Düzgüneş:[10] fusion of vesicles with monolayers, a decrease in the surface tension of the monolayer was observed. The last two experimental methods, (4) and (5), and their experimental results are described in this chapter and in [8] in this volume.

Materials and Methods

Chromatographically pure phospholipids are used to form spherical membranes. Phospholipids may be purchased from Avanti Polar Lipids (Alabaster, AL). Phospholipids dissolved in chloroform are kept below −15°. Before use, the purity of phospholipids should be checked using thin-layer chromatography; the lipids should display a single spot. The membrane-forming solution is prepared by first evaporating the chloroform component of a phospholipid solution with nitrogen gas or *in vacuo,* then dissolving the phospholipids (10 mg) in 0.4 ml chloroform, 0.3 ml methanol, and 0.3 ml *n*-decane. Although slight variations in components and concentrations of the membrane-forming solution are allowed, the concentrations of the individual solvents are extremely important and

[2] D. Papahadjopoulos, G. Poste, B. E. Schaeffer, and W. J. Vail, *Biochim. Biophys. Acta* **352**, 10 (1974).
[3] D. Papahadjopoulos, K. Jacobson, and T. Isac, *Biochim. Biophys. Acta* **311**, 330 (1973).
[4] E. A. Liberman and V. A. Nenashev, *Biofizlca* **15**, 1011 (1970).
[5] E. Neher, *Biochim. Biophys, Acta* **373**, 327 (1974).
[6] M. R. Moore, *Biochim. Biophys. Acta* **426**, 765 (1976).
[7] J. E. Cohen and M. M. Moronue, *J. Supramol. Struct.* **5**, 409 (1976).
[8] N. Düzgüneş and S. Ohki, *Biochim. Biophys. Acta* **467**, 30 (1977).
[9] W. Breisblatt and S. Ohki, *J. Membr. Biol.* **23**, 385 (1975).
[10] S. Ohki and N. Düzgüneş, *Biochim. Biophys. Acta* **552**, 438 (1979).

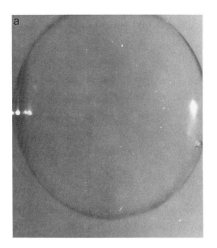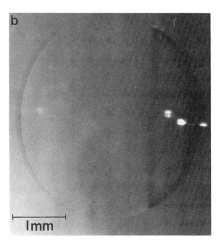

FIG. 1. (a) Spherical membrane prior to bilayer formation. (b) Spherical bilayer (arrow indicates bilayer formation).

have to be kept at the proper ratios in order for the membranes to form properly. Chloroform is the key solvent for a proper membrane-forming solution.

Spherical bilayer membranes are formed in a layered concentration gradient containing a base solution (bottom phase) of 4.5 M NaCl with an upper solution of 10 mM NaCl. The inside solution of the spherical membrane is 0.2 M NaCl. All solutions are buffered with 5 mM HEPES, at pH 6.5. Spherical membranes can also be formed on other salt gradients.[11] The inside solution for the spherical membrane is stored in a microsyringe (Gilmont, Great Neck, NY, 0.2 ml, 2-μl divisions) and is delivered through a polyethylene tube (internal diameter 0.03 inch). The lipid solution described above is drawn into the tube at the tip and then pushed out with the inside solution, forming a bubble. The diameter of the bubble is in the range of a few to several millimeters. Through manipulation, the bubble can be detached into the density gradient, with the bubbles remaining in the upper gradient phase.

Membranes are observed through a low power (30–40×), wide-field microscope while the bubble membranes are illuminated with a microscope illuminator. Bilayer formation is observed as the spherical bilayer membrane starts to thin out. Bilayer formation takes place at the bottom of the bubble (Fig. 1a) as the excess lipid is extruded into a cap at the top of the spherical bilayer membrane (Fig. 1b). Total bilayer formation usually

[11] R. Pagano and T. E. Thompson, *Biochim. Biophys. Acta* **144**, 666 (1967).

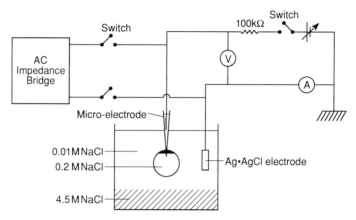

FIG. 2. Schematic diagram of instruments used for current and capacitance measurements.

takes 10–15 min and stops at the membrane cap, which is composed mostly of excess lipid and decanes. Membranes are extremely stable once bilayer formation is completed. Some spherical membranes stay at the same position for several hours. From these observations, the spherical membranes appear to set their vertical positions in the gradient to their isodensity levels.

To confirm that the black membrane in Fig. 1b is a bilayer, membrane resistance and capacitance measurements can be made using the device shown in Fig. 2, in which a microelectrode (tip diameter ~ 1 μm, resistance ~ 1.0 MΩ) penetrates the cap of the spherical membrane, and a reference electrode (e.g., calomel electrode) in the bathing solution is connected to each measuring instrument (Fig. 2). Common values of bilayer resistance and capacitance are 10^8 Ω cm (ref. 12) and 0.4 μF/cm^2, (ref. 13) respectively. The surface area of a spherical bilayer membrane can be estimated from the size of the spherical membrane measured through a microscope with a calibrated reticle. The temperature of the solution can be varied during the experiment by using a water jacket around the trough and a thermoregulator.

Spherical membranes are formed in close proximity (within several millimeters) to facilitate membrane contact. For each study, approximately 10 membranes are made, and, after the bilayer formation is completed, they are maneuvered into contact with each other in order to form the doublet (Fig. 3). To monitor the temperature-dependent fusion of

[12] S. Ohki, *J. Colloid Interface Sci.* **30**, 413 (1969).
[13] S. Ohki, *Biophys. J.* **9**, 1195 (1969).

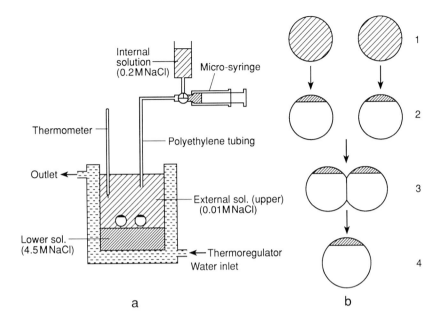

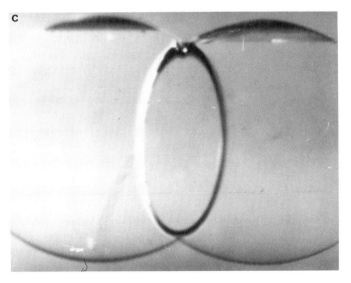

FIG. 3. Schematic diagram of the experimental arrangement for spherical bilayer fusion. (a) Formation of spherical membranes. (b) The sequence for membrane fusion. (c) A doublet of spherical bilayer membranes.

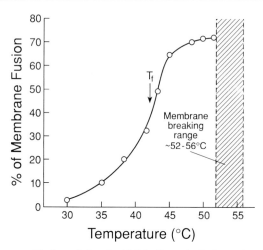

FIG. 4. Percentage of fusion of phosphatidylcholine spherical membranes with respect to temperature.

spherical membranes, the temperature is increased at 5-min intervals and allowed to equilibrate. The temperature scan is started at 25° and completed at 60°. At 25°, the percentage of fusion of the adhered spherical membranes is very small (see Figs. 4 and 5). Diffusion of the concentration gradient has been checked by including a dye in the solution, and the gradient was found to be stable up to 65° and for time intervals of 2 hr or

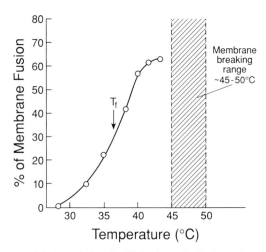

FIG. 5. Percentage of fusion of phosphatidylserine spherical membranes with respect to temperature.

more. Fusion of spherical membranes is observed through a microscope, as two adhering doublets become one spherical membrane. The effect of divalent cations can be measured in a similar manner, but in NaCl solutions (10 mM, 0.2 M, and 4.5 M) containing a certain concentration of divalent salt (Fig. 3).

Experimental Results and Comments

Temperature Effect

Spherical bilayer membranes composed of phospholipids are formed in a concentration gradient in close proximity to each other. Membrane fusion among spherical membrane doublets can be observed with increasing temperature. A spherical bilayer doublet can be defined by the close adhesion of two planar bilayers of two spherical membranes (as is indicated in Fig. 3c). Fusion is defined as the two adhering spherical cells (a doublet) forming one (Fig. 3b).

Figures 4 and 5 show the results obtained for phosphatidylcholine (PC) and phosphatidylserine (PS) membranes with the increase in temperature. The measurement of membrane fusion is based on only those membranes that have come into contact. In the case of PC membranes, approximately 80% of the membranes came into contact at 25° (the remaining 20% either broke or did not achieve contact), and 70% of these ultimately fused. Of the remaining 30% that did not fuse, approximately 8% remained as stable doublets, while the remaining ones (22%) broke. The curve for PC membrane fusion indicates a steep rise in the percentage of fusions at 45° (see Fig. 4). The temperature corresponding to the sharpest increase in the fusion curve is defined as T_f or the fusion threshold temperature. At this temperature PC membranes appear to have undergone some critical structural change due to increased fluidity and not due to a phase transition. The PC curve was compiled from 20 studies and contained results from over 200 membranes. As an alternative method, membrane fusion was scored at specific fixed temperatures. The results were equivalent, however, as the greatest increase in membrane fusion occurred between 44 and 45°. The PC membranes exhibited a breaking point in the range of 52–56°.

For the PS spherical bilayers (Fig. 5), 75% of the membranes came into contact, and approximately 65% of these fused. Of the remaining 25% that did not fuse, only 5% remained as stable doublets, and nearly 20% of the membranes in contact ruptured. A sharp rise in the curve for PS membrane fusion occurs at 38°. For the PS membranes, 20 individual studies were also done, representing the results from close to 200 membranes. The breaking point for PS membranes occurred in the range of 45–50°, somewhat lower than for the PC membranes.

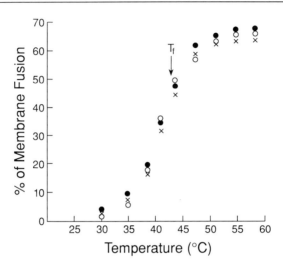

FIG. 6. Effect of divalent cations (1 mM) on the fusion of phosphatidylcholine spherical membranes suspended in 0.1 M NaCl buffer, pH 6.5, as a function of temperature. o, Mg^{2+}; ●, Ca^{2+}; ×, Mn^{2+}.

Divalent Cation Effect

For studies on the effect of divalent cations, equimolar concentrations of the divalent cations (usually 1 mM) are placed on both sides of the spherical bilayer membrane. The effects of Ca^{2+}, Mg^{2+}, and Mn^{2+} on the temperature-dependent fusion of PC and PS membranes are shown in Figs. 6 and 7, respectively. It is clear from Fig. 6 that none of the divalent cations at 1 mM affects the fusion of PC membranes. The T_f for PC membrane fusion remains unchanged from that in the absence of divalent cations. For PS membranes, on the other hand, the divalent cations (Ca^{2+}, Mg^{2+}) affect fusion by rendering it relatively independent of temperature, and the fusion temperature (T_f) is abolished. Ca^{2+} exerts a stronger effect over Mg^{2+} for PS membrane fusion. Both divalent cations enhance PS membrane contact, causing approximately 90% of the spherical membranes to form doublets and, in many cases, aggregates of more than two spherical membranes.

Effect of Other Lipids

The addition of cholesterol into PC and PS bilayer membranes abolishes the T_f and strongly inhibits fusion.[14] It is known that the addition of

[14] W. Breisblatt and S. Ohki, *J. Membr. Biol.* **29**, 127 (1976).

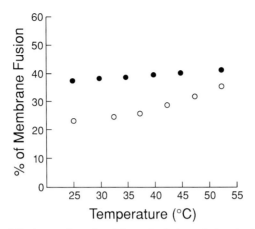

FIG. 7. Effect of divalent cations (1 mM) on the fusion of phosphatidylserine spherical membranes suspended in 0.1 M NaCl buffer, pH 6.5, as a function of temperature. ○, Mg^{2+}; ●, Ca^{2+}.

cholesterol to the membranes decreases membrane fluidity, that is, it makes the membrane (hydrocarbon chain packing) more rigid and constrained.[15] The motional freedom of the hydrocarbon chains is decreased, and the possibility of structural change occurring in the membrane due to increased temperature is suppressed. Therefore, cholesterol incorporation into the membrane should be expected to suppress the membrane fusion reaction induced by temperature.

Further support for the importance of membrane fluidity is given by our studies with some synthetic derivatives of phosphatidylcholine.[14] As evidenced by the different T_f values found for membranes composed of these phospholipids [dioleoylphosphatidylcholine (DOPC), 49°; dimyristoylphosphatidylcholine (DMPC), 59°; dipalmitoylphosphatidylcholine (DPPC), >70°], membrane fluidity appears to play a role in these differences. DPPC, which has the longest saturated hydrocarbon chains of the phospholipids studied, has the highest T_f [it also has the highest phase transition temperature (T_c 41°) and is least fluid with increasing temperature], whereas DOPC, which contains one double bond on each acyl chain, has a lower fusion temperature as well as a lower phase transition. The order of fusion temperature for these lipids is the same as that of their phase transition temperatures.

[15] B. D. Ladbrooke, R. M. Williams, and D. Chapman, *Biochim. Biophys. Acta* **150**, 333 (1968).

The effect of divalent cations on temperature-induced fusion of spherical bilayers relates both membrane fluidity and structural instability to the fusion event. It is clear that none of the divalent cations (Fig. 6) has any effect on the fusion of PC membranes. This is predictable because PC behaves as a neutral species, and divalent ions are not likely to be adsorbed or bound to the membrane strongly.[16-18] On the other hand, PS, which has a negatively charged polar group, will bind divalent cations (more than the neutral species), thus affecting the fusion process. As indicated in Fig. 7, the fusion of PS membranes is indeed affected: the T_f is abolished, and, at lower temperatures, fusion is enhanced.

The relatively temperature-independent process of fusion may be due to stabilization of the acidic phospholipid membrane by divalent cations, which is well documented in the literature.[19] However, the fairly high percentage of fusion for the overall temperature range may be due to another reason. Hauser et al.[20] have shown by NMR that the interaction of divalent cations with polar phospholipid head groups causes a water exclusion effect, which renders the phospholipid–divalent ion complex more hydrophobic. It appears, therefore, that there are two main factors contributing to membrane fusion in this system: instability of the membrane and hydrophobicity of the membrane surface. The latter is related to the increase in surface tension of the membrane surface. Although Ca^{2+} stabilizes the membrane, which in turn decreases membrane fluidity, the acidic phospholipid membrane surface becomes more hydrophobic with Ca^{2+} adsorption at the surface than in the absence of Ca^{2+}, that is, the surface may achieve a higher energy state. On the other hand, the relative temperature independence of fusion may be explained by the low fluidity of the membrane in the presence of Ca^{2+}. This interpretation is also consistent with observations with phospholipid vesicles.[3]

It seems obvious from these experimental results that both structural instability and surface hydrophobicity are closely related to fusion in this model membrane system. It has been shown that as membrane fluidity is increased by increasing temperature, more instability may arise in the membrane bilayer. This has been demonstrated by the effects of temperature and lysolecithin, divalent ions, cholesterol, etc. By increasing membrane fluidity, the freedom of motion of the hydrocarbon chains of lipid molecules can be increased so as to allow the hydrocarbon chains to be

[16] E. Rojas and J. M. Tobias, Biochim. Biophys. Acta 94, 394 (1965).
[17] H. Hauser and R. M. C. Dawson, Eur. J. Biochem. 1, 61 (1967).
[18] T. Seimiya and S. Ohki, Biochim. Biophys. Acta 298, 546 (1973).
[19] S. Ohki and D. Papahadjopoulos, in "Surface Chemistry of Biological Systems" (M. Blank, ed.), p. 155. Plenum, New York, 1970.
[20] H. Hauser, M. C. Phillips, and M. D. Barratt, Biochim. Biophys. Acta 413, 341 (1975).

exposed at the membrane surface. This increase in the hydrophobic surface area would cause an instability in the bilayer structure and possibly form what we have termed a "semimicelle" configuration.[9] If this occurs, hydrophobic interactions between apposed membrane surfaces could lead to membrane fusion as postulated.[21]

Some of the reasons why spherical membrane systems have not been utilized much for the study of membrane fusion may be the following. (1) The spherical membrane has a large radius of curvature compared to those of the lipid vesicles that have been used frequently for model membrane fusion studies. A small radius of curvature appears to play a crucial role for the process of lipid membrane fusion.[22,23] (2) The spherical membrane may contain a small amount of organic solvent, making it difficult to assign an unequivocal role to lipid or organic solvent molecules in the region of intermembrane adhesion.

In spite of this ambiguity and difficulty, the obtained fusion results give a qualitatively correct picture of lipid vesicle fusion. Furthermore, spherical membranes are the only system in which adhesion and fusion are visually observed through a microscope. With the recent development of video microscopy, one should be able to monitor the full course of membrane adhesion and fusion processes. Therefore, the spherical membrane fusion system can provide qualitative but reliable information about fusion with respect to external parameters such as temperature, pressure, or fusogenic substances.

[21] S. Ohki, in "Molecular Mechanisms of Membrane Fusion" (S. Ohki, D. Doyle, T. D. Flanagan, S. Hui, and E. Mayhew, eds.), p. 123. Plenum, New York, 1988.
[22] S. Ohki, in "Cell and Model Membrane Interactions" (S. Ohki, ed.), p. 267. Plenum, New York 1991.
[23] J. Wilschut, N. Düzgüneş and D. Papahadjopoulos, Biochemistry 20, 3126 (1981).

[8] Surface Chemical Techniques in Membrane Fusion

By Shinpei Ohki

In this chapter an experimental method to detect the fusion of lipid vesicles to lipid monolayers[1] is described, as well as some results obtained using this method. Various methods used to monitor the fusion of phospholipid membranes have been outlined in [7] in this volume.

[1] S. Ohki and N. Düzgüneş, Biochim. Biophys. Acta 552, 438 (1979).

Materials and Experimental Methods

Chromatographically pure phospholipids used in this experiment can be obtained from Avanti Polar Lipids (Alabaster, AL) or other companies. Each lipid should form a single spot on TLC (thin-layer chromatography). The phospholipids are dissolved in *n*-hexane [purity >99% as judged by gas chromatography (GC)] at approximately 2×10^{-4} M for the monolayer spreading solution. The exact concentration of the phospholipids may be determined by phosphate analysis (Fiske and Subbarow method).[2] Subphase solutions are 100 mM NaCl, 5 mM HEPES, 0.01 mM ethylenediaminetetraacetic acid (EDTA) ("NaCl buffer"). The pH is adjusted to 7.4 with NaOH. The salts used (reagent grade) should be heated at approximately 500° for 1 hr in order to remove organic contaminants. Water should be doubly distilled in a glass apparatus using alkaline permanganate in the first step. Divalent cation salts (all chloride salts of reagent grade) can be used without further treatment because the concentrations used are small compared with that of the monovalent salt.

Vesicle Preparation

Unilamellar phospholipid vesicles can be prepared according to published methods.[3] Phospholipids are suspended at a concentration of approximately 10 μmol phospholipid/ml in the buffer solution, vortexed for 10 min, and sonicated for 1 hr in a bath-type sonicator (Heat Systems-Ultrasonics, Farmingdale, NY) at room temperature and under an N$_2$ atmosphere. The vesicle suspensions are then centrifuged at 100,000 g for 1 hr, and the supernatant is used as a vesicle stock suspension.

Surface Tension Measurements

Monolayers are formed at the air–water interface of a fixed area (30–65 cm^2) in a round glass dish. Surface tension is measured using an electronic microbalance (e.g., Beckman) with a Teflon plate (11 $\times$ 11 $\times$ 1 mm) as a Wilhelmy plate (see Fig. 1). For these experiments the hydrophobic Wilhelmy plate gives nonerratic and reproducible results, and is superior to the hydrophilic (e.g., glass) plate. For each experiment, water surface tension is first measured to ensure that the aqueous surface is clean. The surface tension of water should also be checked using a glass Wilhelmy plate.

The depth of the dipped plate is kept constant at about 1.0 mm from the water surface, which is monitored by a microscope. For the monolayer

[2] C. H. Fiske and Y. Subbarow, *J. Biol. Chem.* **116**, 375 (1926).
[3] D. Papahadjopoulos, S. Nir, and S. Ohki, *Biochim. Biophys. Acta* **226**, 561 (1972).

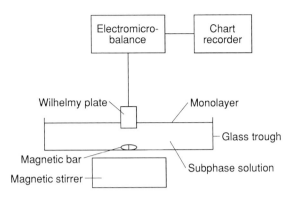

FIG. 1. Schematic diagram of the apparatus for measuring surface tension of the vesicle–monolayer system.

surface tension measurement, the plate is redipped after a monolayer of a given area/molecule is formed on the water surface. The monolayer exerts a net upward force on the Wilhelmy plate relative to the water surface. Certain amounts of concentrated lipid vesicle (10 μmol lipid/ml) and divalent ion (1 M) solutions are then injected into the subphase solution of the monolayer by way of microsyringes (Hamilton, Reno, NV), and the subphase solution is stirred well using a magnetic stirrer. Normally, the vesicles are first introduced and stirred for 1 min, and then, after waiting for 2 min, divalent ion concentrations are raised successively (injected at 1 or 2 mM increments, stirred for 1 min, and incubated for 2 min before the next injection). However, near the critical concentration at which the surface tension starts to decrease sharply, the increment of divalent ion concentration is reduced by 0.33 to 0.25 mM for each injection. The manner of increasing the divalent ion concentration (e.g., injecting amounts of concentrated solutions) as well as the incubation (waiting) time influence the experimental results slightly. However, the deviation in these results should fall within experimental error.

Force – Area Curve Measured by Teflon Wilhelmy Plate

Figure 2 shows the force – area curve for phosphatidylserine monolayers obtained with the Teflon Wilhelmy plate method. Because the contact angles between the Teflon plate and the hydrocarbon phases of lipid monolayers may be different for monolayers at different areas per molecule, and are not known, the surface tension of the monolayer is not shown in Fig. 2. When the area per molecule is 100 Å^2, the upward force acting on the plate is about 40 dyne/cm; when the area per molecule is below 50 Å^2,

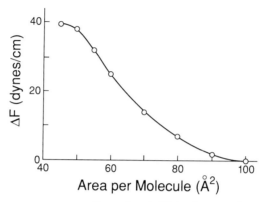

Fig. 2. Force versus area curve for a phosphatidylserine monolayer at the air–water interface. The difference in the upward force on the Wilhelmy plate between that at 100 Å² and that at the indicated area/molecule, $\Delta F = F(100 \text{ Å}^2/\text{molecule}) - F(X \text{ Å}^2/\text{molecule})$, is plotted against the area/molecule.

the force is nearly zero. The difference of about 40 dyne/cm between the forces at 100 and 50 Å²/molecule is slightly smaller than the equilibrium pressure (45 dyne/cm) obtained by use of other surface tension methods for phosphatidylserine monolayers.[4,5] The smaller value obtained in this study may be due to the nonzero contact angle, Θ, between the hydrocarbon phase of the lipid film and the Teflon plate. For an upper limit contact angle of 25°,[6] and a maximum film pressure γ, of 45 dyne/cm (the equilibrium surface pressure), $\Delta F = \gamma \cos \Theta = 41$ dyne/cm. This value compares well with the measured value of 40 dyne/cm.

Fusion Experiments and Comments

When divalent ion concentrations are increased in the presence of a certain amount of phosphatidylserine/phosphatidylcholine (1 : 1) vesicles, the surface tension of the phosphatidylserine monolayer does not show any appreciable change until a certain concentration of the divalent ion is attained. At or above this concentration which we call the threshold concentration, the surface tension reduces sharply toward a value of zero for the upward force (Fig. 3). This sharp decrease in surface tension is not observed when only phospholipid vesicles or divalent ions are present in

[4] S. Papahadjopoulos, *Biochim. Biophys. Acta* **163**, 240 (1967).
[5] T. Seimiya and S. Ohki, *Biochim. Biophys. Acta* **274**, 15 (1972).
[6] W. A. Zisman, *in* "Contact Angle, Wettability and Adhesion" (F. M. Fowkes, ed.), p. 1. 1964. American Chemistry Society, Washington, D.C.

TABLE I
THRESHOLD CONCENTRATIONS OF DIVALENT IONS THAT INDUCE FUSION[a]

NaCl buffer solution	Threshold concentration (mM)				
	Mg^{2+}	Sr^{2+}	Ca^{2+}	Ba^{2+}	Mn^{2+}
100 mM NaCl buffer	16 ± 2	7 ± 1	6 ± 1	3.5 ± 0.7	1.75 ± 0.5
1/10 diluted buffer solution	6.5 ± 1	4 ± 1	3 ± 0.7	1.5 ± 0.5	0.75 ± 0.3

[a] Phosphatidylserine monolayers (100 Å^2 molecule) fused with phosphatidylserine/phosphatidylcholine (1:1) vesicles (0.067 μmol phospholipid/ml) in 100 mM NaCl buffer solution and 1/10 diluted buffer solution.

the subphase of the monolayer. It should be noted that divalent ions reduce the surface pressure (or increase the surface tension) of acidic phospholipid monolayers to a small extent[7] (e.g., several dynes/cm for $1 mM Ca^{2+}$) which is the opposite of that when there are both vesicles and divalent cations present.

It must be emphasized that a Teflon Wilhelmy plate is necessary in order to achieve reproducible results for these measurements. A glass Wilhelmy plate often results in abrupt and erratic behavior of the output of the microbalance, perhaps because the vesicles interfere with proper adhesion of the monolayer to the glass surface.

The threshold concentrations for various divalent cations are given in Table I. Above their threshold concentrations, Mn^{2+}, Ba^{2+}, and Ca^{2+} caused sharp decreases in the surface tension of the monolayer, whereas with Sr^{2+} and Mg^{2+} the rate of decrease in surface tension was slower by severalfold compared with Ca^{2+} and Mn^{2+}. The value of the threshold concentration was affected slightly by the different ways of raising the divalent ion concentration in the subphase solution. For example, the longer the incubation time, the lower the threshold concentration that was obtained. However, the deviation of the threshold concentration obtained using different injection methods was not appreciably large, but fell within the experimental error obtained with a particular method of injection (see legend to Fig. 3). When divalent ions were injected first and then a certain amount of phospholipid vesicles introduced into the subphase solution, no change in surface tension was observed unless the concentration of divalent ions exceeded the threshold concentration. The threshold concentration also depended on the ionic strength of monovalent ions in the subphase. When the NaCl buffer was diluted 1:10 (v/v), the threshold

[7] S. Ohki, Biochim. Biophys. Acta 689, 1 (1982).

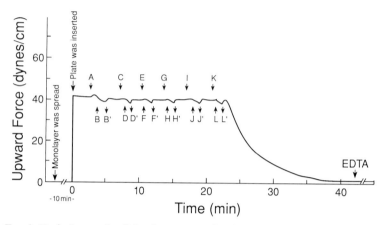

FIG. 3. Typical example of the time course of surface tension decrease for the case of a phosphatidylserine monolayer (100 Å²/molecule) in the presence of phosphatidylserine/phosphatidylcholine (1 : 1) vesicles (0.067 μmol lipid/ml) and various concentrations of Ca^{2+} in 100 mM NaCl buffer solution in the subphase. Here, instead of film surface tension, the upward force exerted on the Wilhelmy plate by the monolayer is plotted with time. After a complete monolayer is formed at the air–water interface, a certain amount of the concentrated vesicle suspension (10 μmol lipid/ml) is injected into the subphase solution (A), which is then stirred well for 1 min (started at B and stopped at B'), and the system is left for 2 min (incubation time) to observe any change in surface tension. Subsequently, an aliquot of CaCl solution (1 M) is added to bring the Ca^{2+} concentration of the subphase up to 2 mM (C), and this is followed by stirring for 1 min (started at D and stopped at D') and incubation for 2 min. Similar procedures are followed for each injection: 4 mM Ca^{2+} (E) and stirring (started at F and stopped at F'); 5 mM Ca^{2+} (G) and stirring (H to H'); 5.5 mM Ca^{2+} (I) and stirring (J to J'); 6.0 mM Ca^{2+} (K) and stirring (L to L'). In some cases, near the threshold concentration, the increment of divalent ion concentration is reduced by 0.33 to 0.25 mM for each injection. The added EDTA is equivalent to the divalent ion concentration in the subphase.

concentration was reduced to about one-half of that for the nondiluted NaCl buffer solution for each divalent ion (see Table I). It has been demonstrated by radioisotope tracer,[8] surface potential,[9] and electrophoresis studies[10] that the amounts of divalent ions bound to phosphatidylserine membranes depend on the concentration of monovalent ions in the subphase solutions.

With the same divalent ion, the values of the threshold concentration also depend on the area per molecule for the monolayer. The smaller the

[8] E. Rojas and K. Tobias, *Biochim. Biophys. Acta* **94**, 394 (1965).
[9] S. Ohki and R. Sauve, *Biochim. Biophys. Acta* **511**, 377 (1978).
[10] S. McLaughlin, N. Mulrine, T. Gresalfi, V. Gerard, and A. McLaughlin, *J. Gen. Phys.* **77**, 445 (1981).

TABLE II

THRESHOLD CONCENTRATIONS OF DIVALENT IONS AT VARIOUS AREAS/MOLECULE[a]

Ion	Threshold concentration (mM) at area/molecule (Å²/molecule)					
	110	100	90	80	70	60
Mg^{2+}	—	16 ± 2	—	—	12 ± 2	—
Sr^{2+}	—	7 ± 1	—	—	5 ± 1	—
Ca^{2+}	6.5 ± 1	6 ± 1	5.5 ± 1	5 ± 0.8	4 ± 0.7	3.5 ± 0.6
Ba^{2+}	—	3.5 ± 0.7	—	—	2 ± 0.4	—
Mn^{2+}	—	1.75 ± 0.5	—	—	1 ± 0.3	—

[a] Phosphatidylserine monolayers in the presence of phosphatidylserine/phosphatidylcholine (1:1) vesicles (0.067 μmol phospholipid/ml) in 100 mM NaCl buffer solution.

area per molecule, the lower the threshold concentration (Table II), indicating that this interaction is strongly related to the net negative charge density on the membrane surface and to the binding of the divalent cation to these negative charge sites.

The concentration of phospholipid vesicles in solution greatly affected the rate of reduction of the surface tension of the monolayer but only slightly influenced the magnitude of the threshold concentration. Higher vesicle concentrations enhanced the rate of reduction of the surface tension and lowered slightly the magnitude of the threshold concentration of divalent ions.

In all the cases mentioned above, the order of the threshold concentrations of divalent cations was unchanged: $Mn^{2+} > Ba^{2+} > Ca^{2+} \geq Sr^{2+} > Mg^{2+}$. This order agrees well with the ability of these ions to induce the fusion of small unilamellar phosphatidylserine vesicles.[7,11] The order is also the same as that of the association constants of divalent cations interacting with phosphatidylserine membranes, obtained from electrophoretic measurements of phosphatidylserine vesicles,[10] and surface potential measurements of phosphatidylserine monolayers.[12] The order also agrees well with that obtained from turbidity measurements on phospholipid vesicle suspensions[1] and those from direct measurements of divalent ion binding to phosphatidylserine molecules.[13] These agreements also support the hypothesis that the divalent ion-induced vesicle–monolayer interaction is strongly related to the degree of binding of divalent ions to the surface of the membrane.

[11] J. Wilschut, N. Düzgüneş and D. Papahadjopoulos, *Biochemistry* **20**, 312 (1981).

[12] S. Ohki and R. Kurland, *Biochim. Biophys. Acta* **654**, 170 (1981).

[13] S. Nir, C. Newton and D. Papahadjopoulos, *Bioelectrochem. Bioenerg.* **5**, 116 (1978).

It is important to mention that the divalent ion-induced lowering of surface tension of the monolayer in the presence of vesicles was not reversed or altered by the addition of EDTA in quantities equivalent to the divalent ion concentrations in the subphase solution. When EDTA was added while the surface tension of the monolayer was decreasing with time, no further reduction or increase in surface tension was observed.

Experiments similar to those described above can be performed by varying the phospholipid components of the monolayer and the vesicles, keeping all other conditions the same. For example, when a phosphatidyl-serine/phosphatidylcholine (1 : 1) monolayer and vesicles of the same molecular composition were used, higher values for the threshold concentrations of Mn^{2+} and Ca^{2+} were obtained (8 ± 1 and 20 ± 2 mM, respectively for a monolayer of 100 Å^2/molecule; 5 ± 0.5 and 13 ± 1.3 mM, respectively for 65 Å^2/molecule) than those with a phosphatidylserine monolayer (Table I). In this case Mg^{2+} caused no change in surface tension up to 25 mM. For a phosphatidylcholine monolayer and phosphatidylserine/phosphatidylcholine (1 : 1) vesicles, similarly high threshold concentrations were observed for each divalent ion. It is interesting to note that in this case the dependence of the threshold concentration on the area per molecule of the monolayer was reversed from that observed with phosphatidylserine monolayers. For Mn^{2+}, the threshold concentration was 9 ± 1 mM at 100 Å^2/molecule, 12 ± 1.3 mM at 80 Å^2 molecule, and 15 ± 1.5 mM at 65 Å^2/molecule. For Ca^{2+} no change in surface tension was observed up to 25 mM. When a phosphatidylcholine monolayer and phosphatidylcholine vesicles were used, no surface tension change was observed up to 25 mM of any divalent ion used (Mn^{2+}, Ca^{2+}, and Mg^{2+}).

In the case of a phosphatidylserine monolayer and phosphatidylserine vesicles, the critical concentrations of Ca^{2+} were about 1.5 mM for monolayers of 100 Å^2/molecule and about 1.2 mM for monolayers of 70 Å^2/molecule in the presence of a vesicle concentration of 0.067 μmol phospholipid/ml in the 100 mM NaCl buffer solution. The threshold values for Mg^{2+} were about 10 and 7 mM, respectively. Other values are given in Table III. An area of 70 Å^2/molecule corresponds to that deduced from X-ray diffraction studies of lipid bilayer membranes.[14] It is interesting to note that the observed threshold concentration of Ca^{2+} is about the same magnitude at which fusion among phosphatidylserine vesicles was observed in an ionic environment similar to our experiments.[7,15] These experiments were particularly difficult because the surface tension, recorded

[14] V. Luzzati and F. Husson, *J. Cell Biol.* **12**, 207 (1962).
[15] N Düzgüneş, S. Nir, J. Wilshut, J. Bentz, C. Newton, A. Portis, and D. Papahadjopoulos, *J. Membr. Biol.* **59**, 115 (1981).

TABLE III
THRESHOLD CONCENTRATIONS OF DIVALENT IONS FOR FUSION OF PHOSPHATIDYLSERINE
MONOLAYERS AND PHOSPHATIDYLSERINE VESICLES[a]

Ion	Threshold concentration (mM) at area/molecule (Å^2/molecule)				
	100	90	80	70	60
Mg^{2+}	12 ± 1	—	—	8 ± 1	—
Ca^{2+}	1.5 ± 0.5	1.4 ± 0.5	1.3 ± 0.5	1.2 ± 0.5	1.1 ± 0.5
Mn^{2+}	1.2 ± 0.5	—	—	0.9 ± 0.5	—

[a] System consists of a phosphatidylserine monolayer of various areas/molecule and phosphatidylserine small unilamellar vesicles (0.04 μmol phospholipid/ml) in 100 mM NaCl buffer.

as the output of the microbalance, occasionally showed erratic behavior, which was confirmed to be unrelated to the instrument. Above 1 mM Ca^{2+}, the subphase solution became turbid, which indicated the formation of large aggregates among phospholipid vesicles and of fusion products.

The observed sharp decrease in the surface tension of certain monolayers in the presence of both phospholipid vesicles and divalent ions is interpreted as a consequence of fusion of vesicles with monolayer membranes. The reasons for this are the following. (1) The presence of phospholipid vesicles only in the subphase solution, without divalent ions, did not give any appreciable change (less than a few dynes/cm) in the surface tension of the monolayer, at least within a few hours, up to a vesicle concentration of 0.5 μmol phospholipid/ml. (2) Concentrations up to 50 mM of divalent ions without the vesicles did not result in any large change in surface tension of the monolayer. It has been observed that divalent ions reduce the surface pressure (or increase the surface tension) of acidic phospholipid monolayers to a small extent (a few dynes/cm).[7] (3) When the monolayer was left by itself, the surface tension was fairly stable, at least for a few hours. It has been found that temperature-dependent molecular dissolution occurs from the monolayer into the bulk phase.[16] However, phosphatidylserine as well as phosphatidylcholine monolayers at an experimental temperature of 24° do not show significant molecular dissolution,[16] and the effect of dissolution on surface tension is an increase in its magnitude, which would not account for the decrease in surface tension observed here. (4) Adhesion of vesicles to a monolayer may contribute to the reduction of the film tension because the surface tension of phospholipid vesicles may be smaller than those of monolayers at relatively

[16] E. Shapiro, Ph.D. Thesis, State University of New York at Buffalo (1974).

large areas per molecule (100 $Å^2$/molecule). However, irreversibility of decreased surface tension by the addition of EDTA at the final (equilibrium) stage, as well as during the surface tension decrease, suggests that vesicle adhesion is not a predominant cause of the observed phenomenon. The incorporation of lipid molecules from the vesicles into the monolayer is most likely responsible for this event. (5) It is possible that when the monolayer and vesicles are in close contact, molecular exchange[17] occurs between the membranes, mainly from the vesicles to the monolayer. The rate of reduction of the monolayer, however, is relatively rapid (change of 40 dynes/cm occurring with 10–30 min, depending on the concentration of phospholipid vesicles, the type of phospholipid vesicle, divalent ion concentration, etc.), so that molecular exchange, which is a rather slow process,[18] especially at the temperature used in our experiments, could not account for the observed large change in surface tension of the monolayers.

There are significant correlations between the present experimental results and those obtained in other systems used to study membrane fusion. The threshold concentration for Ca^{2+} of 1.2 mM for phosphatidylserine monolayers (at 70 $Å^2$/molecule) and phosphatidylserine vesicles corresponds well to the concentration (1.0 mM) at which fusion among small unilamellar phosphatidylserine vesicles occurs in an ionic environment similar to that of our experiments.[7,11,15] The slight difference in threshold concentrations between the two membrane fusion systems may indicate that the geometric difference in membrane structures between planar phospholipid bilayer and vesicle membranes does contribute to a slight difference in the degree of divalent ion-induced fusion among these membrane systems. The possible presence of organic solvents in the monolayer may affect the fusion process, but probably only to a small degree, judging from the good correlation between the two systems. Another correspondence is seen between two experimental systems in which phosphatidylserine/phosphatidylcholine (1:1) vesicles interact either with phosphatidylserine planar bilayers or phosphatidylserine monolayers. The threshold concentration for Ca^{2+} of 4 mM for phosphatidylserine monolayers at 70 $Å^2$/molecule in the presence of vesicles corresponds well to the Ca^{2+} concentration (~4 mM) at which phosphatidylserine bilayers in a similar ionic environment showed a large increase and discrete fluctuations of membrane conductance, which was interpreted to be a consequence of vesicle fusion with the bilayer membrane.[19]

[17] F. J. Martin and R. C. MacDonald, *Biochemistry* **15**, 321 (1976).
[18] J. M. Kremer and P. P. Wiersema, *Biochim. Biophys. Acta* **671**, 348 (1977).
[19] N. Düzgüneş and S. Ohki, *Biochim. Biophys. Acta* **467**, 301 (1977).

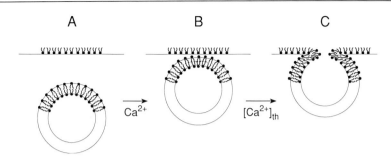

FIG. 4. Schematic diagram of the possible molecular events during the interaction of phosphatidylserine/phosphatidycholine vesicles with a phosphatidylserine monolayer. (A) No Ca^{2+} in the subphase. (B) A certain amount of Ca^{2+} in the subphase brings two membranes to close contact, possibly by Ca^{2+} binding to both membranes. (C) At the threshold concentration of Ca^{2+}, vesicle membranes fuse with the monolayer.

A possible fusion process is illustrated in Fig. 4. First, the two membranes should be brought into close contact. Divalent metal ions would mediate this process, especially for two acidic phospholipid membranes, by reducing surface charges by both screening and binding and/or bridging the two membrane surfaces. Second, the surface properties of the membranes should be altered for the two membranes to be able to fuse. The interaction of divalent cations with phospholipid polar head groups causes water exclusion from the membrane surfaces, which renders the surface of the phospholipid–divalent ion complexes more hydrophobic in nature.[20] It has been suggested[21] that this increased hydrophobicity of the two membrane surfaces would bring two membranes to a closer contact. Finally, a membrane deformation would be created at the boundary of the region of contact of the two membranes and act as the fusion site of the two membranes.

Since the study of Ohki and Düzgüneş,[1] a few reports using the fusion system have been published. One was the study on the effect of osmotic pressure on membrane fusion, where large unilamellar vesicles and monolayers were used,[22] and the other was the examination of the effect of

[20] H. Hauser, M. C. Phillips, and M. D. Barratt, *Biochim. Biophys. Acta* **413**, 341 (1975).
[21] S. Ohki, *in* "Molecular Mechanisms of Membrane Fusion" (S Ohki, D. Doyle, T. D. Flanagan, S. Hui, and E. Mayhew, eds.), p. 123. Plenum, New York, 1988.
[22] N. G. Stoicheva, I. C. Tsonova, and D. S. Dimitrov, *Comptes Rendus Acad. Bulgare Sci.* **37**, 1107 (1984).

Ca^{2+}-binding proteins and other membrane proteins on the fusion of vesicles to monolayers.[23]

Although a monolayer is half of a bilayer, and its physical properties may be different from those of a bilayer, some of the membrane properties which are relevant to membrane fusion are the same for both membranes; for example, the nature of the surface is more important to membrane fusion than that of the bulk phase of the membrane. The advantageous features of a monolayer are that it is mechanically stable and its area per molecule can be varied at will, whereas the area per molecule of a bilayer is difficult to control experimentally. This system may be one of the few systems which study fusion between dissimilar membranes, both in terms of geometry and membrane composition. This system is especially attractive for the study of the mechanisms of exocytotic membrane fusion where the two interacting membranes are always dissimilar to each other.

[23] S. Ohki and K. Leonards, *Chem. Phys. Lipids* **31**, 307 (1982).

[9] Fusion of Semispherical Membranes

By LEONID V. CHERNOMORDIK

Introduction

The physiologically important phenomenon of biomembrane fusion, namely, the simultaneous joining of contacting membranes and the volumes bound by them, appears to be a relatively simple process. At certain stages of the process, the lipid matrixes of the membranes in contact have to fuse, which necessitates some disturbance of the continuity of lipid bilayers. For this reason, the interaction between lipid bilayers has been investigated using different model systems.[1-3] The experimental model in which two semispherical lipid bilayers are brought into contact was proposed by Liberman and Nenashev in 1968.[4] In the golden age of other experimental systems developed to study lipid bilayer fusion (liposomes–

[1] G. Poste and G. L. Nicolson (eds.), "Membrane Fusion." Elsevier North-Holland Biomedical Press, Amsterdam, 1978.
[2] D. Evered and S. Whelan (eds.), "Cell Fusion." Pitman, London, 1984.
[3] A. E. Sowers (ed.), "Cell Fusion." Plenum, New York, 1987.
[4] E. A. Liberman and V. A. Nenashev, *Biofizika* **13**, 193 (1968).

liposomes[5,6] and liposomes–planar lipid bilayers[6a,7,8]), this model was practically forgotten. However, owing to some significant properties of the semispherical lipid bilayer model, the system proved to be very promising in clarifying the molecular mechanisms of certain stages of intermembrane interaction.[9] The model (also referred to as the model of two interacting planar lipid bilayers[9]) is quite simple and easily observed. Two semispherical lipid bilayers are brought into contact, usually by hydrostatic pressure. The stages of interaction are monitored with the aid of ordinary microscopes, as well as by the measurement of the capacitance and conductance of the interacting membranes, and the region of their contact. It is important to stress that this experimental model deals with the interaction of only two bilayers, that is, a single interaction event is investigated.

This chapter considers the experimental techniques applied to investigate the interaction of semispherical lipid bilayers. The main results obtained are discussed along with the advantages and limitations of the model.

Experimental System

Figure 1 shows the experimental cell used to form two planar lipid bilayers and bring them into contact.[10-12] (Somewhat different cells are described in Refs. 4 and 13–18.) Using the Mueller–Rudin technique, membranes are formed on openings 1 and 2 from decane or squalene

[5] S. Nir, J. Bentz, J. Wilschut, and N. Düzgüneş, *Prog. Surf. Sci.* **13**, 1 (1983).
[6] N. Düzgüneş, *in* "Subcellular Biochemistry" (D. B. Roodyn, ed.), Vol. 11, p. 195. Plenum, New York, 1985.
[6a] N. Düzgüneş, and S. Ohki, *Biochim. Biophys. Acta* **640**, 734 (1981).
[7] F. S. Cohen, M. H. Akabas, and A. Finkelstein, *Science* **217**, 458 (1982).
[8] F. S. Cohen, M. H. Akabas, J. Zimmerberg, and A. Finkelstein, *J. Cell Biol.* **98**, 1054 (1984).
[9] L. V. Chernomordik, G. B. Melikyan, and Yu. A. Chizmadzhev, *Biochim. Biophys. Acta* **906**, 309 (1987).
[10] G. Melikyan, I. Abidor, L. Chernomordik, and L. Chailakhyan, *Biochim. Biophys. Acta* **730**, 395 (1983).
[11] G. Melikyan, M. Kozlov, L. Chernomordik, and V. Markin, *Biochim. Biophys. Acta* **776**, 169 (1984).
[12] L. Chernomordik, M. Kozlov, G. Melikyan, I. Abidor, V. Markin, and Yu. Chizmadzhev, *Biochim. Biophys. Acta* **812**, 643 (1985).
[13] G. N. Berestovsky and M. Z. Gyulkhandanyan, *Stud. Biophys.* **56**, 19 (1976).
[14] S. A. Badzhinyan, S. A. Kovalev, and L. M. Chailakhyan, *Biofizika* **17**, 705 (1972).
[15] S. N. Viryasov and V. V. Perelygin, *Biofizika* **29**, 220 (1984).
[16] E. Neher, *Biochim. Biophys. Acta* **373**, 327 (1974).
[17] G. J. Brewer and P. D. Thomas, *Biochim. Biophys. Acta* **776**, 279 (1984).
[18] L. R. Fisher and N. S. Parker, *Biophys. J.* **46**, 253 (1984).

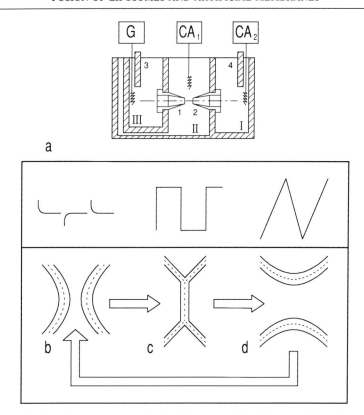

Fig. 1. Design of an experimental cell and its setup (a) and the principle of capacitance monitoring of the stages of membrane interaction (b–d). (a) I, II, and III denote separate cell compartments; 1 and 2, openings in the Teflon walls; 3 and 4, calibrated Teflon rods connected to micrometer screws. G is a generator; CA_1 and CA_2 are current amplifiers. (b–d) Cycle of interaction of bilayers (oscillograms of characteristic signals recorded at the output of CA_2 are displayed at top). (b) Membranes in contact. (c) Formation of a single bilayer in the contact region. (d) Membrane tube.

solutions of different lipids.[19] According to specific capacitance measurements, the membranes formed from squalene solutions can be regarded as solvent-free.[20,21] In contrast to bilayers made by opposing two lipid monolayers,[22] the solvent-free membranes of squalene–lipid solutions, like the usual solvent-containing membranes, are surrounded by a meniscus which

[19] P. Mueller, D. U. Rudin, H. T. Tien, and W. C. Wescott, *Nature (London)* **194,** 979 (1962).
[20] S. H. White, *Biophys. J.* **23,** 337 (1978).
[21] L. Chernomordik, G. Melikyan, N. Dubrovina, I. Abidor, Yu. Chizmadzhev, *Bioelectrochem. Bioenerg.* **12,** 155 (1984).
[22] M. Montal and P. Mueller, *Proc. Natl. Acad. Sci. U.S.A.* **69,** 3561 (1972).

enables one to bulge out such membranes, increasing their area by twice their original size.

By lowering or raising calibrated Teflon rods 3 and 4 (Fig. 1), which are connected to micrometer screws, the hydrostatic pressure in compartments I and III of the cell can be controlled. Increasing the level of solutions in these compartments to a certain extent brings the membranes into contact. In addition, by shifting the third compartment within the second it is possible to change the distance between openings 1 and 2.

Figure 1a shows a diagram of the cell, which is encased in a compartment with three glass windows in its walls to make it possible, using a microscope, to monitor visually both perpendicular to (in reflected light) and parallel (in transmitted light) to the contact plane of the membranes. To carry out various electric measurements Ag/AgCl electrodes are immersed in all the cell compartments. This simple setup includes a generator, G, and two current amplifiers, CA_1 and CA_2 (Fig. 1a), that can be readily used to monitor a number of electrochemical properties of contacting membranes.

Capacitive measurements hold a special place in examining the interaction of bilayer membranes. First, the specific capacitance value characterizes the thickness of membranes and their solvent content.[23] The increase of the contact region capacitance during monolayer fusion ("semifusion") reflects the growth in the area of "contact bilayer." [24] The capacitive measurements are used to record changes in the boundary potentials of the bilayers.[9,25] This method is based on the dependence of bilayer capacitance on the inner membrane electric field, which is the superposition of the external electric field, and the difference of boundary potentials. During the interaction of a negatively charged bilayer with a neutral bilayer, the exchange of lipids between the membranes is reflected in changes in their surface charges and, correspondingly, in the boundary potentials. The use of this technique in the system of interacting bilayers makes it possible to monitor the passage of lipids between different membranes, and between monolayers belonging to the same membrane.[9]

Finally, the simple method of capacitive current monitoring proposed by Neher[16] enables one to separate three possible states of the system (Fig. 1b–d): (1) bilayers before being brought into contact; (2) formation of a single bilayer in the contact region (monolayer fusion); and (3) appearance of a membrane tube following complete fusion of the contacting mem-

[23] H. T. Tien "Bilayer Lipid Membranes (BLM): Theory and Practice." Dekker, New York, 1974.

[24] G. Melikyan, L. Chernomordik, and I. Abidor, Biol. Membr. 2, 1048 (1985).

[25] I. Abidor, S. Aityan, V. Cherny, L. Chernomordik, and Yu. Chizmadzhev, Dokl. Akad. Nauk SSR 245, 977 (1979).

branes. Transitions between these stages can also be determined. The linear sweep of voltage is applied to the electrode in compartment III (Fig. 1a), and the current is recorded by the current amplifier connected to an electrode in compartment I, with the electrode in compartment II being grounded (Fig. 1a). When membranes are brought into contact, a small current, which is the doubly differentiated input signal, appears (Fig. 1b). This signal corresponds to the combination of two capacitors in series shunted to ground via a low impedance path. When monolayer fusion occurs, a purely capacitive current (rectangular) appears (Fig. 1c). Finally, the large conductive current of the membrane tube corresponds to the complete fusion of the membranes (Fig. 1d).

Measuring the conductance of bilayers can reveal local changes in the membrane structure, such as the appearance and development of pores with small radii (i.e., 0.5 nm). Conductance measurements have also been used to control the redistribution among the contacting bilayers of such membrane markers as oleic acid,[14] gramicidin,[16] cholesterol, and polyene antibiotics.[26,27]

The experimental setup shown in Fig. 1 is also used to determine the electromechanical stability of bilayers.[12] Analysis of the voltage dependencies of the mean lifetime for bilayers of different lipid compositions can be used to determine the linear tension of the hydrophilic pore in membranes (the work of formation of the unit of membrane area).[28] The value of this parameter provides quantitative information on the spontaneous curvature of monolayers comprising the membrane.[12]

Main Stages of Intermembrane Interaction: Typical Experiment

Drops of lipid solution (i.e., 10 mg of azolectin/ml of decane) are applied using a thin glass pipette on openings 1 and 2 of the cell (Fig. 1) and filled by electrolyte (usually $0.01 - 1\ M$ KCl). The process of spontaneous blackening (or thinning) of membranes is controlled visually and by monitoring the capacitive current of both membranes by a linear sweep of the applied voltage (usually 100 V/sec). When the specific capacitance reaches a constant value (varying for bilayer membranes of different composition in the range of 3 to 8 mF/m^2),[12,23] rods 3 and 4 (Fig. 1) are lowered to press the membranes toward each other.[12,23] A gradual decrease in the

[26] E. A. Liberman and V. A. Nenashev, *Biofizika* **17**, 1017 (1972).
[27] G. Melikyan, L. Chernomordik, I. Abidor, L. Chailakhyan, and Yu. Chizmadzhev, *Dokl. Akad. Nauk SSSR* **269**, 1221 (1983).
[28] I. Abidor, V. Arakelyan, L. Chernomordik, Yu. Chizmadzhev, V. Pastushenko, and M. Tarasevich, *Bioelectrochem. Bioenerg.* **6**, 37 (1979).

thickness of the water layer between them is monitored visually. When membrane contact is observed, the membranes tend to be flattened at the contact region.

Expulsion of the electrolyte from the region of interaction can continue until plane-parallel contact is established, under which conditions the hydrostatic pressure and molecular attraction (bringing the bilayers closer together) are equalized with the hydration and (for membranes with electrical charge of identical sign) the electrical repulsion force (which causes the bilayers to move apart).[29,30] The time it takes for the bilayers to establish such an equilibrium distance is determined by hydrodynamic factors[31] and depends on the composition of membranes and electrolytes.[4,13] The process of coming together is very nonuniform and is often accompanied by the appearance of dimples.[13] To determine the mean distance between the bilayers at the contact stage, one can use the measurement of a relaxation current flowing in a three-electrode system at a linear voltage sweep.[32] Optical methods have been applied to measure the distance between membranes with the aid of laser interferometry.[15]

The transition from plane-parallel contact to the next stage of membrane interaction is reflected by qualitative changes in practically any controlled parameter. Some examples are the appearance of black spots on the yellow background of the contact region observed "full face" and the change of the electrical equivalent schemes, corresponding to interacting membranes, owing to the disappearance of current leaks through the contact region, which can be monitored by means of the Neher method,[16]

The duration of time from the establishment of contact to the transformation of the system, along with the kinetics of this transformation, depends on the composition of membranes and electrolytes.[9,12,13] In bilayers containing solvent (decane) the transition has a very fast all-or-nothing character: many black spots appear simultaneously in the contact region, quite often before the establishment of plane-parallel contact in all the area of interaction. For solvent-free membranes, this transformation proceeds gradually from one black point that appears, as a rule, after waiting for some time following contact. Liposome–planar lipid bilayer interaction was shown also to be affected by the presence of solvent in planar lipid bilayers.[32a,32b]

[29] R. P. Rand, *Annu. Rev. Biophys. Bioenerg.* **10,** 277 (1981).

[30] R. P. Rand and V. A. Parsegian, *Biochim. Biophys. Acta* **988,** 351 (1989).

[31] D. S. Dimitrov, I. V. Ivanov, and T. T. Traikov, *J. Membr. Sci.* **17,** 79 (1984).

[32] I. Abidor, V. Pastushenko, E. Osipova, G. Melikyan, P. Kusmin, and S. Fedotov, *Biol. Membr.* **4,** 67 (1986).

[32a] J. Zimmerberg, *Bioscience Rep.* **7,** 251 (1987).

[32b] M. S. Perin and R. C. MacDonald, *J. Membr. Biol.* **109,** 221 (1989).

The structure of the contact region at this stage of interaction is quite stable. However, under certain conditions, one can observe transition of the system to a new state: a membrane tube is formed connecting openings 1 and 2 (Fig. 1). Once again, the transition is accompanied by drastic changes in all the controlled parameters. The conductance of the system increases significantly (Fig. 1d), and the membrane tube is quite visible through light microscope monitoring "in profile." [9,11]

The simplest, most universal way to trigger the transition of the system from interacting bilayers to the membrane tube is to apply a voltage pulse of sufficient amplitude and duration to the contact region (0.7 V, 0.1 msec pulse for phosphatidylcholine bilayers in 0.1 M KCl).[10] Spontaneous formation of the membrane tube is observed during the interaction of membranes of certain compositions (e.g., cardiolipin-containing bilayers) in the presence of Ca^{2+} ions.[9,27]

The membrane tube fissions into two separate bilayers when the electrolyte level in compartments I and III is decreased by raising rods 3 and 4 or when the distance between the bases of the membrane tube is increased by a critical amount. So, the system returns to its initial state. This cycle can be repeated dozens of times.

Investigation of Fusion Mechanisms in Model Systems

It is necessary to correlate the stages of membrane interaction observed for the model of two interacting bilayer lipid membranes with the stage corresponding to complete fusion in other model systems, as well as *in vivo*. It is apparent that the simultaneous joining of the membranes and the volumes bounded by them, that is, complete fusion of membranes, occurs during the formation of the membrane tube. Tube fission can be considered as an analogy to the membrane rearrangements in the course of membrane fusions involved in endocytosis and cell division. To clarify the mechanism of membrane fusion, it is necessary to understand the intermediate stages of the process. Thus, the structure of the contact region at the stage between plane-parallel contact and complete fusion is an essential one.

A number of different experimental approaches have been used to demonstrate the formation of a single bilayer from two apposed bilayers (monolayer or semifusion, or trilaminar structure formation) in the contact region at this stage of interaction.[9] The most direct proof that this structure is formed is the correspondence of the values of specific capacitance and electromechanical stability of the contact region with that of a conventional membrane of the same composition.[10,12,18] Experiments have also been conducted with the channel-former nystatin.[26,27] Nystatin causes a considerable increase in the conductance only when it is added symmetrically to a membrane (i.e., on both sides). Introducing nystatin into compartments I and III (Fig. 1) does not lead to an increase of conductance through two membranes. However, after the membranes came into contact, a drastic increase of conductance occurs, reflecting the formation of a single bilayer in which both monolayers are modified by nystatin.

It is important to note that the stage of monolayer fusion, revealed and thoroughly explored in the model of two interacting lipid bilayers, does not appear to be a specific feature of just this experimental system, but is probably observed in certain fusion processes in other model systems, including suspensions of liposomes[6,33-37] as well as the interaction of liposomes with planar lipid bilayers.[38] The local formation of trilaminar structures can also occur in cell membrane fusion,[40-41a] but has been a topic of controversy.[39]

The model under consideration has been used to investigate the mechanism of monolayer fusion.[12] In studying the interaction of solvent-free membranes, there is an opportunity to measure the waiting time of monolayer fusion, t_{mf}, that is, the time interval from the instant of formation of close plane-parallel contact to the initiation of monolayer fusion. The dependence of t_{mf} on the lipid composition of the membranes observed is qualitatively and quantitatively described in terms of the stalk model of membrane fusion proposed by Kozlov and Markin.[42] According to this model, bulging defects, representing local ruptures, appear in the contacting monolayers. Once close to each other, these defects form a stalk; the increase in the diameter of the latter gives rise to a contact bilayer (semifusion). The monolayers forming a stalk are considerably bent, and, consequently, they possess a considerable bending energy. The smaller the polar head group in the lipid molecule, compared with the width of the hydrocarbon tail (so-called nonbilayer lipids),[43] the lower the energy. Therefore, the stalk energy and, correspondingly, the waiting time of monolayer

[33] J. Rosenberg, N. Düzgüneş, and C. Kayalar, *Biochim. Biophys. Acta* **735**, 173 (1983).

[34] H. Ellens, J. Bentz, and F. C. Szoka, *Biochemistry* **24**, 3099 (1985).

[35] J. Bondeson and R. Sundler, *FEBS Lett.* **190**, 283 (1985).

[36] N. Düzgüneş, R. M. Straubinger, P. A. Baldwin, D. S. Friend, and D. Papahadjopoulos, *Biochemistry* **24**, 3091 (1985).

[37] J. Wilschut and D. Hoekstra, *Chem. Phys. Lipids* **40**, 145 (1986).

[38] I. N. Babunashvili and V. A. Nenashev, *Biofizika* **27**, 441 (1982).

[39] R. L. Ornberg and T. S. Reese, *J. Cell Biol.* **90**, 40 (1981).

[40] P. Pinto da Silva and B. Kachar, *Cell Biol. Int. Rep.* **4**, 625 (1980).

[41] J. A. Lucy, *Biochem. Soc. Trans.* **17**, 623 (1989).

[41a] L. Song, Q. F. Ahkong, G. Georgescauld, and J. A. Lucy, *Biochim. Biophys. Acta* **54**, 1065 (1991).

[42] M. Kozlov and V. Markin, *Biofizika* **28**, *242 (1983)*.

[43] A. J. Verkleij, J. Leunissen-Bijvelt, B. de Kruijff, M. Hope, and P. R. Cullis, *in* "Cell Fusion, Ciba Foundation Symposium 103" (D. Evered and S. Whelan, eds.), p. 45. Pitman, London, 1984.

fusion are determined by the effective shape of lipid molecules in a bilayer.[9,12]

For experimental verification of the stalk hypothesis,[12] solvent-free phosphatidylethanolamine membranes modified by different concentrations of lysophosphatidylcholine (added to the aqueous medium between membranes) are used. These membranes can be regarded as having two components, each with a different spontaneous curvature, and their relative concentration can be varied continuously. The influence of the waiting period of monolayer fusion on the lysophosphatidylcholine concentration is compared with predictions of the theoretical model. To avoid other unknown parameters in the comparison, the influence of lysophosphatidylcholine on the linear tension of the hydrophilic pores in membranes of the same composition is studied in parallel experiments, in order to ascertain if the spontaneous curvature of the membrane monolayer would influence not only the formation of the stalk, but also another reorganization of the bilayer, namely the formation of hydrophilic pores. The signs of curvature of the monolayer in a pore and in a stalk are opposite. The experimental dependencies of t_{mf} and linear tension on the lysophosphatidylcholine concentration prove to be quantitatively in agreement with theoretical predictions of the stalk model.[12] Also, structures similar to stalks have been observed in multilamellar liposomes of a certain composition.[44] Lysolipids inhibit not only the fusion of purely lipid bilayers, for these lipids also have been found to inhibit Ca^{2+}, H^+, GTP, and GTPγS triggered fusion of biological membranes.[44a]

The monolayer fusion stage has been shown to be the necessary intermediate step between contact and complete fusion of two interacting bilayer lipid membranes.[9,10,27] Applying an electric field to the membranes at the plane-parallel contact stage causes rupture of both membranes, but never leads to their complete fusion. Also, the Ca^{2+}-induced fusion of bilayers is always preceded by the local formation of a single bilayer in the contact region.

The experimental model under consideration has also been used to investigate the mechanism of the electrofusion of lipid bilayers.[10] The voltage dependence of the duration of a square pulse required for the induction of complete membrane fusion and the voltage dependence of the lifetime of a conventional bilayer are identical, suggesting that the same mechanism (the development of hydrophilic pores) underlies electrofusion and electroporation phenomena. Similar relationships between the ampli-

[44] V. Markin, M. Kozlov, and V. Boroviagin, *Gen. Physiol. Biophys.* **5**, 361 (1984).
[44a] L. V. Chernomordik, S. S. Vogel, E. Leikina, and J. Zimmerberg, *Biophys. J.* **61**, A499 (1992).

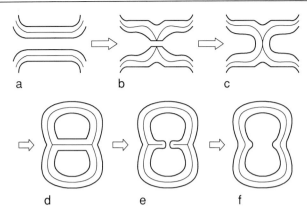

FIG. 2. Successive stages of membrane fusion according to the stalk mechanism.[42,47,48] (a) Membranes at the equilibrium distance. (b) Local approach and rupture of contacting monolayers. (c) Formation of a stalk. (d) Increase of the area of the contact bilayer. (e) Rupture of the contact bilayer, which completes the fusion process (f).

tude and duration of electric field pulses inducing fusion and electroporation is also observed for biological membranes.[45,46]

The results obtained by the experimental system described above have been discussed by Leikin *et al.* and Kozlov *et al.*[47,48] as background for a proposal concerning the mechanism of fusion of lipid membranes: fusion is considered as a process requiring successive local bending deformations in opposite directions (Fig. 2). Theoretical estimates[47] show that the thermal fluctuations of bilayers can provide the energy for overcoming the hydration barrier[29] and for the approach of membranes to a short distance of about 0.5 nm. As a result of the competition of the hydration repulsion and hydrophobic interactions,[47] the structure of contacting monolayers is disturbed, and a stalk is formed (Fig. 2b,c). Increase in the stalk diameter may lead to the formation of a single contact bilayer (Fig. 2d). The contact bilayer developed at this stage can occupy a small part of the area of intermembrane contact and may be a short-lived structure. To complete

[45] D. V. Zhelev, D. S. Dimitrov, and P. Doinov, *Bioelectrochem. Bioenerg.* **20,** 155 (1988).
[46] A. Sowers, *in* "Charge and Field Effects in Biosystems–2" (M. J. Allen, S. F. Cleary, and F. M. Hawkridge, eds.), p. 315. Plenum, New York, 1989.
[47] S. L. Leikin, M. M. Kozlov, L. V. Chernomordik, V. S. Markin, and Yu. A. Chizmadzhev, *J. Theor. Biol.* **129,** 411 (1987).
[48] M. M. Kozlov, S. L. Leikin, L. V. Chernomordik, V. S. Markin, and Yu. A. Chizmadzhev, *Eur. Biophys. J.* **17,** 121 (1989).
[49] D. Papahadjopoulos, S. Nir, and N. Düzgüneş *J. Bioenerg. Biomembr.* **22,** 157 (1990).

fusion, the contact bilayer has to be ruptured via the development of the overcritical hydrophilic pore (Fig. 2e).[48]

The monolayer fusion of membranes is promoted by the negative spontaneous curvature of the monolayers in contact. This may be achieved by the presence of nonbilayer lipids (which have a tendency to form the hexagonal phase), such as unsaturated phosphatidylethanolamines or cardiolipin, in the presence of Ca^{2+}. Complete monolayer fusion is promoted by the positive spontaneous curvature of the internal monolayers of liposomes. Such a spontaneous curvature can be determined, for example, by the electrostatic repulsion between the lipid polar head groups which, in contrast to the lipids in the external monolayer, make no contact with Ca^{2+}. The general picture of lipid bilayer fusion described above appears to be in agreement with those based on fusion phenomena in liposome suspensions[49] and lipid bilayers formed on the surface of mica cylinders.[50,51] The modified-stalk model has been suggested for virus influenza mediated fusion.[52]

Conclusions

The experimental model of two interacting semispherical lipid bilayers is, of course, only one of a number of models successfully developed for the investigation of membrane fusion. Naturally, any one of these models has its own limitations and advantages. Some of the limitations of the model are presented here. (1) The bilayer lipid membranes do not include proteins of cell membranes and, consequently, simulate only their lipid matrix. (2) It is necessary to remember that any bilayer lipid membrane certainly contains some amount of solvent. Even for the "solvent-free" squalene membranes, one cannot rule out the possibility of the existence in bilayers of a small number of solvent inclusions.[21] (3) The contact areas during the interaction of biological and liposomal membranes are significantly smaller than in the case of the semispherical lipid bilayers. In contrast to bilayer lipid membranes, cell and liposome membranes exist in closed (vesicular) form and have no menisci, which create tension of membranes and serve as a store of lipid molecules. This circumstance makes it difficult to use the semispherical bilayer model to study the osmotic effects in fusion.[7,53] However, this model has its unique advan-

[50] C. A. Helm, J. N. Israelachvili, and P. M. McGuiggan, *Science* **246,** 919 (1989).
[51] C. A. Helm, J. N. Israelachvili, and P. M. McGuiggan, *Biochemistry* **31,** 1794 (1992).
[52] D. P. Siegel, *in* "Viral Fusion Mechanisms" (J. Bentz, ed.), CRC, Boca Raton, in press.
[53] J. Zimmerberg, M. Curran, F. S. Cohen, and M. Broderick, *Proc. Natl. Acad. Sci. U.S.A.* **84,** 1585 (1987).

tages. The results presented here show the possibility of investigating single fusion events with easy to interpret and sensitive electrical measurements, simultaneously with visual control of the two semispherical bilayer system. Prospects of this system for studying the molecular mechanisms of various stages of membrane fusion are great.

Acknowledgments

I express my gratitude to the Department of Bioelectrochemistry of the Frumkin Institute of the Russian Academy of Sciences, where the studies described here were performed. In particular, I am very obliged to my friends and colleagues, Drs. M. Kozlov, G. Melikyan, S. Leikin, V. Markin, and Yu. Chizmadzhev, for fruitful discussions.

[10] Measurement of Interbilayer Adhesion Energies

By DAVID NEEDHAM

Introduction

Certain surfactant molecules, such as lipids, spontaneously form bilayer membrane-bound capsules in aqueous media. As a first approximation, they resemble the plasma and organelle membranes of biological cells and therefore serve as models of natural membranes. These bilayers are condensed states of matter that exhibit solidlike or liquidlike material behavior and interact nonspecifically via long range electrostatic, electrodynamic, and solvation forces. Basic knowledge of their mechanical and interactive properties is necessary in order to propose realistic mechanisms that underlie instability phenomena and interbilayer fusion. Experimental difficulties, however, arise in carrying out micromechanical and adhesion tests on lipid vesicles and cells owing to the small size range of these capsules ($5-50$ μm) and membranes (~ 5 nm) and the small stresses involved in mutual adhesion ($10^{-2}-10^{-1}$ dyne/cm), membrane deformation and destruction ($10^{-4}-30$ dyne/cm). Micromechanical methods have been developed that allow direct control over the application of such small forces and the microscopic measurement of the resulting deformations. These methods have been used to study certain aspects of the mechanical and adhesive properties of vesicles having a wide range of membrane composition as well as of many types of cells.[1-5]

---

[1] E. Evans and R. Skalak, "Mechanics and Thermodynamics of Biomembranes." CRC Press, Boca Raton, Florida, 1980.

In this chapter, methods and equipment are described that are concerned with the formation and micropipette manipulation of giant (30 μm) lipid bilayer vesicles and the measurement of their mutual adhesion energies. This is followed by some examples of lipid vesicle systems in which adhesion phenomena have been investigated. Finally, a brief description is given of some recent experiments in which the micromanipulation technique has been applied to electropermeabilization and electrofusion of lipid vesicles.

Methods and Equipment

Lipid Vesicle Preparation

The simple addition of water to a dried phospholipid such as stearoyloleoylphosphatidylcholine (SOPC) results in the formation of a vesicular suspension. A variety of structures are produced, most of which are not suitable for the precise measurement of bilayer mechanical properties and mutual adhesion. Much of the suspension is composed of multilamellar vesicles and nonvesicular hydrated lipid. A key requirement in these experiments is that the vesicles selected for adhesion tests should be closed and comprised of a single bilayer so that, on aspiration into a micropipette, the total membrane area and vesicle volume are accurately known and the applied stresses are supported by only a single continuous membrane. In appearance, these vesicles are the most transparent and do not contain any internal bilayer structures. Even vesicles that "appear" to be single-walled may be double-walled and can contain internal vesicles that are connected to the outer membrane. The only sure way of testing for "hidden" membrane area and the presence of more than one wall is to prestress the vesicle and carry out a measurement of the elastic area expansion modulus, the value of which groups with the number of walls stressed.[6]

In preparation procedures, it is therefore necessary to maximize the formation of giant single-walled vesicles that can be subsequently selected for experiment from the greater proportion of unsuitable lipid structures. For neutral lipids that hydrate to give a lamellar phase, the conditions that

[2] E. Evans and D. Needham, *J. Phys. Chem.* **91**, 4219 (1987).

[3] E. Evans, this series, Vol. 73, p. 3.

[4] E. Evans, *in* "Physical Basis of Cell–Cell Adhesion" (P. Bongrand, ed.), pp. 91, 173. CRC Press, Boca Raton, Florida, 1988.

[5] D. Berk, R. M. Hochmuth, and R. E. Waugh, *in* "Red Blood Cell Membranes" (P. Agre and J. C. Parker, eds.), p. 423. Dekker, New York and Basel, 1989.

[6] R. Kwok and E. Evans, *Biophys. J.* **35**, 637 (1981).

promote the formation of single-walled, giant vesicles include gentle rehydration in nonelectrolyte, from thin layers of prehydrated lipid, at a temperature above the gel-to-liquid crystalline phase transition of the lipid. A small amount of negatively charged lipid (1 mol %) of matching hydrocarbon chain composition may be added to a neutral lipid preparation, in order to introduce interbilayer repulsion and so enhance lamellae separation and vesicle formation. Subsequent adhesion measurements in 100 mM NaCl solution are unaffected by such a small amount of surface charge.[2]

The preparation of giant vesicles from lipid solutions and the procedures for carrying out thermomechanical studies on a range of lipid and mixed lipid systems are documented in several recent publications concerning the micromanipulation of such structures.[2,6-9] A large selection of natural and synthetic lipids can be purchased from a number of suppliers (e.g., Avanti Polar Lipids, Alabaster, AL; Sigma, St. Louis, MO) and are delivered either lyophilized or in organic solvent solution (chloroform/methanol). Lipids and lipid mixtures are made up at a suitable working concentration (e.g., 10 mg/ml) in a chloroform/methanol solution. The lipid solutions can be conveniently stored under an argon atmosphere in Parafilm-sealed Eppendorf tubes at −20°. For lipid mixtures it is necessary to sonicate freshly prepared solutions in order to ensure complete mixing of the lipids in the organic solvent prior to vesicle formation.

When the lipid is dried from chloroform/methanol solution onto a glass substrate, thick layers or mounds of lipid are formed, and many multilamellar and nonvesicular structures are subsequently produced on rehydration. By observing the rehydration process under an optical microscope, it is clear that unilamellar vesicles are best produced from flat, thin layers. A roughened Teflon disk provides a suitable substrate from which to form a greater number of unilamellar vesicles, because capillary action of the grooves keeps the solution spread over as large an area as possible during the short time required for evaporation of the solvent. A circular Teflon disk is cut from a 2-mm-thick sheet, such that it will fit into the bottom of a 50-ml beaker. The disk is roughened with emery paper to form tiny grooves covering the whole surface. The disk must be thoroughly cleaned by washing in detergent, tap water, and distilled water, and then dried, before a final rinse in chloroform/methanol. Thirty microliters of lipid solution (10 mg/ml) is then applied to the clean dry disk from a 50-μl glass syringe (Hamilton Company, Reno, NV) and is quickly spread over

[7] D. Needham and E. Evans, *Biochemistry* **27**, 8261 (1988).
[8] D. Needham, T. J. McIntosh, and E. Evans, *Biochemistry* **27**, 4668 (1988).
[9] D. Needham and R. S. Nunn, *Biophys. J.* **58**, 997 (1990).

the whole surface with the syringe needle. The solvent rapidly evaporates at room temperature to leave the lipid spread over the disk surface. Final traces of solvent can be removed by a brief (1–2 h) evacuation under reduced pressure.

To produce a lipid vesicle suspension from dried lipid, the lipid must be hydrated at a temperature above its gel-to-liquid crystalline transition temperature, T_C. Fully hydrated SOPC has a T_C of 5° and so must be rehydrated above this temperature. The lipid-containing disk is first prehydrated in the loosely Parafilm-sealed 50-ml beaker for approximately 15 min, with water vapor that is delivered to the beaker via an inert carrier gas such as argon. This procedure allows the closely stacked and dried lamellae to swell and hydrate as much as possible prior to the addition of aqueous solution. Full hydration of the lipid is achieved by placing the beaker in an oven at approximately 40° and gently delivering 10 ml of 160 mOsm sucrose solution down the side of the beaker. The sucrose solution is delivered from a 10-ml syringe through a 0.2-μm filter, which removes suspended particles. The beaker is flushed with argon, sealed with Parafilm, and left undisturbed in the oven overnight. The beaker is then taken out of the oven and inspected for the presence of a fluffy cloud of vesicles suspended in solution. If no such cloud is initially observed, two or three heating/cooling cycles may induce cloud formation by lifting the hydrated lipid off the disk by gentle convection. Just prior to the experiment, the vesicle cloud is gently aspirated by a Pasteur pipette into an Eppendorf tube. Vesicles are essentially resuspended in glucose solution by diluting approximately 0.2 ml of the sucrose-containing vesicle sample with 0.4 ml of a 160 mOsm (isosmotic) glucose solution.

Micropipette Preparation

An essential part of the manipulation system is a glass micropipette of desired internal diameter (~8–10 μm) and having a flat tip. The micropipette not only applies the force to the aspirated vesicle, it also acts to produce a well-defined geometry for vesicles with excess area compared to a sphere of the same volume and as a sensitive transducer of area change in vesicle deformation tests. Thus, pipette geometry must be well defined and accurate.

Pipettes are formed from 1-mm outer diameter, 0.7-mm inner diameter glass capillary tubing (A-M Systems, Inc., Everett, WA). The tubing is mounted in a pipette puller, and heater and solenoid settings are adjusted to give a long gradual taper to the formed pipette. The flat pipette tip is made by mounting the pipette in a microforge so that it is aligned on a horizontal axis with a heated glass bead of lower melting temperature than the glass of the pipette. The pipette is inserted into the molten glass bead by moving the bead horizontally with respect to the pipette until the desired diameter of the taper of the pipette is reached. The bead is then cooled while very slight horizontal motion is maintained in order to produce a small dimple in the solidifying bead. After a few seconds to allow further cooling, the bead is slowly withdrawn, and the pipette is broken by a quick fracture to leave a flat tip (as shown later in Fig. 2).

Pipettes are filled with the desired aqueous solution (usually salt buffer that is isosmotic with the solution in which the adhesion test is performed) by boiling several pipettes held in a sealed Büchner flask under low vacuum, or, individually, by back-filling. The pipettes must be completely filled with buffer and must contain no air bubbles. Tiny air bubbles can be removed from the untapered end by holding the pipette, tip down, and gently tapping the pipette to dislodge the bubble. The bubble can then be removed from the pipette by using a diamond knife to break off the small (~2 mm) portion that contains the bubble at the end of the pipette.

In addition to the measuring and holding micropipettes, a larger transfer pipette (50 μm diameter) is necessary in order to span the air gap during transfer of vesicles from a suspension chamber into an adjacent, but separate, test chamber, This pipette is pulled to a long taper, and the relatively flat tip is made by simply breaking the taper at some point with a quick fracture, after etching with a diamond knife. The pipette is then back-filled to within 1 mm of the tip with a low-viscosity oil, just prior to the experiment. On insertion of the oil-filled transfer pipette into the aqueous solutions of the microchambers, a small air space is maintained which separates the oil from the working solutions.

Micromanipulation System

The micromanipulation system is centered around an inverted microscope that has three micromanipulators mounted directly on the microscope stage. The small micromanipulators (Research Instruments Inc., Durham, NC) are pneumatically controlled by hand-operated joysticks that allow smooth motion in all three axes, zero drift, and control of position on the micron scale. The optics essentially consist of a long working length (>1 mm) objective (20X) of high numerical aperture, followed by large empty magnification (16X eyepieces in binocular; 25X eyepiece in monocular to camera, with additional empty magnification by elevating camera) in order to give a large video image with good optical resolution. A mercury vapor lamp with narrow-band interference filters provides the best image quality for video recording, although a more convenient 100 W, 12 V halogen lamp can give acceptable quality. An interference contrast system (Modulation Optics, Inc., Greenvale, NY) is necessary in order to form suitable images of vesicles and vesicle projections inside the micropipette. A video recording is made of the progress of an experiment, including important data such as pipette suction pressure, time, and temperature, with the use of video multiplexing (Vista Electronics, La Mesa, CA). A series of experiments can then be carried out, and geometrical analyses (vesicle and pipette dimensions) are made subsequently using a video caliper system (Vista Electronics).

Control of micropipette suction pressure is achieved by coupling the pipette, via a small chuck, to a continuous water system comprising a micrometer-positioned water manometer and sensitive in-line pressure transducer (Validyne Engineering Corp., Northridge, CA). The complete absence of air bubbles is absolutely necessary in this system in order to obtain reproducible, stable pressures. The lowest suction force that can be measured with a pressure transducer having a resolution of 1 dyne/cm^2 is about 10^{-8} dyne. This is approximately the pressure required to bend a single fluid bilayer membrane into a pipette. Thus, membrane deformations, such as area expansion, and adhesion measurements in the strong adhesion region ($> 10^{-3}$ erg/cm^2) are readily measured with a pressure range of $1-10^5$ dyne/cm^2. This range of suction pressures is achieved by using the micrometer-positioned manometer for the low ($1-1000$ dyne/cm^2) pressures and a well-sealed 1-, 3-, or 5-ml syringe acting on the air space above the manometer reservoir to give the volume displacements necessary for the high ($1000-50,000$ dyne/cm^2) range. The water-filled manometer comprises a series of taps to allow each section of the system to be equilibrated at atmospheric pressure prior to operation and to isolate the pipette side of the manometer/pressure transducer in operation.

Double microchambers for operation at room temperature can easily be fabricated out of a microscope slide and thin strips of coverslip glass as shown in Fig. 1a. A temperature controlled microchamber (Fig. 1b) is constructed from glass (microscope slide and coverslips) and fabricated steel parts and comprises two adjacent but separate chambers which have open ports for entry through air–liquid interfaces.

Vesicle–Vesicle Adhesion Experiment

The microchambers are first filled with aqueous buffer solutions. One chamber (suspension chamber, Fig. 1a) is filled with an isosmotic (160 mOsm) glucose solution (with respect to the sucrose-containing vesicle suspension), and the other chamber (test chamber) is filled with a slightly hyperosmotic (180 mOsm) NaCl solution. Both solutions contain bovine serum albumin ($\sim 0.1\%$, w/w) in order to precoat glass surfaces prior to the introduction of the vesicle suspension.

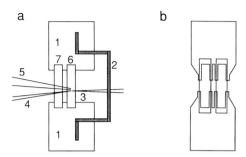

FIG. 1. Microchambers and micropipette arrangement. (a) The simplest double micro-chamber is constructed from two pieces of cut microscope slide (1) joined by a metal handle (2) that is fixed with epoxy onto each piece. The chambers are then simply made from four thin strips of cut cover glass. Two strips are positioned adjacent to each other and form the top surfaces, and two strips are positioned on the other side to form the bottom surfaces. The strips of cover glass are attached by small amounts of silicone grease and can be disposed of at the end of an experiment and replaced with fresh coverslips for the next experiment. The measuring pipette on the right (3) is aligned with the holding pipette (4) on the left. The larger pipette is a transfer pipette (5) an allows vesicles to be transferred from the suspension chamber (6) to the measuring test chamber (7) across the separating air gap. (b) Plan view of a temperature-controlled double microchamber that can be constructed from a sandwich of steel and glass parts so that an equilibrating bath fluid circulates above the chambers. Air–aqueous interfacial areas are kept to a minimum while still allowing working entry ports for the pipettes. Experiments may then be carried out at a temperature just above the dew point, where evaporation losses are minimized, or at selected temperatures within a relatively limited range of 4–40° when it is desired to study the effects, for example, of related lipid phase transitions.

The measuring, holding, and transfer pipettes are mounted in the manipulators on the microscope stage and arranged as shown in Fig. 1a. They are first positioned in the glucose solution (suspension chamber) so that the transfer pipette is just off the bottom of the chamber and the measuring and holding pipettes are on a level with the middle of the transfer pipette lumen. Even with a long taper, the measuring pipette must be at a slight downward angle so that its tip can reach vesicles on the bottom of the chamber. The other two pipettes are horizontally aligned.

With the pipettes aligned so that their relative positions are not disturbed on stage translation, the microscope stage is moved so that they are positioned in the salt solution (test chamber), and the pressure in the measuring and holding pipettes is zeroed with respect to fluid flow at the pipette tips. It is essential that the pressure system be zeroed accurately and that this reference state of zero pressure be maintained and frequently checked throughout the adhesion experiments. Typically, suction pressures need to be accurate to less than 5 dyne/cm^2. However, suction pressures can change by several hundred dynes/cm^2 in these thin chambers due to alterations in curvature of the air–water interfaces through which the micropipettes pass. Changes in curvature occur readily on slight evaporation of solution. Such curvature changes are minimized by not completely filling the room temperature chamber (Fig. 1a) so that any slight evaporation of solution occurs at almost constant mean curvature of the menisci, or better still by working at a low enough temperature, just above the dew point, by using a temperature-controlled chamber (Fig. 1b) when evaporation losses are negligible.

Pipettes are zeroed by observing the motion of a small particle inside the pipette, within 10–20 μm of the pipette tip. With all taps in the water manometer system open, the height of the water level in the manometer reservoir is carefully adjusted by means of a coarse screw drive until the motion of the particle ceases. If the particle stays in this position without drifting, the pipette can be considered to be at zero pressure. If after a few seconds the particle rapidly moves out of, or into, the pipette, the system usually contains a bubble. Likely positions for bubbles are in the pipette, where the taper starts, or at the back end of the pipette in the chuck, in the chuck itself, or at sharp corners, grease, or small tap apertures in the manometer system. Bubbles adhering to the metal surfaces in the transducer chamber itself can be particularly troublesome and difficult to remove. Because of the detrimental effect of the presence of bubbles, the manometer system is initially filled with freshly boiled/evacuated water.

Once zeroed, the holding and measuring pipettes are set with a small suction pressure, in order to draw in salt solution and to precoat their surfaces with albumin, for approximately 10 min. The pipette and microchambers are also electrically grounded. Without electrical grounding and the presence of albumin in the working solutions, vesicles are attracted to the pipette tips and frequently break on gentle aspiration. The glass surfaces appear to be charged, and the presence of albumin reduces this static charge effect. After the albumin pretreatment, most of the glucose solution in the suspension chamber is carefully withdrawn, using a twisted tissue paper, and the vesicle suspension is added

The vesicles are prepared in sucrose. Their subsequent resuspension in a glucose solution not only improves visible detection, owing to the difference in refractive index, but also causes the vesicles to sink to the bottom coverslip of the chamber, close to the objective lens, for ease of identification and capture. Within about 2 min vesicles accumulate on the bottom of the microchamber. The measuring pipette is inserted into the protective solution space of the transfer pipette and the stage is translated to position the three pipettes in the vesicle suspension chamber. Although few in number, giant (20–30 μm in diameter) single-walled vesicles can be identified and captured for transfer into the salt-containing test chamber.

Two vesicles of comparable size are identified; one is placed in the aqueous lumen of the transfer pipette and allowed to float freely, while the other is held in the transfer pipette by the measuring pipette. The stage is then translated to position the pipette in the second, salt solution chamber. The presence of oil in the transfer pipette restricts motion of the fluid, and the pipettes and vesicles should remain in relative position during this transfer procedure. The measuring pipette is withdrawn from the transfer pipette and the vesicle (which will be the test surface) is handed to the waiting holding pipette that has been preset with a large suction pressure. The second vesicle (adherent vesicle) is retrieved from the transfer pipette, which is then moved out of the field of view. The vesicles rapidly deflate (by ~ 10%) to new equilibrium volumes in the hyperosmotic solution, so it is important to carry out the above manipulations as fast as possible, in order that the excess membrane area of the adherent vesicle is always supported by pipette suction once it starts to deflate.

In Fig. 2, the vesicle on the right, in the holding pipette, is under high suction pressure (10,000 dyne/cm^2) and forms a rigid, spherical test surface, while the vesicle on the left, in the measuring pipette, is under lower (starting pressure of ~ 1000 dyne/cm^2), variable suction pressure and remains deformable. The vesicles are aligned and maneuvered into close proximity (Fig. 2a), and the pipettes are maintained in these fixed positions so that no axial force is exerted during the adhesion test. In the experiment, the extent of adhesion is controlled via the tension in the left-hand adherent vesicle membrane, which in turn is controlled by pipette suction pressure. The adherent vesicle is allowed to spread on the test surface in discrete equilibrium steps by reducing the pipette suction pressure (Fig. 2b, c). The suction pressure is then increased in discrete steps, and the adherent vesicle is peeled off the test surface in order to test for reversibility and true equilibrium, as well as constancy of vesicle area and volume.

The low internal pressures, the limited permeability of the membranes, and the presence of trapped solutes ensure that vesicle volumes do not change during the experiment because of water filtration. Also, the low membrane area compressibility ensures constant vesicle area. The vesicle projection in the measuring pipette should therefore return to the same length as at the beginning of the adhesion test (Fig. 2a).

At the end of the test, the adherent vesicle is expelled from the pipette, and the suction pressure is adjusted to give no flow in the pipette in order to test for the zero pressure condition. For the pipettes and vesicles shown in Fig. 2, typical suction pressures to control equilibrium adhesion (Fig. 2c) are relatively low (~ 100 dyne/cm^2), so it is important that the pipette zero should not change during the adhesion test. The adhesion measurement represents the difference between spreading of the adherent vesicle on the test vesicle surface and on the measuring glass pipette. This is why every

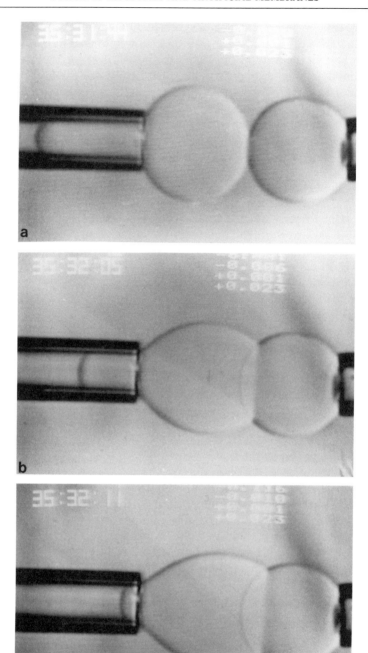

effort must be made to prevent adhesion to glass surfaces (e.g., by coating with albumin).

A control experiment must be performed in the absence of the rigid test vesicle. The suction pressure holding the nominally adherent vesicle is reduced, and the pressure at which the vesicle comes out of the pipette is measured. This threshold pressure must be much lower than the pipette pressure required to balance vesicle–vesicle adhesion. In the absence of vesicle self-adhesion, any displacements of the aspirated length of the vesicle membrane in the pipette, as a function of pipette suction in this control experiment, should be due only to an entropy-driven tension and membrane bending stiffness.[10] Several pipette coatings (e.g., silanization) have been tested in the past, but a completely satisfactory inert surface for every experimental condition has not yet been found.

At the end of a series of adhesion tests, a recording is made of a micrometer graticule (2 mm, 200 divisions) in order to calibrate the video caliper system used for subsequent geometrical analyses.

Analysis

The reversible work of adhesion, ω_a, to assemble bilayer surfaces can be considered to be the reduction in free energy per unit area of membrane–membrane contact formation. It is the cumulation of the action of densely distributed interbilayer forces from large separation (infinity) to intimate contact. Mechanical equilibrium at stable contact is the balance between the free energy required to increase contact area ω_a and the mechanical work to deform the vesicle contour, and it is represented by the Young–Dupre equation:

$$\omega_a = \tau_m(1 - \cos\,\theta_c) \tag{1}$$

[10] E. Evans and W. Rawicz, *Phys. Rev. Letts.* **64**, 2094 (1990).

Fig. 2. Adhesion test. Video micrographs of vesicle–vesicle adhesion test between neutral lipid bilayers in 0.1 *M* NaCl. Vesicle membrane tensions are controlled by pipette suction pressures, and vesicle areas and volumes remain constant throughout the adhesion test. Vesicle and pipette diameters are 30 and 8 μm, respectively. (a) The vesicle pictured at right is held under high tension ($\sim$3 dyne/cm) so that it forms a rigid spherical test surface. The vesicle at left is held under lower, variable suction pressure and is deflated in the hyperosmotic solution of the test chamber so that it has enough excess area to give an appreciable coverage of the test surface. The vesicles are shown maneuvered into close proximity but not forced into contact. The pipettes are then left in these fixed positions. (b, c) Spontaneous adhesion is allowed to progress in discrete, equilibrium steps controlled by pipette suction pressure applied to the left-hand vesicle. Finally, the procedure is reversed (not shown), and the adherent vesicle is peeled off the test surface by increasing the pipette suction pressure to return to the original position.

where τ_m is the membrane tension in the adherent vesicle and θ_c is the included angle between the adherent and test surface membranes exterior to the contact zone. Thus, for constant adhesion energy, a unique relation exists between membrane tension and contact angle for the membrane assembly experiment. The contact angle can be derived from measurements of the extent of encapsulation x_c of the spherical test surface, because there exists a precise relation between x_c and θ_c for fixed vesicle area and volume, determined prior to adhesion.[11,12,14] Membrane tension is calculated from the pipette suction pressure ΔP and vesicle and pipette geometry:

$$\tau_m = \Delta P R_p / 2(1 - R_p c) \tag{2}$$

where R_p is the pipette radius and c is the mean curvature [$c = (1/R_1 + 1/R_2)/2$]. A first approximation would take the vesicle surface as a perfect sphere of radius R_o, and c would then equal $1/R_o$. Otherwise, mean curvature must be established from the geometrical requirements imposed by the fixed distance between the pipettes and the fixed surface area and vesicle volume.[11]

Thus, for a given vesicle pair both adhesion and separation can be represented by a plot of extent of encapsulation x_c versus the reciprocal of pipette suction pressure multiplied by pipette radius $1/\Delta P R_p$, as shown in Fig. 3a. These measurements can be converted to contact angle and reciprocal membrane tension for the adherent vesicle membrane as shown in Fig. 3b. A single value of the adhesion energy can be used to fit the experimental data. Figure 3 shows results and analyses for adhesion of phosphatidylcholine bilayer vesicles produced by two different long-range forces: van der Waals attraction ($\omega_a = 10^{-2}$ erg/cm^2) and depletion attraction in a concentrated solution of nonadsorbent polymer ($\omega_a = 10^{-1}$ erg/cm^2).[2,13,14]

The above analysis relates to a regime where the adhesion energy is strong enough that edge energy effects arising from membrane bending and random thermal undulations are insignificant.[10,12] This strong adhesion regime starts at the very limit of experimental measurement, using the micropipette method, that is, for adhesion energies greater than 10^{-3} erg/cm^2. The technique has been successfully used to measure a whole range of interbilayer interactions (van der Waals attraction, short-range hydration repulsion, electrostatic double-layer repulsion, and attraction

[11] E. Evans, *Biophys. J.* **31**, 425 (1980).
[12] E. Evans, *Colloids Surf.* **43**, 327 (1990).
[13] E. Evans and D. Needham, *Macromolecules* **21**, 1822 (1988).
[14] E. Evans and D. Needham, *in* "Molecular Mechanisms of Membrane Fusion" (S. Ohki, D. Doyle, T. D. Flanagan, S. W. Hui, and E. Mayhew, eds.), p. 89. Plenum, New York and London, 1988.

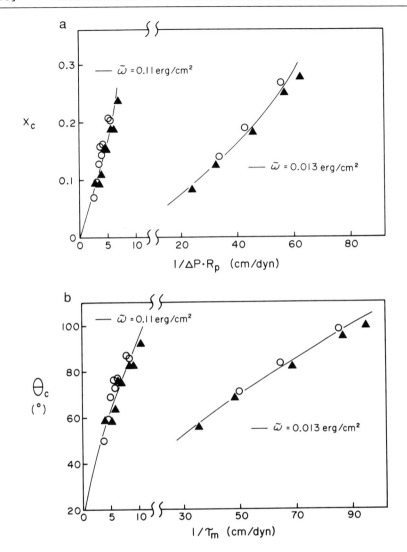

FIG. 3. Results and analyses of adhesion and separation. (a) At each discrete, equilibrium step in the adhesion and peeling experiment (triangles represent contact formation and circles represent separation of contact) the necessary geometrical measurements and mechanical analysis yield a plot of extent of encapsulation of the rigid vesicle surface, x_c, versus reciprocal suction pressure ΔP multiplied by pipette radius R_p. (b) With values for vesicle area and volume determined prior to the adhesion experiment, plus mechanical analysis of membrane geometry, measurements in (a) can be converted to precise contact angles and plotted versus the reciprocal of membrane tension for the adherent vesicle. Solid curves are predictions for uniform fixed values of free energy reduction per unit area of contact formation. Two cases are shown representing the operation of two different long-range forces,[2,13]: van der Waals attraction ($\omega_a \sim 10^{-2}$ erg/cm²) and depletion–attraction in a concentrated solution of nonadsorbent dextran polymer ($\omega_a \sim 10^{-1}$ erg/cm²).

induced by the presence of nonadsorbing polymer) that result in stable adhesive contact.[2,13,14]

One example is shown in Fig. 4, where SOPC vesicle membranes have been assembled in the presence of various volume fractions of aqueous soluble dextran and polyethylene oxide (PEO) polymers of different molecular weights.[13,14] A threshold level of adhesion energy of approximately 0.015 erg/cm² exists for the neutral bilayers in salt solution, in the absence of dextran polymer. This free energy potential for adhesion represents the intrinsic colloidal attraction/repulsion (van der Waals attraction and short-range repulsion) between the membranes. With increasing dextran concentration, the membranes adhere much more strongly, and the addition of approximately 12% (w/v) dextran produces an order of magnitude

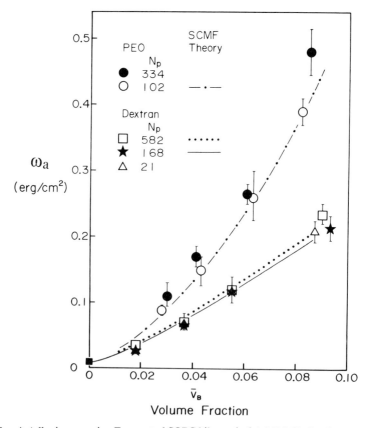

FIG. 4. Adhesion energies. For neutral SOPC bilayers in 0.1 M NaCl plus dextran or PEO polymer, the adhesion energy ω_a increases with increasing volume fraction of the polymer. The curves are predictions for the dextran and PEO polymers obtained from self-consistent mean-field theory (SCMF).[13,14]

increase in the adhesion energy. Further increases in polymer concentration continue to raise the level of adhesion energy and to decrease the gap separation between the membranes. Polyethylene oxide polymer at concentrations of 0.1 volume fraction increases the adhesion energy to values on the order of 0.5 erg/cm^2 and leads to smaller bilayer separations. This is an important result when considering the action of PEO in membrane fusion. Adhesion in concentrated solutions of nonadsorbing polymers is promoted by interaction of depletion layers due to polymer exclusion from the bilayer surfaces, and it is independent of molecular weight for these large polymers. The curves are predictions obtained from self-consistent mean-field (SCMF) theory, in which the only fitting parameter is the choice of polymer segment length.[13,14] For these large polymers the adhesion-enhancing effect is essentially independent of polymer molecular weight.

Membrane Deformation, Electropermeabilization, Adhesion, and Electrofusion

Quantitation of membrane material properties and interbilayer adhesion energies are essential in order to arrive at a full description of instability phenomena such as mechanical and electric field-induced membrane breakdown and interbilayer fusion. Experiments have been reported in which the micropipette technique is extended to lipid vesicle studies involving electropermeabilization and electrofusion.[15,16] Methods are being developed that are ideally suited to experiments in which control over membrane tension and controlled assembly of membranes is combined with the application of electropermeabilizing and electrofusing pulses to single vesicles or adherent vesicle pairs.

A double microchamber is simply modified to include two strip electrodes in the test chamber. In the electropermeabilization experiment, a single lipid vesicle is selected from the suspension chamber and is transferred into the adjacent electrode chamber by means of the shielding transfer pipette. The vesicle is then positioned between the electrodes (parallel to the pipette) for the permeabilization test (Fig. 5a). Vesicle membrane tension is set by pipette suction pressure, and the pulsed (60 μsec, dc) electric field strength is sequentially increased until the vesicle is permeabilized. Permeabilization is easily assayed. At high suction pressure the vesicle is rapidly drawn up the pipettes; at low suction pressure permeabilization causes a release of internal pressure, and the flaccid membrane is slowly aspirated (Fig. 5b). At lower suction pressures, which produce boundary tensions of approximately 0.03 dyne cm, permeabiliza-

[15] D. Needham and R. M. Hochmuth, *Biophys. J.* **55**, 1009 (1989).
[16] D. Zhelev and D. Needham, *Biochim. Biophys. Acta* in press (1993).

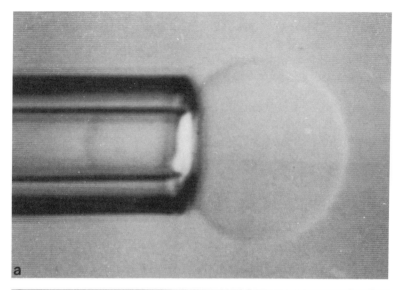

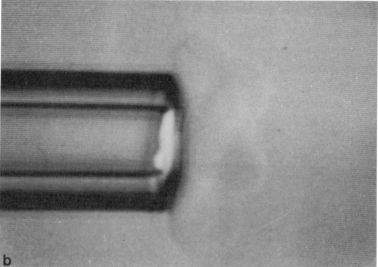

FIG. 5. Vesicle electropermeabilization. (a) A single-walled lipid vesicle (20 μm diameter) is aspirated by micropipette (8 μm diameter) under a low suction pressure, less than that required to break the membrane under conditions of zero voltage. For very low membrane tension (~0.03 dyne/cm), a permeabilizing electric field pulse causes the pressurized vesicle to lose a small amount of internal volume before resealing. The small change in volume of the repressurized vesicle can be measured from the change in the projection length in the pipette. (b) At membrane tensions above this threshold, electropermeabilization of the vesicle causes larger volume displacements, and the flaccid vesicle is aspirated up the suction pipette.

tion causes a release of internal pressure followed by membrane resealing.[15,16] Small volume displacements can therefore be measured by tracking small changes in the length of the projection in the pipette (Fig. 5a), and information regarding postpulse hole geometry and kinetics is obtained.[16] Critical electric field strengths for membrane failure have been determined as a function of applied membrane tension for membranes of different composition and compressibility.[15,16] Concomitant increases in membrane area and decreases in membrane thickness can be made to occur by either tensile or electrocompressive stresses. Both store elastic energy in the membrane and eventually cause breakdown.

In the electrofusion test, two vesicles are transferred into the electrode chamber, where they can be made to adhere as shown in Fig. 6a and are axially positioned by micropipette between the electrodes. The application of a short (60 μsec) dc pulse of critical field strength causes the closely proximal membranes (gap separation ~ 28 Å, adhesion energy ~ 0.02 erg/cm^2) to fuse in the contact zone (Fig. 6b). Excess membrane, over a sphere of the same volume as the two daughter vesicles, forms an adherent flap, and on rotation of the fusion product, this flap is seen to contain a large hole that connects the two halves. Adhesional spreading in the flap causes a tension in the surface membrane that results in the preferred spherical geometry. This fusion product can now be manipulated with a suction pipette (Fig. 6c). The application of suction pressure causes an unpeeling of the membrane in the flap. This membrane can be pulled into the surface, showing that the fusion of the two adherent vesicles produces a single capsule with one continuous membrane. Furthermore, measurement of vesicle and pipette geometries shows that the original total membrane area and internal volume of the initial adherent vesicle pair are conserved in the membrane fusion process. The critical field strengths for fusion are found to be the same as those required to cause membrane permeabilization of a single lipid membrane.

Conclusions

The micropipette manipulation technique offers many features that make it extremely versatile and powerful in studies of the thermomechanical and interactive properties of vesicles, cells, and other small particles, such as emulsion droplets. These features include the following: the investigator is able to select and manipulate single objects or pairs of objects for study from a heterogeneous suspension and to transfer them into a new suspending medium; vesicle and cell geometries are well-defined and excess membrane area is supported by pipettes; well-characterized forces may be directly applied, and the resulting deformations are readily observed

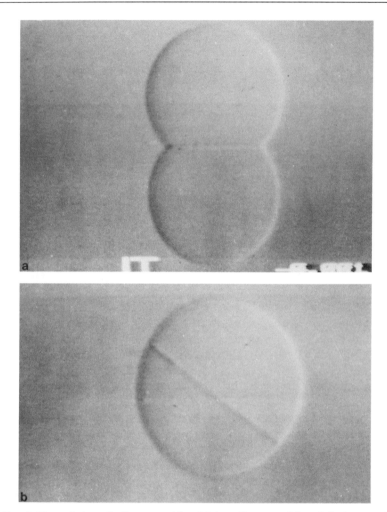

FIG. 6. Electrofusion of adherent vesicles. (a) An adherent vesicle pair is shown axially aligned between the electrodes (with the contact zone parallel to the electrodes). Vesicles are pulsed (60 μsec, dc) at increasing electric field strengths until they fuse in the contact zone. (b) The fusion product is spherical, and excess membrane area is taken up in an adherent flap that contains a hole connecting the two halves. The spherical geometry is a direct result of adhesional spreading in the flap, which causes a tension in the surface membrane. (c) Application of a low suction pressure results in an unpeeling of the membrane in the flap, showing that the original total membrane area and internal volumes of the adherent vesicle pair are conserved in the fusion process and that fusion produces a single capsule with one continuous membrane.

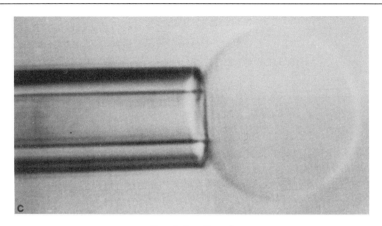

FIG. 6. *(continued)*

and measured; and finally, the whole history of the experiment is recorded on video tape for subsequent detailed analysis.

With this micropipette technique, it is expected that many more aspects of intermembrane interactions will be studied. For example, it is possible to examine the role of certain instability and defect phenomena, such as spontaneous curvature and lamellar-to-hexagonal phase transitions for assembled bilayers containing mixtures of lipids. Asymmetric vesicle–vesicle and vesicle–cell adhesion and fusion tests can be carried out owing to the ease of sample selection. Interactions may also be studied that involve the adhesion of vesicles and cells with other surfaces of physical, medical, and biotechnological importance.

Acknowledgments

The author gratefully acknowledges support in part by the National Institutes of Health, through Grant GM40162, and the Hunt Faculty Scholarship of Duke University.

[11] Forces between Phospholipid Bilayers and Relationship to Membrane Fusion

By CHRISTIANE A. HELM and JACOB N. ISRAELACHVILI

Introduction

The adhesion and fusion of model membranes, that is, surfactant and lipid bilayers of known composition, can be quite complex since they interact with each other through a number of different forces. Thus, apart from the expected attractive van der Waals and repulsive electrostatic forces, which are satisfactorily described by the Derjaguin–Landau–Verwey–Overbeek (DLVO) theory,[1] some unexpected and fascinating repulsive steric/hydration forces[2,3] and an attractive hydrophobic force[4,5] have been discovered. The origin of these forces is still not clear, but they appear to dominate the interactions at small separations, below 10–20 Å. Also, repulsive "undulation forces" due to macroscopic thermal undulations of the elastic membranes, which become restricted when membranes approach each other, were theoretically predicted and experimentally verified.[6] A similar theoretical approach has been used to explain the steric/hydration force by considering the molecular scale undulations arising from the protrusion of amphiphilic groups from bilayers into water.[7]

The nature of intermembrane forces is still under discussion, and the relative importance of the different forces for adhesion and fusion is still far from clear.[8,9] There are examples of strong adhesion without fusion[8,10] and of fusion with no or little adhesion.[8,11] In other cases, fusion has been

[1] J. N. Israelachvili, "Intermolecular and Surface Forces." Academic Press, New York, 1985.

[2] D. M. LeNeveu, R. P. Rand, V. A. Parsegian, and D. Gingell, *Nature (London)* **259**, 601 (1976).

[3] J. Marra and J. Israelachvili, *Biochemistry* **24**, 4608 (1985).

[4] H. Christenson, *in* "Modern Approaches to Wettability" (M. Schnader and G. Loeb, eds.), in press, Plenum, New York, 1992.

[5] R. M. Pashley and J. N. Israelachvili, *Colloids Surf.* **2**, 169 (1981).

[6] R. M. Servuss, and W. Helfrich, *in* "Physics of Complex and Supermolecular Fluids" (S. A. Safran and N. A. Clark, eds.), p. 85. Wiley, New York, 1987.

[7] J. N. Israelachvili and H. Wennerström, *Langmuir* **6**, 873 (1990); *J. Phys. Chem.* **96**, 520 (1992).

[8] L. V. Chernomordik, G. B. Melikyan and Yu. A. Chizmadzhev, *Biochim. Biophys. Acta* **906**, 201 (1986).

[9] N. Düzgüneş, K. Hong, P. A. Baldwin, J. Bentz, S. Nir, and D. Papahadjopoulos, *in* "Cell Fusion" (A. E. Sowers, ed.), p. 241. Plenum, New York, 1987.

[10] J. Marra, *Biophys. J.* **50**, 815 (1986).

[11] R. G. Horn, *Biochim. Biophys. Acta* **778**, 224 (1984).

induced without apparently altering the interbilayer forces simply by stressing bilayers by an electric field[8] or by creating an osmotic gradient.[8,12]

Various models of fusion have been suggested, mainly based on electron micrographs of small vesicles or membranes.[8,13] The principal problems of conventional freeze–fracture electron microscopy (and rapid freezing techniques in general) are the rapidity of the fusion process and the very small area over which it occurs. It is very difficult to "trap" or "freeze" the various stages of fusion, since the whole process appears to be over within 0.1–1 msec.[14] Thus, the precise molecular rearrangements accompanying the fusion process are still unknown, and it is also not known whether there is one general fusion mechanism or whether many different pathways exist, each specific to a particular system.

With the aim of answering some of these questions, we describe how the surface forces apparatus technique can be used for measuring interbilayer forces while monitoring the deformations of fusing bilayers supported on solid surfaces. The technique allows one to follow the molecular rearrangements in real time during the fusion process, and to identify the most important forces involved.[15]

Experimental Methods

Surface Forces Apparatus

The surface forces apparatus (SFA) measures the force F between two cylindrically curved, molecularly smooth surfaces, of radii R_1 and R_2, as a function of their separation D.[16] The distance D between the two surfaces can be controlled and measured to within $1-2$ Å by an optical interference technique using fringes of equal chromatic order (FECO).[17] This technique also allows determination of the index of refraction of molecules adsorbed on the surfaces. One of the surfaces is mounted on a spring with spring constant K. From the measured deflection ΔD of the spring, the force between the surfaces F can be determined. The corresponding interaction energy per unit area E between two *flat* surfaces is simply related to the

[12] M. H. Akabas, F. S. Cohen, and A. Finkelstein, *J. Cell Biol.* **98,** 1063 (1984); L. R. Fisher and N. S. Parker, *Biophys. J.* **46,** 253 (1984).

[13] A. J. Verkleij, L. Leunissen-Bijvelt, B. De Kruijff, M. Hope, and P. R. Cullis, *in* "Cell Fusion," p. 355. Pitman, London, 1984.

[14] R. L. Ornberg and T. S. Reese, *J. Cell Biol.* **90,** 40 (1981).

[15] C. A. Helm, J. N. Israelachvili and P. M. McGuiggan, *Science* **246,** 919 (1989).

[16] J. N. Israelachvili and G. E. Adams, *J. Chem. Soc. Faraday Trans. 1* **74,** 975 (1978).

[17] J. N. Israelachvili, *J. Colloid Interface Sci.* **44,** 259 (1973).

force F between the two *curved* surfaces by the so-called Derjaguin approximation[1]

$$E = F/2\pi R \tag{1}$$

where $R = \sqrt{R_1 R_2}$.

Static Measurements of Forces

For repulsive forces, the surfaces are simply pushed together and the spring deflection ΔD is measured optically, from which the force is obtained[16]:

$$F = \Delta DK \tag{2}$$

In the case of attractive forces, the system becomes unstable if the gradient of the force $\partial F/\partial D$ exceeds the spring constant K, and the surfaces jump to the next stable point. The point of instability is therefore[16]

$$\partial F/\partial D = -K \tag{3}$$

By varying the spring constant during an experiment one can thus determine the slope of the force law, from which the force is then obtained by integration.[10]

Dynamic Measurements of Forces

In the dynamic "drainage" method, one surface is driven toward the other at a constant speed.[18] The deflection of the spring ΔD now has two contributions: the equilibrium force between the surfaces, as above, and an additional hydrodynamic force due to the viscosity η of the medium. Thus, at any time, t, the force between the surfaces is

$$F = \Delta DK - 6\pi RR_h\eta v/D \tag{4}$$

where the velocity is $v = \partial D/\partial t$, and where $R_h = 2R_1 R_2/(R_1 + R_2)$. To determine the velocity of the surfaces, it is necessary to record the moving optical fringes during a drainage experiment using a video camera (Model SIT 66, Dage-MTI Inc., Michigan City, IN) and a video cassette recorder (Sony, Model VO5800H, McBain Instruments, Chatsworth, CA). Later, the drainage runs are analyzed using a video micrometer (Model 305A, Colorado Video, Boulder, CO) with a time resolution of 0.02 sec.

Measurements of Pressure and Adhesion

Under very large compressive forces the two initially curved surfaces flatten elastically (the glue supporting the mica sheets becomes com-

[18] D. Y. C. Chan and R. G. Horn, *J. Chem. Phys.* **83**, 5311 (1985).

pressed). This deformation can be directly observed from the shapes of the fringes, which represent a cross section of the contact zone. For example, Fig. 1 (top) shows fringes of two curved surfaces not in contact, which is reflected in the curved shapes of the fringes. Figure 1A,B shows the fringes when the surfaces are in close contact, where the flattening is clearly seen. From such fringes one can measure the contact diameter (to within 1 μm) and, hence, the contact area, A. Note that the 1 Å resolution normal to the surfaces is a factor of 10^4 higher than the 1 μm resolution parallel to the surfaces, the latter being essentially the same as for an optical microscope.

When one observes flattening, the Derjaguin approximation [Eq. (1)] no longer applies. Instead of the energy per surface area, one can now determine the mean pressure from $P_{mean} = F/A$. However, the maximum pressure occurs at the center of the contact area. If the surfaces deform elastically and do not mutually adhere, the "Hertz theory" [19,20] predicts that $P = 0$ at the edges and that it rises to a maximum at the center, where

$$P_{max} = 1.5F/A = 1.5P_{mean} \qquad (5)$$

The elastic deformation of adhering surfaces is more complicated.[20] Even under no applied load the adhesion causes a finite contact area. This was first treated by the Johnson–Kendall–Roberts (JKR) theory,[21] which predicted that at the edges (the contact boundary) the pressure is negative and infinite. This was because the JKR theory assumes an infinitely short-range attractive force. More realistic theories[20] do not give infinite stresses but still very high values; however, the solutions require complex numerical calculations. The effects of adhesion on the deformation of contacting bilayers can also be seen on the shapes of the fringes in Fig. 1. In Fig. 1A the adhesion is low, and the edges are rounded. In contrast, in Fig. 1B a high adhesion causes sharp edges.

According to the JKR theory the adhesion energy γ can be determined from the adhesion or "pull-off" force F_0 using

$$\gamma = F_0/3\pi R \qquad (6)$$

whereas the maximum pressure is given by [cf. Eq. (5)]

$$P_{max} = 1.5F/A + 2(F_0/A)[1 + (1 + F/F_0)^{1/2}] \qquad (7)$$

All of these parameters can be directly measured in the SFA experiments. More elaborate theories suggest that the factor 3 in Eq. (6) should be higher, but not above 4.[20]

[19] H. J. Hertz, *Reine Angew. Math.* **92**, 156 (1881).
[20] R. G. Horn, J. N. Israelachvili, and F. Pribac, *J. Colloid Interface Sci.* **115**, 480 (1987).
[21] K. L. Johnson, K. Kendall, and A. D. Roberts, *Proc. R. Soc. London A* **324**, 301 (1971).

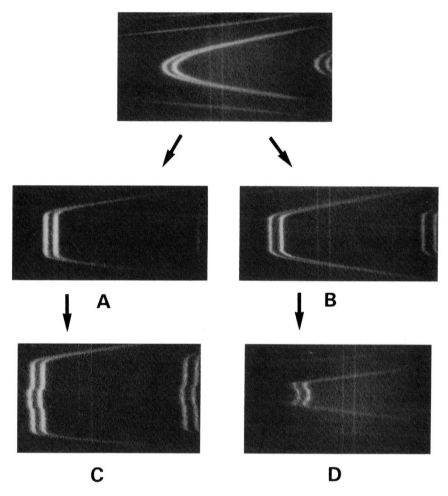

FIG. 1. Odd-order interference fringes, showing typical deformations during hemifusion. *Top:* Surfaces separated. *Middle:* Contact flattening dominated by the applied load (A) or high adhesion energy (B). *Bottom:* Fusion of weakly adhering DMPC bilayers (C) and strongly adhering CTAB bilayers (D). Vertical lines are spectral lines from a mercury lamp, used for calibration.

The above equations were used to compute pressures and adhesion energies shown in the subsequent figures. Pressures up to about 150 atm can be attained when using mica sheets with an approximate radius R of 1 cm, and pressures as high as 5000 atm have been attained by using surfaces of smaller radii.[22]

Observation of Fusion

Fusion can start at any point within the contact zone. Between nonadhering bilayers under an applied pressure, fusion usually starts at the center, where the compressive pressure is highest and where the bilayers are probably at their closest point, and then spreads *outward* (Fig. 1C). Between strongly adhering bilayers fusion often starts at the edges, where the stress is highest, and then spreads *inward* (Fig. 1D). All these events can be followed and recorded in real time with the video system, and they are described more fully in the section on results.

Preparation of Bilayers and Control of Thickness

Transparent, molecularly smooth surfaces are employed with the SFA. Mica is the one most often used, but recently silica[23] and sapphire surfaces[24] have also been used successfully. In the experiments reported below, various lipid bilayers were adsorbed on mica surfaces, which were allowed to equilibrate with the monomers in solution during the experiments.[5] For monomer concentrations down to about 10^{-6} M, the solution can be treated as an infinite reservoir since the bathing chamber has a volume of 350 ml and the total area of the mica surfaces is about 2 cm². Two methods have proved successful for coating mica surfaces with monolayers or bilayers: adsorption from solution and Langmuir–Blodgett (L-B) deposition.

Adsorption from Solution. For soluble surfactants the monomer concentration in the solution can be varied systematically so as to alter the equilibrium thickness of the adsorbed bilayers or monolayers.[5,25,26] Thus, for CTAB (see Materials), a "fully developed" bilayer is usually observed at concentrations above twice the critical micelle concentration (CMC), whereas at approximately 10% of the CMC there is only a monolayer.[26] Bilayers can also be adsorbed from vesicle dispersion if the chains are in the

[22] B. J. Briscoe and D. C. Evans, *Proc. R. Soc. London A* **380,** 389 (1982).
[23] R. G. Horn, D. T. Smith, and W. Haller, *Chem. Phys. Lett.* **162,** 404 (1989).
[24] R. G. Horn, D. R. Clarke, and M. T. Clarkson, *J. Mater. Res.* **3,** 413 (1988).
[25] P. Herder, *J. Colloid Interface Sci.* **134,** 336 (1990); **134,** 346 (1990).
[26] P. Kékicheff, H. K. Christenson, and B. W. Ninham, *Colloids Surf.* **40,** 31 (1989).

fluid state.[11] However, adhesive vesicles tend to get trapped between the surfaces, which is inconvenient. The adsorption from solution method fails for highly insoluble, nonvesicle-forming surfactants and lipids. In such cases, the Langmuir–Blodgett deposition method is preferred.

Langmuir–Blodgett Deposition. The L-B trough allows for controlled deposition of insoluble surfactants or lipids at known surface coverage.[27] The method has been found to be suitable for depositing a whole range of monolayers and bilayers on mica.[3,4,10,15,28]

Lipid monolayers can be spread on water from a 3 : 1 (v/v) chloroform/ methanol solution and then compressed to the desired surface pressure, which is equivalent to a given area per molecule. Each lipid exhibits a specific isotherm, which shows the surface pressure as a function of the molecular area.[27] The monolayer is deposited at a controlled speed onto the solid support. Asymmetric bilayers can thus be prepared, where the first (inner) monolayers are different from the second (outer) monolayers, between which the forces are measured.[3,4,10,15] For the outer monolayer to remain stable it must not be removed from the aqueous solution.

The lipids in the outer monolayer are in equilibrium with the mono- mers in the aqueous solution. If the monomer concentration is above the CMC the bilayers stay "fully developed" (the CMC is normally the critical micelle concentration, but for lipids it is understood to be the critical aggregate or vesicle concentration and can be as low as $10^{-10} M$). By diluting the solution of lipids, a slow desorption of lipids from the bilayers results, until a new equilibrium bilayer thickness is reached. This "deple- tion" method provides a reliable and quantitative way of controlling the thickness of bilayers.

Materials

In the experiments reported below the following materials are used. The lipids DLPC (L-α-diaurylphosphatidylcholine), DMPC (L-α-dimyris- toylphosphatidylcholine), and DPPE (L-α-dipalmitoylphosphatidylethan- olamine) are from Sigma (St. Louis, MO) or Avanti (Alabaster, AL) and are used without further purification. The single-chain surfactant CTAB (hexadecyltrimethylammonium bromide) from Sigma is recrystallized in 9 : 1 (v/v) ethanol/ether. The water is distilled and additionally filtered in a water purification unit (Model 90004, Labconco Corp., Kansas City, MO). The NaBr from Mallinckrodt (Paris, KY) is analytical grade.

[27] O. Albrecht, H. Gruler, and E. Sackmann, *J. Phys.* **39**, 301 (1978).
[28] Y. L. E. Chen, M. L. Gee, C. A. Helm, J. N. Israelachvili, and P. M. McGuiggan, *J. Phys. Chem.* **93**, 7057 (1989).

During experiments, the temperature of the apparatus should be controlled to $\pm 0.1°$. This is achieved by heating or cooling the whole room over the range $10-32°$. In this way, with lipids such as DMPC (T_c 24°) it is possible to make measurements both above and below T_c.[3]

Results

We describe here our measurements of interbilayer forces and the fusion process, including the intermediate stages, on a variety of lipid bilayer systems.[15] The thicknesses of "fully developed" bilayers are determined from refractive index measurements, which compare very well with the known (hypothetically anhydrous) thickness of L-B deposited layers.[3] With "fully developed" bilayers no fusion is observed up to applied pressures of 150 atm. Fusion takes place only between "depleted" bilayers, that is, when the solution is diluted below the CMC. Depletion also changes the force law between bilayers.

Figure 2 shows results obtained for the force laws between phosphatid-

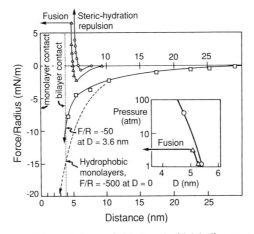

FIG. 2. Force versus distance between fluid phosphatidylcholine monolayers (each on a solid DPPE monolayer) with decreasing monomer concentration in the aqueous solution at 22°. (○) Forces between two fully developed DLPC bilayers in water showing a van der Waals attraction and a steric/hydration repulsion. The van der Waals attraction causes the two surfaces to jump into "contact" from a bilayer separation of 4.2 nm, and no fusion occurs even at a pressure of 40 atm. (△) Forces between two depleted DLPC bilayers. The two surfaces now jump into contact from a bilayer separation of 6.2 nm, and fusion occurs at a pressure of 2 atm. (□) Strong hydrophobic attraction between even more depleted DMPC bilayers which spontaneously fuse into one bilayer under a very weak force. The dashed curve shows the hydrophobic force between two hydrophobic monolayers.

ylcholine bilayers in the fluid state, where for convenience $D = 0$ refers to monolayer–monolayer contact. Fully developed DLPC bilayers have an anhydrous thickness of 36Å (i.e., bilayer–bilayer contact would occur at $D = 36$ Å, for DMPC at $D = 38$ Å). The measured force law (O points, Fig. 2) exhibits a van der Waals attraction and at short range, below 25 Å, a strong steric/hydration repulsion which prevents the bilayers from coming into close contact. Depletion of the solution causes the bilayers to thin: the attractive force (△ points, Fig. 2) is stronger and of longer range than for the "fully developed" bilayers, and the thinned bilayers now also fuse under a slight compression (see inset, Fig. 2). On further dilution, the attractive force continues to increase (□ points, Fig. 2), and the force law begins to approach an exponential distance dependence, characteristic of the hydrophobic interaction between two purely hydrophobic monolayers (dashed line, Fig. 2). In addition, these very depleted bilayers fuse almost immediately after they jump into contact.

Figure 3 shows the changing fringe pattern during the fusion of two DMPC bilayers, and Fig. 4 shows the most likely bilayer deformations corresponding to these fringe patterns. As shown, the crucial first breakthrough step starts by a lateral parting of the head groups on opposite sides of the bilayers, thereby exposing hydrophobic hydrocarbon regions that had previously been shielded from the aqueous phase. Owing to the long-range nature of the hydrophobic interaction (Fig. 2), these opposing hydrophobic regions spontaneously jump together across the gap and fuse. The two outer monolayers of the locally hemifused bilayers now slide radially outward from the fusion site until only one bilayer remains. As can be seen in the inset of Fig. 2, the two bilayers do not have to first come into contact to fuse, but can fuse spontaneously when the bilayer surfaces are more than 10 Å apart.

For the phosphatidylcholine, the initial breakthrough always takes place close to the center of the contact area, where the pressure is highest. The fusion process, as outlined in Fig. 4 (steps a–e), appears to be quite general for all the bilayers we have studied, with one exception. With bilayers of CTAB, fusion starts at the boundary/edge of the contact zone rather than at the center (Fig. 1B–D). This may be correlated with the fact that CTAB bilayers exhibit a large adhesion energy γ of approximately 7 mJ/m^2 (erg/cm^2), compared to 0.2 mJ/m^2 of PC bilayers in Fig. 2.

The second stage, from hemifusion to full fusion or complete fusion, involves the total removal of the central bilayer. This is shown in Fig. 4 (steps f–h) and is characterized by a second critical "breakthrough" step. It is only observed between supported hydrophilic monolayers in humid air or in water–wet organic solvents, when the binding of the head groups to the mica surfaces becomes weak.[28]

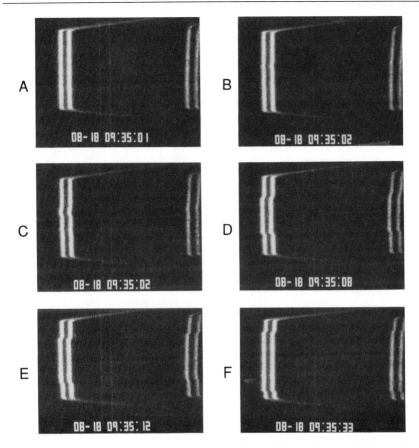

FIG. 3. Fringe patterns showing the progress of hemifusion of two partially depleted DMPC bilayers in the fluid state. (A) Flattening under a pressure of 50 atm with a circular contact area of 95 μm diameter (given by the length of the vertical parts of the fringes). (B) Breakthrough, followed by fast spreading which slows with time: (C) a fraction of a second later, (D) 6 sec later, (E) 10 sec later, and (F) 21 sec later. The thickness difference between unfused and fused bilayers is 3.3 ± 0.15 nm. Same magnification is in Fig. 1.

Conclusions and Implications

The dilution of lipid monomers in the aqueous solution below the CMC leads to a slow desorption ("depletion") of the lipids from the outer monolayers of adsorbed bilayers, that is, to a thinning of the bilayers. This results in a larger number of hydrocarbon chain segments, or hydrophobic area per lipid molecule, exposed to the aqueous phase. Thus, dilution increases the "hydrophobicity" of depleted bilayers, and the effects of

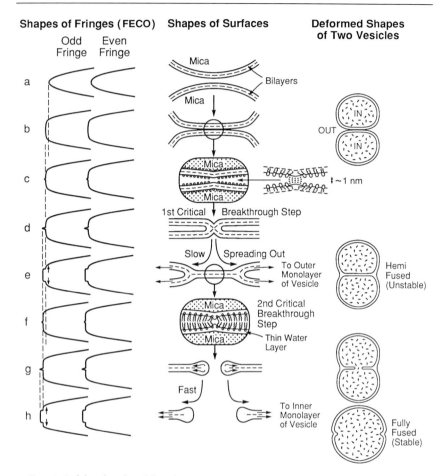

FIG. 4. *Left hand and middle columns; steps (a)-(h):* Intermediate stages in fusion of two adsorbed (supported) bilayers on mica in aqueous solutions (middle column) as monitored from the shapes of the FECO fringes (left-hand column). The "critical breakthrough" step (c) is followed by the sliding (spreading out) of the lipids of the fused outer monolayers, leaving a single hemifused bilayer (e). The second breakthrough step (f) completes the full fusion process. *Right-hand column:* Schematic illustration of fusion of two vesicles. In the case of vesicles, the hemifused state is "unstable" or "highly stressed" because the outer monolayer now contains a higher density of lipid than the inner monolayers. These stresses are relieved on full fusion.

hydrophobic interactions on adhesion and fusion can then be studied under controlled conditions.

For adsorbed bilayers fusion always occurs in two distinct stages. The first is the hemifusion of two bilayers into one bilayer, shown by Fig. 4

(steps a–e). This is essentially the same as found by Horn[11] for the fusion of egg phosphatidylcholine bilayers. Hemifused bilayers have previously been observed as stable or metastable states during the fusion of large unilamellar vesicles in solution or with black lipid membranes (BLMs).[8] Thus, the molecular mechanism we describe appears to apply to both supported and free bilayers.

Concerning the forces, we find that the occurrence of fusion is not simply related to the force law between bilayers. First, the internal stresses within bilayers which determine how easily a bilayer can locally deform also play a major role in fusion (no fusion is observed between defect-free bilayers in the frozen state). Second, fusion does not require that some repulsive (e.g., steric/hydration) force barrier be "overcome," nor that two bilayers first have to be able to come into adhesive contact. Rather, these forces are "bypassed" via molecular rearrangements so that the fusion process can start spontaneously when the bilayers are still a finite distance apart. It is also concluded that the most important force leading to the direct fusion of bilayers is the hydrophobic interaction, which attracts the interiors of membranes (in contrast to other types of attractive forces, such as ion-binding forces, which only act to bring the exterior surfaces of membranes together and adhere, but not necessarily fuse).

Our results relate to a number of previous reports on fusion-inducing effects, some of which are illustrated in Fig. 5. First, both osmotic and electric field stresses are often necessary for fusion to occur.[8,12] Although the popular explanation for this is that these stresses cause vesicles to rupture before they fuse,[8] we propose that fusion is more likely to be caused by the bilayer thinning induced by these effects,[2,29] and thus by an increase in the hydrophobic area exposed between two adjacent bilayers (Fig. 5C,D). As we have seen here, an increase in the area by only a few percent above the equilibrium (unstressed) value can initiate fusion.

Perhaps our most important conclusion is that only the attractive hydrophobic interaction between bilayers leads to *direct* bilayer–bilayer fusion. There have been many reports of fusion occurring because of the exposure of hydrophobic "domains," "pockets," or "segments" within proteins or membranes,[30–33] again suggestive of mechanisms that may be basically the same as those found here (Fig. 5A). Finally, ionic effects that

[29] O. Alvarrez and R. Latorre, *Biophys. J.* **21,** 1 (1978).

[30] J. D. Andrade, *in* "Surface and Interfacial Aspects of Biomedical Polymers, Volume 2: Protein Adsorption" (J. D. Andrade, ed.), p. 1. Plenum, New York, 1985.

[31] S. Ohki and N. Düzgüneş, *Biochim. Biophys. Acta* **552,** 438 (1979).

[32] N. Düzgüneş, J. Paiement, K. B. Freeman, N. G. Lopez, J. Wilschut, and D. Papahadjopoulos, *Biochemistry* **23,** 3486 (1984).

[33] D. Papahadjopoulos, S. Nir, and N. Düzgüneş, *J. Bioenerg. Biomembr.* **22,** 157 (1990).

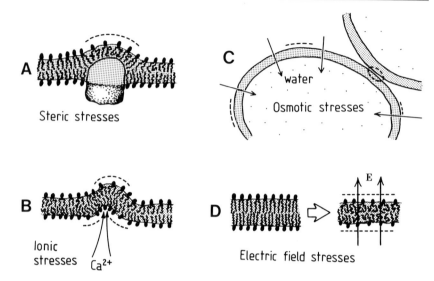

Fig. 5. Opening of hydrophobic areas (----) arising from protein packing stresses (A), ionic stresses (B), osmotic stresses (C), and electric field stresses (D). These areas become likely sites for the initiation of fusion.

do not induce fusion between adsorbed bilayers but do between free membranes are postulated to act via a more complex mechanism from the direct hydrophobic interaction, for example, by destabilizing free bilayers or membranes (Fig. 5B).

Advantages and Disadvantages of Method

The SFA allows direct measurements of both the attractive and repulsive regions of the force laws, and other interfacial properties, between supported bilayers. Such studies identified and quantified most of the fundamental interactions occurring between bilayers and other surfaces in aqueous solutions.[3-5,10,11,16,17,25,26,34,35] We may also note the good agreement in the measured forces between lecithin bilayers using the SFA and the osmotic pressure technique.[35] The SFA is the only device which allows the observation of the fusion process directly (in real time) with a resolution on the molecular level, at least in one dimension.

The major drawback is that the bilayers or membranes must be supported, which changes the viscoelastic properties within the bilayers. Also,

[34] J. N. Israelachvili and P. M. McGuiggan, *Science* **241**, 795 (1988).
[35] R. G. Horn, J. N. Israelachvili, J. Marra, V. A. Parsegian, and R. P. Rand, *Biophys. J.* **54**, 1185 (1988).

the force law between the bilayers is changed because the repulsive undulation forces are suppressed.[6,35] In addition, supported bilayers may be less fluid than free bilayers, a factor which is important for determining the breakthroughs in the fusion process.[8,36] The properties of the inner monolayer are obviously more affected by its interaction with the mica surfaces, which determines its molecular area, fluidity, etc.[28] Accordingly, the SFA technique currently does not allow all aspects of the fusion process to be probed, although it is hoped that future developments with other supporting surfaces may circumvent some of these limitations and allow measurements to be made between much more weakly supported bilayers and biological membranes.

Acknowledgments

We are very grateful to Patricia McGuiggan for much good advice and contributions to the experiments reported here as well as the exploratory experiments preceding them. This work was funded by the National Science Foundation under Grant CBT87-21741. Christiane Helm is grateful to the Deutsche Forschungsgemeinschaft for a postdoctoral research scholarship.

[36] J. Wilschut, N. Düzgüneş, D. Hoekstra, and D. Papahadjopoulos, *Biochemistry* **24**, 8 (1985).

[12] Effects of Fusogenic Agents on Membrane Hydration: A Deuterium Nuclear Magnetic Resonance Approach

By KLAUS ARNOLD and KLAUS GAWRISCH

Introduction

Hydration of surfaces is a common property of all glycerophospholipid bilayers. The phospholipid head groups, especially the phosphate, choline, carboxyl, and carbonyl groups, are mainly involved in the interaction with water molecules. Because the hydration of lipids is generally determined in the lamellar phase, the concept of hydration of phospholipids includes both bound and possibly trapped water. The thickness of the layer of trapped water depends on the strength of attractive and repulsive forces between the bilayers. Swelling of a lamellar phospholipid/water dispersion ceases if the repulsive forces are counterbalanced by attractive forces.

An important consequence of the binding of water is the creation of water-mediated repulsive forces between phospholipid bilayers. These hydration forces result from the structuring of water at the phospholipid surface. The bound water molecules exert an effect on the arrangement of adjacent water molecules, and this effect is carried over into the next layer of water. An approximately exponential decay with the separation between bilayers is observed for the hydration forces.[1,2] The hydration of lipids also influences thermotropic phase transitions and the formation of nonlamellar phases such as the inverted hexagonal phase.

Experiments have shown that vesicle–vesicle fusion is strongly dependent on the phospholipid head groups and their hydration properties. The inclusion of phosphatidylethanolamine (PE) and phosphatidylserine (PS) enhances cation-induced fusion, whereas phosphatidylcholine (PC) has an inhibitory effect.[2a] It was concluded that fusion can only occur after the breakdown of the hydration repulsion. This is possible either by a change in the properties of the membrane surface to reduce the magnitude of the repulsive force or by forcing out the intervening water by applied stress. Some representative examples of such processes are discussed in this chapter. The modulation of surface hydration due to the action of fusogenic agents such as cations leads to a reduction in the hydration repulsion. The fusogenic polymer polyethylene glycol (PEG) exerts its influence on membrane fusion by causing an efflux of water from the gap between membranes.

There are different experimental approaches to the study of water hydration. The methods have been reviewed.[2,3] Water adsorption isotherms specify the work necessary to transfer water from the bulk phase to the region between the membranes. Calorimetry detects the changes in thermodynamic properties of the interbilayer water as a shift in the freezing temperature of water. From dielectric relaxation measurements the dipole relaxation times of water molecules are derived; from X-ray and neutron diffraction the thickness of the water layer and the location of water molecules are determined. Hydration was also studied by measuring the quantity of water included by inverted micelles formed in organic solutions. Spectroscopic methods such as nuclear magnetic resonance (NMR), infrared, and Raman spectroscopy are sensitive to the orientational and dynamic properties due to the motion of the water.

[1] V. A. Parsegian, N. Fuller, and R. P. Rand, *Proc. Natl. Acad. Sci. U.S.A.* **76,** 2750 (1979).
[2] R. P. Rand and V. A. Parsegian, *Biochim. Biophys. Acta* **988,** 351 (1989).
[2a] N. Düzgüneş, J. Wilschut, R. Fraley, and D. Papahadjopoulos, *Biochim. Biophys. Acta* **642,** 182 (1981).
[3] K. Gawrisch, V. A. Parsegian, and R. P. Rand, *in* "Biophysics of the Cell Surface" (R. Glaser and D. Gingell, eds.), pp. 61–73. Springer-Verlag, Berlin and New York, 1989.

We focus our discussion on the application of ^{2}H NMR of ^{2}H$_2$O.[4,5] ^{2}H NMR has some significant advantages because this method can provide information about both the ordering of water molecules, that is, the structure of hydration layers, and the dynamics of the water molecules (molecular correlation times) in the lamellar phase. The purpose of this chapter is to show how ^{2}H NMR can be used to obtain a more detailed picture of lipid hydration and the effects of some fusogenic agents.

^{2}H NMR Spectroscopy of Water

Sample Preparation and Instrumentation

The natural abundance of ^{2}H$_2$O in H$_2$O is approximately 0.02%, and the NMR sensitivity is relatively low. For measurements at low water content it is therefore necessary to replace the water in phospholipid dispersions partly or completely by ^{2}H$_2$O to ensure sufficient sensitivity. Because of the differences in dynamic properties of hydrogen bonds in H$_2$O and ^{2}H$_2$O, deuteration can conceivably influence hydration; however, so far there has been no indication that deuteration changes structure and hydration properties.

Several methods are used for the preparation of phospholipid/water mixtures having different water contents and a homogeneous distribution of water in the sample: (1) A well-defined amount of dry phospholipid is weighed with about the same amount of water. The sample is thoroughly mixed at elevated temperature. If necessary the sample is partially dehydrated in a desiccator to adjust the concentration. The amount of water is determined by weighing. (2) Dry phospholipid is transferred to a sample ampoule containing a suitable weight of ^{2}H$_2$O. The ampoule is sealed, and complete mixing is achieved by centrifugation back and forth through a constriction having a diameter of about 0.5 mm.

The morphology of the system under study is important because, as shown below, ^{2}H NMR spectral parameters are sensitive to morphology-dependent motional averaging processes.[6] The lateral diffusion of water molecules over the surface of vesicles with diameters smaller than 2 μm influences the line shape of the ^{2}H NMR spectrum significantly. A decrease in the diameter of multilamellar vesicles is often found in vesicles having a range of higher water content.[6]

[4] N. J. Salsbury, A. Darke, and D. Chapman, *Chem. Phys. Lipids* **8**, 142 (1979).

[5] E. G. Finer and A. Darke, *Chem. Phys. Lipids* **12**, 12 (1974).

[6] K. Gawrisch, W. Richter, A. Möps, P. Balgavy, K. Arnold, and G. Klose, *Stud. Biophys.* **108**, 5 (1985).

^{2}H NMR spectra of ^{2}H$_2$O can be recorded in a conventional high-resolution Fourier transform (FT) NMR spectrometer. Typical parameters for recording are a 50 kHz sweep width and 8K data points. In the case of relatively broad spectra (>2 kHz), distortions of the line shape can occur due to incomplete recording of the free induction decay (FID) that results from a time delay between excitation and acquisition. The results are phase errors in the spectra and baseline distortions. This effect can be avoided by using the on-resonance quadrupole echo and collecting the FID from the top of the echo.

^{2}H NMR Spectrum of ^{2}H$_2$O

The ^{2}H NMR experiment is sensitive to the characteristics of motion of the O–^{2}H bond. Because ^{2}H has a spin quantum number $I = 1$, an electric quadrupole moment of the nucleus, eQ, occurs which interacts with an electric field gradient, eq, of mainly intramolecular origin.[7] The NMR experiment can give information about the anisotropy and rate of motion due to the modulation of the quadrupole interaction by molecular motion.

Every residual electric field gradient at the deuterium nucleus, not averaged out by fast isotropic motion, splits the ^{2}H NMR signal into a doublet. The resulting splittings provide one of the most direct ways of measuring the anisotropy of the motion of water. The quadrupole splitting of ^{2}H$_2$O of a macroscopically oriented bilayer system, whose symmetry axis of molecular reorientation is oriented at an angle θ with respect to the static magnetic field, is[8,9]

$$\Delta\nu_Q(\theta) = \left| \sum_i p_i \nu_{Qi} S_i (3 \cos^2 \theta - 1) \right| \tag{1}$$

where p_i is the fraction of deuterons at site i, S_i is the order parameter which represents the extent of the motional averaging of water at site i, and $\nu_{Qi} = (3/4)\chi$, where $\chi = e^2 qQ/h$ is the quadrupole coupling constant of the solid water (222 kHz). It is assumed that the motional rates of rotations and exchange processes are faster than the maximal quadrupole splitting ($> 10^5$ sec^{-1}) and that, consequently, the observed splitting is an average over the splittings of every site. Comparison of experimental data with Eq. (1) shows that the symmetry axis of the molecular reorientation of water is parallel to the bilayer normal.[9,10]

[7] B. Halle and H. Wennerström, *J. Chem. Phys.* **75**, 1928 (1981).
[8] H. Wennerström, G. Lindblom, and B. Lindman, *Chem. Scr.* **6**, 97 (1974).
[9] J. M. Pope and B. A. Cornell, *Chem. Phys. Lipids* **24**, 27 (1979).
[10] B. A. Cornell, J. M. Pope, and G. J. F. Troup, *Chem. Phys. Lipids* **13**, 183 (1974).

For a nonoriented sample, Eq. (1) can be written as

$$\Delta v_Q = \left| \sum_i p_i \, \Delta v_{Qi} \right| \tag{2}$$

with $\Delta v_{Qi} = v_{Qi} S_i$, the splitting in site i. The resulting spectrum is termed the powder spectrum. Quadrupole splitting is the difference in frequency units between the major maxima in the spectrum. The summations in Eqs. (1) and (2) are dominated by sites close to the surface because the splittings are largest for water molecules close to the head groups. An equation for the calculation of the quadrupole splitting as a function of two order parameters and the asymmetry parameter of the electric field was derived by Halle and Wennerström.[7]

Typical ^2H NMR spectra for egg lecithin/^2H_2O dispersions having different water contents are given in Fig. 1. On addition of low quantities of water (up to about 23 water molecules per lipid), all the water is arranged between the lamellae, and a spectrum similar to Fig. 1a is observed. The splitting decreases with higher water content. For more than about 23 water molecules, free water appears and is characterized by a narrow signal component in the center of the spectrum (Fig. 1b). This signal is superimposed on the broad spectrum of the water incorporated between the lamellae. The existence of the narrow signal is indicative of free water arranged in a separate phase and exchanging slowly with trapped or bound water.

The highest value of quadrupole splittings measured for water in phospholipid multilayers was only 6.9 kHz.[5] This is much less than the maximum value of about 166 kHz of immobilized water molecules. These data

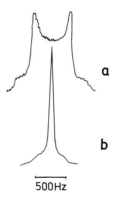

500Hz

FIG. 1. ^2H NMR spectra of ^2H_2O in egg lecithin/^2H_2O dispersions at molar ratios of water to lipid of 15.0 (a) and 32.5 (b).

are indicative of water ordering near the bilayer surface. However, relatively low order parameters would result from Eqs. (1) and (2). From this result we cannot conclude that the ordering of water is low at the bilayer surface; the problem is more complex. Low values of the average splitting can occur because of partial cancellation of contributions, for example, due to an exchange of water molecules between sites having order parameters of opposite signs. In a recent discussion of the apparent contradiction between the ordering of water at bilayer surfaces and the measured low quadrupolar splittings, a model of motion of water molecules was favored that assumes rapid motions of water molecules around their symmetry axis.[10a] This motion reduces the measured quadrupolar splitting to nearly zero.

^2H NMR Relaxation Times

A large proportion of the dynamic data on water molecules in lamellar systems comes from ^2H NMR relaxation studies.[10,11] The spin–lattice relaxation time T_1 is determined from the recovery of the magnetization after a 180° pulse. The dominant relaxation mechanism is quadrupole relaxation, which is a purely intramolecular process.

The experimental data can be well fitted by a two-site exchange model.[11] A primary hydration shell comprises n_L water molecules whose motional properties are influenced by the phospholipid surface. The molecules of this shell have an average relaxation time T_{1L}. A second shell is formed from additional water (n_T molecules) trapped between the lipid lamellae, and a relaxation time T_{1T} close to the T_1 of free water is expected. These water molecules are in rapid exchange with the primary hydration shell. For $n > n_L$ the average relaxation time observed in the experiments is described by Eq. (3):

$$\frac{1}{T_1} = \frac{1}{T_{1T}} + \frac{n_L}{n}\left(\frac{1}{T_{1L}} - \frac{1}{T_{1T}}\right) \tag{3}$$

with $n = n_L + n_T$.

Equation (3) shows that the number of water molecules n_L of the primary hydration shell can be determined from a plot of $1/T_1$ versus $1/n$. Effective correlation times (τ_{ce}) of the motion of these water molecules can

[10a] K. Gawrisch, D. Ruston, J. Zimmerberg, V. A. Parsegian, R. P. Rand, and N. Fuller, Biophys. J. **61**, 1213 (1992).
[11] F. Borle and J. Seelig, Biochim. Biophys. Acta **735**, 131 (1983).

be calculated from the relaxation times using an equation derived for the case of extreme motional narrowing:

$$\frac{1}{T_1} = \frac{3}{2}\,\pi^2\chi^2\tau_{ce} \tag{4}$$

The following motions can influence the quadrupole splitting and relaxation times: fast rotations around the $O-^2H$ bonds of those water molecules which are connected to the phosphate group by only one hydrogen bond, fast rotation of the head group–water complex around the director, angular fluctuations of the rotational axes, rapid exchange of water molecules between the binding sites on one phosphate group, exchange between water layers, diffusion along the surface of the bilayer or between the opposite layers, and rapid exchange of deuterons between water molecules located at different sites or in the surface.

Phospholipid Hydration

The different species of phospholipids have different limiting hydrations, which have been related to the water–lipid interaction, the strength of lipid–lipid interaction via hydrogen bonding between adjacent molecules, and the electrical charge of the lipid. The charges of the lipid molecule play a minor role for the direct binding of the water molecules. However, they increase the bilayer separation, resulting in a larger amount of water incorporated between bilayers. Hydrogen bonds between lipids influence the ability of the head group to bind water molecules.

In Fig. 2 the ^2H_2O quadrupole splittings for a sample consisting of egg lecithin with varying amounts of heavy water are given as a function of the water content. A quantitative analysis of such data has been attempted by Finer and Darke[5] and Södermann et al.[12] They have shown that the plot of the measured Δv_Q versus $1/n$ consists of a series of linear regions with different slopes and that every straight line in such a representation corresponds to a hydration shell, each with a characteristic quadrupole splitting, Δv_{Qi}, and number of water molecules, n_i. The water molecules exchange rapidly between the lowest occupied shells, and therefore one average quadrupole splitting occurs for all water molecules. The number of water molecules, n_i, per shell can be calculated from the projection of the breakpoints of the curve on the $1/n$ axis. The intersections of the straight lines

[12] O. Södermann, G. Arvidson, G. Lindblom, and K. Fontell, *Eur. J. Biochem.* **134**, 309 (1983).

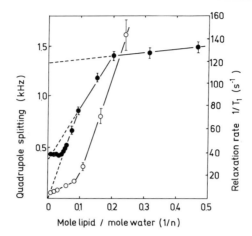

FIG. 2. ^{2}H NMR quadruple splitting (●) and spin lattice relaxation rate (○) of egg lecithin/^{2}H$_2$O dispersions as a function of the inverse molar ratio of water to lipid. (The relaxation data were kindly provided by Dr. F. Volke.)

with the $\Delta\nu_Q$ axis indicate the intrinsic quadrupole splittings. In many cases it was possible to get a good fit of the data by using this model.[13]

From Fig. 2 the existence of two hydration shells with nonzero quadrupole splitting (1.35 and 0.35 kHz) and a water shell with zero quadrupole splitting (trapped water) can be determined. The single shells contain about 5, 7, and 11 water molecules. This means that about 12 water molecules distributed over two shells are partly oriented (bound water) and about 11 water molecules with motional characteristics similar to those of free water are trapped in the interbilayer region. At water concentrations higher than 23 water molecules per lecithin, the thickness of the water layer remains unchanged, and the excess water forms a separate phase.

With a surface area of 60–70 Å^2 per lipid molecule and an average cross-sectional area of 7–9.6 Å^2 for water, 12 water molecules correspond to approximately one to two layers of bound water.[11] At a water concentration of about 23 water molecules per lipid the minimum value of splitting is measured, indicating that at this water content the incorporation of water between lipid bilayers is completed.

In Fig. 2 the spin–lattice relaxation rates are given for egg lecithin/ ^{2}H$_2$O dispersions as a function of water content. As predicted by Eq. (3), these measurements are well described by a two-site exchange process. A qualitatively similar behavior is observed for PE and phosphatidylglycerol

[13] K. Gawrisch, K. Arnold, K. Dietze, and U. Schulze, in "Electromagnetic Fields and Biomembranes" (M. Markov and M. Blank, eds.), p. 9. Plenum, New York, 1988.

(PG), suggesting common properties of phospholipid hydration.[11] A minimum of 11–16 water molecules is needed to form a primary hydration shell. Effective correlation times are calculated from the relaxation rates [Eq. (4)]. Compared to the correlation time of 3.0 psec for pure water, correlation times of about 140 psec are found for the first five water molecules and about 16 psec for the other water molecules of this hydration shell. These values are consistent with ^1H relaxation measurements[14] and dielectric relaxation measurements.[15]

Other quantities such as the chemical potential of water and the dipole correlation times vary quasi-continuously with water content and argue strongly against any division of water in layers of trapped, bound, and strongly bound water. For comparisons of results obtained by different methods, the sensitivity to different types of structural changes and to motions on different time scales for each method must be considered. The quadrupole splitting should be very sensitive to structural properties of the head group region because the first five water molecules are attached to the head group, as Monte Carlo and molecular dynamics calculations have shown.[16] These water molecules affect the distance between head groups and thus the mobility of the head groups.[17] The second discontinuity of the quadrupole splitting observed at about 12 water molecules coincides with the increase in the lamellar repeat spacing owing to a rapid increase in the thickness of the water layer as measured by X-ray diffraction.[18] These structural changes in the lamellar phase may influence the quadrupole splitting to a larger extent than other quantities such as chemical potential.

The ^2H NMR measurements of the hydration of lipids agree with those obtained using other methods. The hydration of PC bilayers in excess water as determined by ^2H NMR studies was always lower than estimates from other methods. A number of studies indicate that the water content in fully hydrated PE is less than that in fully hydrated PC.[5] The egg PE lamellar phase incorporates no more than 12 water molecules per lipid molecule. A maximum hydration of about 8–9 mol of water per mole of lipid was found for mono- and diglucodiacylglycerols.[19] These glucoglycerolipids have several physical properties similar to those of PE, such as low hydration and the formation of the inverted hexagonal H_{II} phase. One interesting feature of hydration studies is the large difference in hydration

[14] D. A. Wilkinson, H. J. Morowitz, and J. H. Prestegard, *Biophys. J.* **20,** 169 (1977).

[15] G. Nimtz, A. Enders, and B. Binggeli, *Ber. Bunsenges. Phys. Chem.* **89,** 842 (1985).

[16] G. Klose, K. Arnold, G. Peinel, H. Binder, and K. Gawrisch, *Colloids Surf.* **14,** 21 (1985).

[17] F. Volke, K. Arnold, and K. Gawrisch, *Chem. Phys. Lipids* **31,** 179 (1982).

[18] M. J. Janiak, D. M. Small, and G. G. Shipley, *J. Biol. Chem.* **254,** 6068 (1979).

[19] A. Wieslander, J. Ulmius, G. Lindblom, and K. Fontell, *Biochim. Biophys. Acta* **512,** 241 (1978).

between lipids that form the lamellar phase only (e.g., PC) and lipids that are able to form the H_{II} phase (e.g., PE).[20]

The negatively charged membrane lipids PS, PG, and glycerophosphoryldiglucosyldiglyceride have higher water-binding capacities than uncharged lipids owing to the repulsion of adjacent bilayers.[5,19] A low hydration level was found for the acid form of PS.[5] However, it is possible that the PS used in earlier studies had some trace amounts of calcium. A water-binding capacity of 16 water molecules was found using inverted micelles and purified lipid.[20]

In summarizing the results of measurements on different lipids, it can be concluded that the spectroscopic behavior of particular numbers of water molecules per lipid molecule is influenced by the bilayer surface. These water molecules show restricted motion, are partially oriented, and are classified as "bound" water. Besides these water molecules, additional water is trapped between the bilayers to minimize the interbilayer force. All water molecules exchange rapidly between the different water sites, and averaged quadrupole splittings and relaxation rates are observed. Even at low water content, the reorientation of the water molecules is very rapid, and an almost complete averaging of the quadrupole splitting occurs. Additional water molecules are arranged outside the bilayers in a separate phase, and this leads to a narrow signal resulting from isotropic rotation of the water molecules.

Effect of Fusogenic Agents

Polyethylene Glycol

The quadrupole splitting of ^2H_2O can be used for the determination of the water content of lamellar lipid systems if a calibration curve similar to that in Fig. 2 is known for the system.[13,21] The accuracy of this method is greater where the quadrupole splitting is strongly dependent on water concentration (e.g., between 5 and 15 water molecules for the PC lamellar phase). Because the accuracy of the measurement of the quadrupole splitting is usually better than ± 50 Hz, an accuracy of better than ± 0.5 water molecules is possible for the calculation of the water content.

For studies on the influence of polyethylene glycol (PEG) on phospholipid hydration, samples are prepared by mixing dry lipid and an excess of PEG/^2H_2O solution of the appropriate concentration. The samples are

[20] A. Sen and S. W. Hui, *Chem. Phys. Lipids* **49**, 179 (1988).
[21] K. Arnold, L. Pratsch, and K. Gawrisch, *Biochim. Biophys. Acta* **728**, 121 (1983).

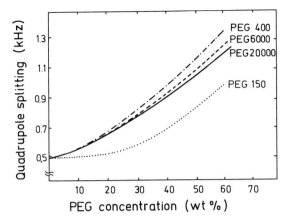

FIG. 3. Quadrupole splitting of ^2H_2O in egg lecithin/^2H_2O dispersions as a function of PEG concentration for polymers of different molecular weights. (For clarity, experimental points are not given.)

homogenized as described above. Before the NMR measurements are completed, the lipid phase is concentrated by gentle centrifugation.

The addition of PEG of different molecular weights to egg lecithin/^2H_2O dispersions leads to an increase of the quadrupole splitting (Fig. 3). It can be concluded that the amount of water incorporated between the bilayers is decreased. Other experiments have shown that PEG is excluded from the gap between the membranes and that it acts in an osmotic manner on the water layer.[22,23] Because of the relatively high osmotic pressures of PEG solutions [20 MPa at ~50% (w/w) PEG], the water content is effectively reduced.[23] The number, n, of water molecules per lipid molecule derived from the quadrupole splitting (Fig. 3) and the calibration curve (Fig. 2) is given for PEG 20,000 in Fig. 4. At a fusogenic concentration of 50% (w/w) PEG, the number of water molecules is reduced from 23 to approximately 8. This means that the trapped and part of the bound water molecules are extracted from the phospholipid bilayer system. The PEG induces a close apposition of the bilayers.

A significantly lower quadrupole splitting is observed for PEG 150. For such low molecular weight PEG, it is possible that the penetration of the

[22] K. Arnold, O. Zschörnig, D. Barthel, and W. Herold, *Biochim. Biophys. Acta* **1022**, 303 (1990).

[23] K. Arnold, A. Herrmann, K. Gawrisch, and L. Pratsch, in "Molecular Mechanisms of Membrane Fusion" (S. Ohki, D. Doyle, T. D. Flanagan, S. W. Hui, and E. Mayhew, eds.), p. 255. Plenum, New York, 1988.

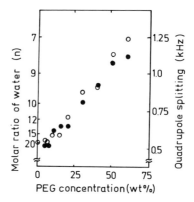

FIG. 4. Quadrupole splittings and number of water molecules per lipid molecule incorporated between bilayers of an egg lecithin/^2H_2O dispersion as a function of the concentration of PEG 20,000. For the calculation of the number of water molecules per lipid molecule, data taken from Figs. 2 and 3 were used. PEG was directly added (●) or separated from the dispersion by a dialysis membrane (○).

polymer into the space between bilayers results in a direct effect on the arrangement of water in the gap and on the osmotic pressure.

The thickness of the water layer can be calculated from the number of water molecules incorporated between the bilayers if the area per lipid molecule and the density of water between lipid layers are known. The area per lipid molecule can be calculated from data given in Lis et al.[24] An exponential relation between the thickness of the water layer calculated in this way and the osmotic pressure exerted by PEG was found for all lamellar systems studied.[13] The repulsive forces determined from these measurements are in a good agreement with the forces calculated from X-ray diffraction studies using an osmotic stress technique.[1,2] X-Ray diffraction has also been used to identify the effect of PEG on the lamellar repeat spacing, and various structural defects in lipid dispersions having high PEG content have been observed by freeze–fracture electron microscopy.[25]

Cations

Owing to their anionic character at physiological pH, phospholipids such as PS, phosphatidic acid (PA), and cardiolipin interact with biologically important cations, and this interaction has been implicated in mem-

[24] L. J. Lis, M. McAlister, N. Fuller, R. P. Rand, and V. A. Parsegian, *Biophys. J.* **37**, 657 (1982).
[25] L. T. Boni, T. P. Stewart, and S. W. Hui, *J. Membr. Biol.* **80**, 91 (1984).

brane fusion processes (for reviews, see Papahadjopoulos *et al.*[26,26a] and Düzgüneş[26b]). It was suggested that interbilayer Ca^{2+} complexes may initially trigger membrane fusion processes.

In 1972 Rand and Sengupta reported that millimolar levels of Ca^{2+} trigger a collapse of PS multilayers, leaving negligible water spacing between bilayers.[27] At such low hydration the water molecules are bound between the bilayers, resulting in a reduced reorientational motion and high quadrupole splittings which are difficult to detect. The change in the measured quadrupole splittings per water molecule increases for reduced water content, but sensitivity problems limit most measurements to water concentrations greater than 2 water molecules per phospholipid.[28]

Lipid bilayers are relatively impermeable to cations. Therefore, suitable means must be found to expose all lipid surfaces to the cations. Similar difficulties occur in the preparation of samples for binding studies,[29] X-ray diffraction, differential scanning calorimetry, and other methods if measurements are performed at a low water content in the presence of cations.[30] Dry phospholipid can be mixed with the cation/^2H_2O solution and homogenized as described above. Alternatively, a sonicated dispersion of the phospholipid in salt solution is precipitated by addition of divalent cation, washed, vacuum dried, and remixed with a small amount of ^2H_2O ($n \approx 20$).

Studies of PC systems in the presence of monovalent cations show that increasing the amount of salt leads to a reduction in the water between the bilayers.[12] A similar effect is observed for PS in the presence of Na^+ and K^+. The dominant effect of these cations is to shield the negatively charged surface, enabling a closer approach of bilayers. In contrast to these cations, Ca^{2+} changes the ^2H NMR spectrum of ^2H_2O completely. The singlet observed is a strong indication that Ca^{2+} displaces the water of hydration, forming a dehydrated or probably anhydrous Ca–PS complex.[28] A comparable effect was found for some trivalent cations, but not for Mg^{2+} where quadrupole splitting is still observed, indicating a residual hydration.[31] Similar differences were found for the decrease in the surface dielectric constant as detected by fluorescence probes.[32] These findings indicate that

[26] D. Papahadjopoulos, A. Portis, and W. Pangborn, *Ann. N.Y. Acad. Sci.* **308**, 50 (1978).

[26a] D. Papahadjopoulos, S. Nir, and H. Düzgüneş, *J. Bioenerg. Biomembr.* **22**, 157 (1990).

[26b] N. Düzgüneş, *Subcell. Biochem.* **11**, 195 (1985).

[27] R. P. Rand and S. Sengupta, *Biochim. Biophys, Acta* **255**, 484 (1972).

[28] H. Hauser, E. G. Finer, and A. Darke, *Biochem. Biophys. Res. Commun.* **76**, 267 (1977).

[29] G. W. Feigenson, *Biochemistry* **25**, 5819 (1986).

[30] H. Hauser and G. Shipley, *Biochemistry* **23**, 34 (1984).

[31] S. Ohki and K. Arnold, unpublished (1989).

[32] S. Ohki and K. Arnold, *J. Membr. Biol.* **114**, 195 (1990).

the interaction of Ca^{2+} with PS leads to a hydrophobic surface. Studies of a $DMPA/Ca^{2+}$ system have also demonstrated that the DMPA molecules are highly dehydrated.[32a] Given the different ability of cations to induce phospholipid bilayer fusion, it can be argued that the formation of an anhydrous cation–phospholipid complex is of importance for fusion processes.

Hydration of Nonlamellar Phases

In the discussion given above the hydration of phospholipids has only been considered for the lamellar phase in the liquid crystalline state. However, it could be shown that phase diagrams can be constructed from the analysis of the quadrupole splitting.[33]

Decreasing the temperature below the liquid crystal–gel transition point, the quadrupole splitting of dimyristoylphosphatidylcholine (DMPC), dipalmitoylphosphatidylcholine (DPPC), and dimyristoylphosphatidylethanolamine (DMPE) always displays a minimum near the phase transition temperature.[4,7,21,33,34] This phenomenon was explained in terms of a spatial modulation in the orientational order induced by the ripplelike structure[35] and as a result of pretransitional fluctuations.[34] No splitting is observed in the gel phase of DMPE, probably because of the small hydration in this phase. For DMPC and DPPC, an increase in the quadrupole splitting with decreasing temperature can be measured in the gel state. The hydration is lower in the gel state compared to the liquid–crystalline state for lamellar phases of PC.

The inverted micelle has been proposed as a possible intermediate in membrane fusion[34a] (however, see Ref. 34b). Such inverted micelles are related to the inverted hexagonal phase H_{II} of phospholipids. In the H_{II} phase, one important additional degree of motional freedom occurs for the water molecules, namely, lateral diffusion around the cylinders formed by the lipid molecules. This diffusion results in a reduction by one-half of the quadrupole splitting compared to the lamellar phase, provided that the interactions can be assumed to be the same and that the length of the tubes is sufficiently great.

[32a] G. Laroche, E. J. Dufourc, J. Dufourc, and M. Pézolet, *Biochemistry* **30**, 3105 (1991).
[33] J. Ulmius, H. Wennerström, G. Lindblom, and G. Arvidson, *Biochemistry* **16**, 5742 (1977).
[34] M. W. Hawton and J. W. Doane, *Biophys. J.* **52**, 401 (1987).
[34a] A. J. Verkleij, *in* "Membrane Fusion" (J. Wilschut and D. Hoekstra, eds.), p. 155. Dekker, New York, 1991.
[34b] E. Bearer, N. Düzgüneş, D. S. Friend, and D. Papahadjopoulos, *Biochim. Biophys. Acta* **693**, 93 (1982).
[35] L. M. Strenk, P. W. Westermann, N. A. P. Vaz, and J. W. Doane, *Biophys. J.* **48**, 355 (1985).

When the lamellar–hexagonal transition temperature is reached for PE, a drastic reduction of the splitting by more than one-half is observed. One interpretation is that the head groups are less exposed to water because they are packed in the inside of the cylinders.[36]

A complex behavior of the ^{2}H NMR spectrum of ^{2}H$_2$O was observed for a mixture of dioleoylphosphatidylcholine (DOPC) and gramicidin.[37] At higher molar ratios of gramicidin, lamellar and H$_{II}$ phases are simultaneously formed, resulting in a superposition of two spectra of different quadrupole splittings. In this case the H$_{II}$ phase is extremely rich in ^{2}H$_2$O. Usually, decreasing the water content promotes H$_{II}$ phase formation, and H$_{II}$ phase-forming lipids have a low head group hydration. What is remarkable is that the addition of gramicidin creates a hexagonal phase that takes up more water than does the lamellar phase. A similar behavior was observed in a $3:1$ (v/v) dioleoylphosphatidylethanolamine (DOPE)/ DOPC mixture after additions of alkanes.[38] It was suggested that the gramicidin-rich lamellar phase is transformed to a hexagonal phase by an interbilayer fusion event involving aggregates of gramicidin molecules.

The biological relevance of these findings is that nonbilayer structures could be stabilized by hydrophobic polypeptides. The experiments discussed above show that ^{2}H NMR spectroscopy of ^{2}H$_2$O can significantly contribute to the elucidation of the hydration properties of nonlamellar phases.

[36] P. L. Yeagle and A. Sen, *Biochemistry* **25**, 7518 (1988).
[37] J. A. Killian and B. de Kruijff, *Biochemistry* **24**, 7890 (1985).
[38] S. M. Gruner, V. A. Parsegian, and R. P. Rand, *Faraday Discuss. Chem. Soc.* **81**, 29 (1986).

Section II

Induction of Cell–Cell Fusion

[13] Chemically Induced Fusion of Erythrocyte Membranes

By JOCELYN M. BALDWIN and JACK A. LUCY

Introduction

In this chapter we describe a number of methods by which chemical agents may be used to induce the fusion of erythrocyte membranes. Erythrocytes do not normally fuse, and their membrane structure is very well characterized and relatively simple. Thus their chemically induced fusion serves as a useful experimental model for naturally occurring processes of membrane fusion.

Some information on the mechanisms involved in chemically induced membrane fusion has come directly from the development of procedures that are described here. For example, the fact that cell swelling accompanies fusion in every case strongly suggests an essential role for osmotic forces in the fusion process. However, we also describe a variety of techniques which are usefully applied in investigations on the mechanisms of erythrocyte fusion induced by chemicals.

Preparation of Cells for Fusion Experiments

Preparation of Washed Erythrocytes

Blood is withdrawn into a sterile solution of citrate anticoagulant[1] and stored at 4°. It is then diluted with buffer (40 mM HEPES, 110 mM NaCl, pH 7.4) and centrifuged at 1000 rpm using an IEC Centra-4X bench-top centrifuge (Damon Ltd., Dunstable, Beds., UK) (200 g) for 10 min. The pelleted cells are washed 3 times with buffer. After each centrifugation, the extreme top layer of cells is taken off by suction to remove the buffy coat, which contains most of the leukocytes. The final packed cell pellet typically contains 5×10^9 cells/ml.

Formation of Monolayers of Erythrocytes

The fusion of erythrocytes in response to treatment with chemical agents may conveniently be studied using monolayers of cells that are attached to Alcian blue-coated surfaces.[2] Plastic petri dishes (35 mm) are

[1] E. L. De Gowin, R. C. Hardin, and J. B. Alsever, *in* "Blood Transfusion," p. 330. Saunders, Philadelphia and London, 1949.
[2] M. V. Nermut and L. D. Williams, *J. Microsc.* **118**, 453 (1980).

treated with 1.0 ml of an aqueous solution of Alcian blue (1 mg/ml) at room temperature for at least 30 min. The Alcian blue solution is then removed, and the dishes are washed thoroughly, first with water and then with buffer. Packed washed erythrocytes are diluted to approximately 5×10^8 cells/ml with buffer (40 mM HEPES, 110 mM NaCl, pH 7.4). It should be noted that too great a dilution of the cells will result in the formation of a patchy or poorly packed monolayer. The buffer is removed from the Alcian blue-coated dish, and 1.0 ml of the cell suspension is added. After 30 min, nonattached erythrocytes are carefully washed off with buffer. It is advisable to add all solutions gently down one side of the dish to avoid detaching the cells. The attached cells are maintained in 1.0 ml of buffer until needed.

It should not be overlooked that, because the cells are firmly attached to the Alcian blue-coated dish, their ability to undergo extensive shape changes is restricted. However, in general, the method provides a convenient way of monitoring fusion reactions by light microscopy.

Cell Fusion by Osmotic Shock and in Progressively Diluted Media

Exposure to a hypotonic medium of a monolayer of erythrocytes, on an Alcian blue-coated surface, results in cells lysis. If, however, steps are first taken to free the erythrocyte membrane from the restraining effects of the cytoskeleton, it is possible to achieve fusion by osmotic swelling.[3] Incubation of hen erythrocytes with ionophore A23187 (Calbiochem Novabiochem Ltd., Nottingham, UK) and Ca^{2+} leads to a proteolytic breakdown of spectrin and goblin.[4] Spectrin may also be broken down by incubating human erythrocytes at 50°,[5] but ionophore treatment is usually more convenient.

Osmotic Fusion of Ionophore A23187-Treated Human Erythrocytes

Reagents

Salt/dextran: 40 mM HEPES buffer (pH 7.4) containing 110 mM NaCl, 80 mg/ml dextran (M_r 60,000–90,000)
Ionophore A23187: stock solution 2 mg/ml in dimethyl sulfoxide
Subtilisin BPN (nagarse, Serva, Heidelberg, Germany): stock solution 1.0 mg/ml in salt/dextran
80 mg/ml Dextran
$CaCl_2$: 1 M stock solution

[3] Q. F. Ahkong and J. A. Lucy, *Biochim. Biophys. Acta* **858**, 206 (1986).
[4] P. Thomas, A. R. Limbrick, and D. Allan, *Biochim. Biophys. Acta* **730**, 351 (1983).
[5] P. Heubusch, C. Y. Jung, and F. A. Green, *J. Physiol. (London)* **122**, 266 (1985).

Procedure. The buffer is removed from a monolayer of human erythrocytes and replaced by 1.0 ml of salt/dextran containing 10 mM $CaCl_2$, 10 μg/ml A23187, and 50 μg/ml subtilisin. The cells are incubated at 37° for 60–90 min. The medium is then diluted with 2.0 ml of 80 mg/ml dextran. Fusion begins to occur after about 10 min. Hen erythrocytes may be fused by the same procedure, but in this case subtilisin is not required.

Fusion of Erythrocytes in Progressively Diluted Media

Although human erythrocytes show greatly restricted volume changes at osmolarities in the physiological range,[5] cell swelling in very dilute media nevertheless results in disruption of the cytoskeleton under appropriate conditions. In this procedure, human erythrocytes are gradually diluted (dextran is included to reduce the rate of swelling) until fusion occurs. $LaCl_3$ is included in the diluent to prevent lysis occurring without fusion.

Reagents

40 mM HEPES (pH 7.4), 110 mM NaCl, 80 mg/ml dextran (M_r 60,000–90,000), 0.5 mM $LaCl_3$
80 mg/ml Dextran, 0.5 mM $LaCl_3$
0.5 mM $LaCl_3$

Procedure. The buffer is removed from a monolayer of erythrocytes and replaced by 1.0 ml of 40 mM HEPES (pH 7.4), 110 mM NaCl, 0.5 mM $LaCl_3$, 80 mg/ml dextran. After 5 min at 37°, 0.5 ml of 0.5 mM $LaCl_3$ containing 80 mg/ml dextran is added; after a further 5 min at 37°, another 1.0 ml of La^{3+}/dextran is then added. After 10 min at 37°, the whole volume of medium (2.5 ml) is removed and replaced by 1.0 ml of 0.5 mM $LaCl_3$. This procedure yields many large irregularly shaped polycells.[3]

Cell Fusion Induced by Permeant Molecules

When human erythrocytes attached to monolayers are exposed to permeant molecules, such as malonamide and polyethylene glycol 300 (PEG 300), in which a stable osmotic equilibrium is not obtainable, the cells swell slowly and continuously and, under the appropriate conditions, then fuse.[6]

Procedure. The buffer is removed from a monolayer of cells, and 1.0 ml of 500 mM PEG 300 (15%, w/v) containing 10 mM $CaCl_2$ and 10 mM

[6] Q. F. Ahkong and J. A. Lucy, *J. Cell Sci.* **91,** 597 (1988).

HEPES, pH 7.4, is added. Extensive cell swelling and fusion are visible after about 15 min at 37°, although cell lysis also occurs. If Ca^{2+} is absent or is replaced by Mg^{2+}, the swelling cells lyse without fusing. Cell swelling, lysis, and fusion are almost completely inhibited by the presence of 60 mg/ml serum albumin or 40 mM HEPES, 110 mM NaCl, pH 7.4. Thus, whereas the addition of salt to arrest cell swelling may be useful in studying the mechanism of PEG 300-induced fusion, care must be taken to ensure that any salt that is added with potential activators or inhibitors of fusion does not itself inhibit cell swelling. PEG 300 may be replaced by 600 mM malonamide or 500 mM PEG 400 (20%, w/v), but swelling and fusion then occur more slowly.

Fusion by Lipid-Soluble Fusogens

A number of simple lipids have been found to be capable to inducing the fusion of hen erythrocytes.[7] These include saturated carboxylic acids (containing 10–14 carbon atoms), longer but unsaturated carboxylic acids, and their monoesters of glycerol. One of the most fusogenic substances is oleoylglycerol, and this has also been shown to fuse mammalian erythrocytes from a number of different species.

Reagents

Salt/dextran solution: see above

Lipid dispersion: Lipids are stored at 4° as 10 mg/ml solutions in hexane. The required volume of lipid (0.1 ml of stock) is transferred to a round-bottomed flask, and all traces of solvent are removed, either by evaporation under reduced pressure in a rotary evaporator at 37° or under a stream of N_2. The dried lipid is dispersed in 6.0 ml salt/dextran solution by sonication for about 1 min. During this process the flask is cooled on ice. The emulsion is used within 5 min of preparation.

Procedure. Erythrocytes are washed in salt/dextran solution, and their concentration is adjusted to between 7×10^8 and 8×10^8 cells/ml. A volume (0.5 ml) of the cells is preincubated with $CaCl_2$ (10 mM) at 37° for about 5 min, before being mixed with 0.8 ml of freshly prepared lipid emulsion. The treated cells are maintained at 37°, and small samples are removed for examination by phase-contrast microscopy. Although all the initial studies with lipid fusogens were carried out with cell suspensions, it is nevertheless possible to fuse monolayers of attached erythrocytes using lipid fusogens.

[7] Q. F. Ahkong, D. Fisher, W. Tampion, and J. A. Lucy, *Biochem. J.* **136,** 147 (1973).

Relatively high concentrations of fusogen are required, and extensive lysis may occur in addition to fusion because lipid-soluble fusogens make erythrocyte membranes permeable to small ions.[8] Lysis is minimized by the inclusion of dextran in the salt solution.

Fusion of Human Erythrocytes by Chlorpromazine

The following method differs from the procedure described above with lipid fusogens in that it does not require Ca^{2+}.[9]

Reagents

Salt/dextran solution: see above
Chlorpromazine: Chlorpromazine hydrochloride is dissolved to a concentration of 15 mM in salt/dextran, and the pH is readjusted to 7.4; the chlorpromazine should form a white milky suspension
Procedure. Washed erythrocytes are suspended in salt/dextran at approximately 3×10^8 cells/ml and warmed to 37°. Chlorpromazine is added to a final concentration of 2 mM, and the progress of the incubation is followed by removing samples for phase-contrast microscopy. It should be noted that chlorpromazine is fusogenic only between pH 6.8 and 7.6. Below pH 6.8 it is inactive, whereas above pH 7.6 the cells lyse rather than fuse. The optimum pH is 7.4.

Fusion of Erythrocytes by Polyethylene Glycol 6000

When erythrocytes are treated for a short period with a high concentration (30–50%) of high molecular weight polyethylene glycol, the dehydrating action of the PEG enables the plasma membranes of adjacent cells to approach extremely closely. This treatment also permeabilizes the membranes to small ions.[8] The addition of isotonic buffer to the cells then results in rehydration, cell swelling, and cell fusion.[3]

Reagents

PEG in buffer: 40 mM HEPES buffer (pH 7.4) containing 110 mM NaCl, 50 μM $LaCl_3$, 40% (w/v) PEG 6000
Buffer: 40 mM HEPES buffer (pH 7.4), 110 mM NaCl, 50 μM $LaCl_3$
Procedure. Washed erythrocytes are attached to an Alcian blue-coated dish as previously described. The cells are incubated for 5 min at 37° in 1 ml of PEG in buffer (see reagents above). The polyethylene glycol solution is then diluted to 13% by the addition of 2 ml of buffer, and the cells

---

[8] A. M. J. Blow, G. M. Botham, and J. A. Lucy, *Biochem. J.* **182**, 555 (1979).
[9] R. D. A. Lang, C. Wickenden, J. Wynne, and J. A. Lucy, *Biochem. J.* **218**, 295 (1984).

are incubated for a further 5 min at 37°. Finally, the whole volume of medium is removed and is replaced by 2 ml of buffer. Fusion begins after a further 10-min incubation at 37°.

In all of the fusion procedures described herein, cell swelling accompanies fusion. It therefore follows that lysis of both fused and unfused cells also occurs. Although inclusion of dextran and $LaCl_3$ (and to some extent $CaCl_2$) in the medium improves the ratio of fused to lysed cells, it is important when considering the mechanism of fusion to remember that (at least in the case of La^{3+} and Ca^{2+}) these substances may actually be promoting fusion in addition to simply inhibiting lysis.

Use of Fluorescent Probes to Study Chemically Induced Fusion

Erythrocytes may be labeled using fluorescent probes specific for either the membrane or the cytoplasm. This enables both membrane fusion and cell fusion to be investigated by fluorescence microscopy. Under certain conditions, it is possible to separate these two processes and to observe that a state of membrane continuity (hemifusion) occurs before cytoplasmic fusion.[10]

Movement of Fluorescent Probes During Fusion by Polyethylene Glycol 6000

Reagents

Cytoplasmic marker: 6-Carboxyfluorescein diacetate (Molecular Probes, Eugene, OR), stock solution 20 mg/ml in acetone

Membrane markers: Rhodamine B chloride (Molecular Probes), stock solution 1 mg/ml in ethanol, or $DiIC_{16}$ (1,1-dihexadecyl-3,3,3′,3′-tetramethylindocarbocyanine perchlorate) (Molecular Probes), stock solution 1 mg/ml in ethanol

Procedure. Erythrocytes are washed and suspended in 40 mM HEPES, 110 mM NaCl (pH 7.4) as described previously. Ten microliters of carboxyfluorescein solution and 10 μl of a solution of membrane probe are added to 0.9 ml of buffer and mixed using a vortex mixer. Then 0.1 ml packed washed erythrocytes is added and incubated for 20 min at 37°. The labeled cells are centrifuged and washed 3 times with buffer. Labeled and unlabeled cells are mixed in the proportion of 1:9. Monolayers of the mixed cell population are prepared as described previously, except that for work involving fluorescence microscopy and photography at high magnification, better results are obtained by using dishes with a thin glass coverslip

---

[10] Q. F. Ahkong, J.-P. Desmazes, D. Georgescauld, and J. A. Lucy, *J. Cell Sci.* **88**, 389 (1987).

glued across a hole in the base.[11] Fusion is carried out using PEG 6000 exactly as described previously.

Using the techniques described above it is found that, although membrane probes rapidly diffuse into the membranes of adjacent unlabeled cells in the presence of 40% PEG 6000, the cytoplasmic probe does not. This indicates that, in 40% PEG 6000, membrane fusion (hemifusion) but not cell fusion is achieved between closely adjacent erythrocytes. Rapid diffusion of the cytoplasmic probe from labeled to unlabeled cells is not observed until the PEG solution is finally replaced by isotonic buffer, indicating that a membranous barrier between the hemifused cells is ruptured by osmotic swelling as the cells rehydrate.[10]

The fusion of erythrocytes associated with cell swelling caused by the entry of malonamide[6] is preceded also by hemifusion, as is demonstrated by the rapid movement of a carbocyanine membrane probe from labeled to unlabeled cells.[11a] Since the movement of fluorescent probes in hemifusion is unaffected by cell lysis, this technique provides a useful way of investigating the mechanism of action of inhibitors of membrane fusion, such as ethanolamine.[11a]

Protease Inhibitors in Cell Fusion

There is considerable evidence that the activation of endogenous proteases, which degrade the erythrocyte membrane skeleton, occurs in at least some instances of chemically induced erythrocyte fusion. Experiments on the effects of various protease inhibitors on chemically induced cell fusion have therefore yielded useful information on the molecular mechanisms involved.[9,12-14]

Selection and Preparation of Protease Inhibitors

For each fusion protocol, it is important to test a number of protease inhibitors with differing specificities since more than one type of protease has been implicated in fusion processes.[9] It is also important to take membrane permeability into account, since a membrane-impermeant substance may not have any effect on the fusion of intact cells.

[11] M. C. Willingham and I. H. Pastan, this series Volume 98, p. 266.
[11a] J. M. Baldwin, R. O'Reilly, and J. A. Lucy, *Biochem. Soc. Trans.* **30,** 326S (1992).
[12] S. J. Quirk, Q. F. Ahkong, G. M. Botham, J. Vos, and J. A. Lucy, *Biochem. J.* **176,** 159 (1978).
[13] Q. F. Ahkong, G. M. Botham, A. W. Woodward, and J. A. Lucy, *Biochem. J.* **192,** 829 (1980).
[14] T. Glaser, and N. S. Kosower, *FEBS Lett.* **206,** 115 (1986).

The following low molecular weight inhibitors have been used successfully: 7-amino-1-chloro-3-L-tosylamidoheptan-2-one (TLCK), 1-chloro-4-phenyl-3-L-toluene-*p*-sulfonamidobutan-2-one (TPCK), *N*-ethylmaleimide, and iodoacetamide. Stock solutions are made as follows: TLCK and TPCK, 0.1 M in methanol; *N*-ethylmaleimide (50 mM) and iodoacetamide (100 mM), in buffer containing 80 mg/ml dextran.

Inhibition of Fusion by Protease Inhibitors

Procedure. Washed erythrocytes (~ 3 × 10⁸ cells/ml) in the salt/dextran solution are preincubated with or without protease inhibitors for 15 min at 37°. CaCl₂, MgCl₂, or ethylene glycol bis (β-aminoethyl ether)-*N*,*N*,*N'*,*N'*-tetraacetic acid (EGTA) are then added to 10 mM, and the incubation is continued for a further 5 min. The fusogen under investigation is subsequently added, and the incubation is continued and monitored as described earlier. We have used protease inhibitors at the following concentrations: TLCK and TPCK, 1 or 7.5 mM; iodoacetamide, 10–100 mM; *N*-ethylmaleimide, 1–50 mM.

Determination of Fusion Index

The inhibition of fusion may be quantified as follows.[12] A sample (1 volume) is fixed at 0° for 1 hr in a solution (2 volumes) of glutaraldehyde (2%, v/v) in 0.1 M sodium cacodylate/HCl buffer, pH 7.4.[15] The occurrence of cell–cell fusion is determined by inspection of cell size and shape. In each sample 800–1000 cells are counted and scored as being either composed of 2, 3, 4, or more cells, or single unfused cells. The total number of single erythrocytes originally present is then obtained from these data. The fusion index may be defined as (number of erythrocytes that participated in fusion)/(total number of erythrocytes originally present) × 100. The fusion index is least accurate when the incidence of fusion is high and multiple fusion events occur, because it is difficult to determine the number of cells involved when more than 4 cells fuse together. However, when samples are recounted they rarely differ by more than 5%.

When assessing the effect of protease inhibitors on fusion, it is necessary to recognize the fact that many of these compounds promote lysis. This is particularly important in experiments with monolayers of cells attached to Alcian blue-coated surfaces. Because the attached cells are not free to move, a protease inhibitor may decrease the number of cells which

[15] A. M. Glauert, *in* "Fixation, Dehydration and Embedding of Biological Specimens" (A. M. Glauert, ed.), p. 43. North-Holland, Amsterdam, 1975.

are close enough to fuse simply by lysing a proportion of the cells, rather than by directly inhibiting fusion.

Electrophoresis of Membrane Proteins

The proteolytic breakdown of specific membrane proteins during the fusion process is of considerable interest, and sodium dodecyl sulfate (SDS)–polyacrylamide gel electrophoresis may be used to correlate changes in the polypeptide composition of membranes with the effect on fusion of substances that activate or inhibit endogenous proteases.[9,12-14]

In experiments of this kind, erythrocyte "ghosts" are prepared by the method of Hanahan and Ekholm,[16] with the modification that EGTA is included in the first two of the hypoosmotic washes in order to prevent proteolysis during the washing procedure.[13] The concentration of protein in the samples is determined by the method of Lowry *et al.,*[17] and phosphate (as P_i) is determined by the method of Baginski *et al.*[18]

A sample of "ghosts" in water is added to an equal volume of modified Laemmli sample buffer, 125 mM Tris-HCl, pH 6.8, 1.25% (w/v) SDS, 40 mM dithiothreitol, 12.5% (v/v) glycerol, 2 mM ethylenediaminetetraacetic acid (EDTA), and 100 μg/ml pyronin Y.[19] The sample is immediately heated in a boiling water bath for 5 min, then either used the same day or stored at $-15°$. In the latter case, it is reheated at $100°$ for a further 5 min and cooled before use.

Samples (30–40 μl) containing 40–50 μg protein (or ~0.03 μmol phosphate) are loaded onto slab gels [containing 5 or 7% (w/v) acrylamide], and electrophoresis is performed using the buffer system of Laemmli.[19] The proteins are detected by staining with Coomassie blue essentially as described by Fairbanks *et al.*[20] Where quantitation of the protein bands is to be carried out by densitometry scanning, it is more accurate to load equal quantities of total phosphate than equal quantities of protein because some preparations of membranes may contain adsorbed cytosolic proteins.[13]

Electron Microscopy

Experiments with protease inhibitors, such as those described above, have indicated that the proteolytic degradation of membrane proteins is

[16] D. J. Hanahan and J. E. Ekholm, this series Volume 31, p. 168.
[17] O. H. Lowry, N. J. Rosebrough, A. L. Farr, and R. I. Randall, *J. Biol. Chem.* **193,** 265 (1951).
[18] E. S. Baginski, P. P. Foa, and B. Zak, *Clin. Chem.* **13,** 326 (1967).
[19] U. K. Laemmli, *Nature (London)* **227,** 680 (1970).
[20] G. Fairbanks, T. L. Steck, and D. F. H. Wallach, *Biochemistry* **10,** 2606 (1971).

involved in the chemically induced fusion of erythrocytes. Ultrastructural studies provide a way of investigating the effect of this degradation on cell structure. It has been proposed that treatments which induce fusion may do so by activating a proteolytic breakdown of the cytoskeleton, leading to an increased lateral mobility of integral membrane proteins.[9,12] In early work it was also suggested that the formation of protein-free areas of lipid bilayer may participate in membrane fusion.[21,22] This hypothesis has been tested by electron microscopy on freeze-fractured specimens of erythrocytes in which fusion has been induced or inhibited.[9,12,13,23] Ahkong and Lucy[6] have presented electron micrographic evidence that fusion may occur at highly localized sites of osmotic swelling in certain circumstances. It is anticipated that the application of rapid freezing techniques[23a] will advance such investigations further.

Phospholipid Asymmetry in Chemically Induced Cell Fusion

Although much work has been done on the possible role in cell fusion of alterations in membrane proteins, less attention has been paid to the possibility that changes in the distribution of membrane phospholipids may be important.

In normal erythrocytes, the outer leaflet of the plasma membrane is composed of phosphatidylcholine and sphingomyelin, while phosphatidylethanolamine and phosphatidylserine are restricted to the inner leaflet.[24] Schlegel and Williamson[25] have suggested that fusion may be controlled in part by the packing of phospholipids in the membrane, and that the tightly packed outer leaflet of the normal erythrocyte membrane is fusion-incompetent. Using the response of treated erythrocytes to the fluorescent dye merocyanine 540 as a measure of phospholipid asymmetry, Tullius et al.[26] have obtained evidence that erythrocytes fuse more readily with PEG 6000 after their normal membrane asymmetry has been lost. The mechanism by which changes in membrane phospholipid asymmetry influence fusion is as yet unknown, but it is possible that a looser packing of the exterior lipids may increase membrane fluidity, facilitate the movement of intramem-

[21] Q. F. Ahkong, D. Fisher, W. Tampion, and J. A. Lucy, Nature (London) 253, 194 (1975).

[22] D. J. Volsky and A. Loyter, Biochim. Biophys. Acta 514, 213 (1978).

[23] J. Vos, Q. F. Ahkong, G. M. Botham, S. J. Quirk, and J. A. Lucy, Biochem. J. 158, 651 (1976).

[23a] D. E. Chandler and C. J. Merkle, this series, Vol. 221 [9].

[24] J. A. F. Op den Kamp, Annu. Rev. Biochem. 48, 47 (1979).

[25] R. A. Schlegel and P. Williamson, in "Molecular Mechanisms of Membrane Fusion" (S. Ohki, D. Doyle, T. D. Flanagan, S. W. Hui, and E. Mayhew, eds.), p. 289. Plenum, New York, 1988.

[26] E. K. Tullius, P. Williamson, and R. A. Schlegel, Biosci. Rep. 9, 623 (1989).

branous particles, and thereby enhance fusion. Alternatively, exposure of the negatively charged phospholipid phosphatidylserine on the erythrocyte surface may be responsible for the enhanced fusion. Hydrated bilayers of phospholipid normally repel each other strongly as a consequence of hydration repulsion arising from solvation of the polar head groups. When Ca^{2+} ions are added to phosphatidylserine, however, the surface water that gives rise to the repulsive forces is completely displaced by the bound Ca^{2+}, and the phospholipid bilayers achieve molecular contact. This is thought to allow the membranes to come sufficiently close for fusion to occur. Until recently, however, little attention has been paid to the possibility that such a phosphatidylserine–Ca^{2+} complex might be required for biological membranes to achieve contact prior to fusion.

The rate of conversion of prothrombin to thrombin by the enzyme complex (factor Xa–factor Va) provides a convenient, sensitive, and semiquantitative way of monitoring the surface exposure of acidic phospholipids. This method was initially employed to monitor the surface exposure of phosphatidylserine in the plasma membranes of platelets[27,28] and sickled erythrocytes,[29] but it was subsequently reported that the assay detected phosphatidylethanolamine almost as well as phosphatidylserine in mixtures with phosphatidylcholine.[30] We further investigated the specificity of the prothrombinase assay with phospholipid vesicles. In our experiments phosphatidylethanolamine was marginally more active in the presence of bovine brain phosphatidylserine than in its absence. Phosphatidylserine was nevertheless many orders of magnitude more effective in catalysing the conversion of prothrombin to thrombin than either dioleoylphosphatidylethanolamine or egg phosphatidylcholine. It thus appears that the procoagulant activity of cells can be used as a semiquantitative measure of phosphatidylserine that is exposed at the cell surface.

Use of Prothrombinase Assay

Reagents for Prothrombinase Reaction

Buffer: 40 mM HEPES (pH 7.4), 110 mM NaCl
$CaCl_2$: stock solution 1 M

[27] E. M. Bevers, P. Comfurius, J. L. M. L. Van Rijn, H. C. Hemker, and R. F. A. Zwaal, *Eur. J. Biochem.* **122**, 429 (1982).

[28] P. F. J. Verhallen, E. M. Bevers, P. Comfurius, and R. F. A. Zwaal, *Biochim. Biophys. Acta* **903**, 206 (1987).

[29] E. Middlekoop, B. H. Lubin, E. M. Bevers, J. A. F. Op den Kamp, P. Comfurius, D. T.-Y. Chiu, R. F. A. Zwaal, L. L. M. Van Deenen, and B. Roelofsen, *Biochim. Biophys. Acta* **937**, 281 (1988).

[30] I. Gerads, J. W. P. Govers-Riemslag, G. Tans, R. F. A. Zwaal, and J. Rosing, *Biochemistry* **29**, 7967 (1990).

Bovine Factor II (prothombin), bovine Factor X_a (both from Sigma, St. Louis, MO)

Bovine Factor V_a is prepared as follows: Bovine Factor V (from Diagnostic Reagents Ltd., Thame, Oxon, UK) is dissolved at a concentration of 35 μg/ml in 10 mM HEPES buffer (pH 7.4) containing 2 mM, CaCl$_2$ and incubated with 0.04 μg/ml bovine thrombin (Sigma) for 30 min at 37°

Reagents for Determination of Thrombin

50 mM Tris-HCl, 120 mM NaCl, 2 mM EDTA, pH 7.5

Chromogenic substrate for thrombin: S2238 (H-D-phenylalanyl-L-pipecolyl-L-arginine-p-nitroanilide dihydrochloride, purchased from KabiVitrum, Stockholm, Sweden) stock solution 1.5 mM

Procedure. Factor X_a, Factor V_a, and CaCl$_2$ are added to cells or membranes in buffer, and the mixture is incubated at 37° for 2 min. The assay is started by the addition of prothrombin. The concentrations in the assay are 5×10^6 to 2×10^7/ml cells, 3 nM Factor X_a, 6 nM Factor V_a, 4 mM CaCl$_2$, and 2 μM prothrombin. At various times, thrombin formation is stopped by diluting a 20-μl aliquot of the incubation mixture into 0.5 ml Tris/NaCl/EDTA buffer (above) at room temperature. The chromogenic substrate S2238 is added to a final concentration of 150 μM, and the quantity of thrombin present is calculated from the rate of change of absorbance at 405 nm using a calibration curve.

It is essential to ensure that the procoagulant surface is the limiting quantity under the conditions of the assay. Because commercial preparations of blood coagulant factors may vary in their concentration, it is advisable to check that none of them are rate-limiting before using the assay to monitor changes in the surface exposure of phosphatidylserine. It is also important to correct the data obtained for the contribution to exposed phosphatidylserine that is made by unsealed erythrocyte membranes. To this end, a sample of the material assayed is centrifuged (13,000 g for 10 min), and the absorbance at 414 nm of the hemoglobin in the supernatant is determined. This is then compared with the corresponding absorbance obtained for a sample which has been totally lysed by freezing and thawing, followed by sonication. (The latter sample is also used to determine the total prothrombinase activity.)

The main problem that we have encountered when using the prothrombinase assay to study the mechanism of chemically induced fusion is the high degree of lysis which results from the use of some methods for inducing fusion. Unless lysis is minimized, the magnitude of the corrections needed to allow for lysis is too great to yield meaningful results. In addition, some of the reagents that are used to promote or inhibit fusion (e.g., LaCl$_3$, PEG 6000, subtilisin) interfere with the assay. Their inhibitory effects may be reduced or abolished in some cases by dilution (e.g., LaCl$_3$ does not inhibit when diluted to 0.5 μM). It is nevertheless important to test the effect on the assay of any new reagent used in a fusion protocol before embarking on determinations of asymmetry by this method.

Assays of prothombinase activity in monolayers of attached cells during fusion experiments have proved unreliable, and we have found it more satisfactory to set up parallel incubations of cells in suspension under equivalent conditions.[31] When we treated cells in suspension with A23187, Ca^{2+}, and subtilisin, we found that they develop extensive procoagulant activity. Without A23187, this activity does not develop. We have also used this assay to test the surface exposure of phosphatidylserine during fusion induced by treatment with the permeable molecule PEG 400. It was observed that cells which swell in the presence of Ca^{2+} and PEG 400 develop procoagulant activity and also fuse, whereas in the absence of Ca^{2+} the cells neither fuse nor develop procoagulant activity, indicating that rupture and resealing of the erythrocyte membrane in swelling cells leads to entry of extracellular Ca^{2+}. Both with A23187 and with PEG 400, it seems that the entry of extracellular Ca^{2+} leads to a change in membrane asymmetry which is linked to membrane fusion.[31] In view of experiments with other methods of inducing fusion (described above), it also seems likely that the entry of Ca^{2+} into erythrocytes results in the proteolytic breakdown of a component of the cytoskeleton which helps to maintain normal phospholipid asymmetry.

The techniques described here also have been used to show that surface exposure of acidic phospholipids is associated with the fusion of human erythrocytes that is induced by electrical breakdown.[32]

[31] J. M. Baldwin, R. O'Reilly, M. Whitney, and J. A. Lucy, *Biochim. Biophys. Acta* **1028,** 14 (1990).

[32] L. Y. Song, J. M. Baldwin, R. O'Reilly, and J. A. Lucy, *Biochim. Biophys. Acta* **1104,** 1 (1992).

[14] Electrofusion

By GARRY A. NEIL and ULRICH ZIMMERMANN

Introduction

Monoclonal antibody (MAb) technology has become an indispensable tool for many aspects of biomedical research and medical diagnostics. Although still in its infancy, the long-awaited therapeutic application of this methodology is now being realized. Although the majority of hybridomas are generated by polyethylene glycol fusion, new methods, notably electrofusion, are increasingly coming into use. Electrofusion offers several advantages over conventional methods including the ability to produce hybridomas with much smaller numbers of fusion partners, higher fusion efficiency, higher hybrid viability, effectiveness with a wider variety of cell types, and a more controlled, reproducible approach. Since this technique was last reviewed in this series,[1] a number of significant improvements in the methodology have been introduced. In this chapter, we underscore some of these developments and outline current electrofusion technology and protocols. Although many of the principles and methods discussed in this review are applicable to fusion of plant protoplasts, yeast, and bacteria, the production of murine and human hybridomas using this methodology is emphasized.

Physical Parameters

Successful application of the electrofusion technique requires an understanding of the rudiments of cell membrane behavior in electrical fields. The fundamental step in electrofusion is reversible membrane breakdown. When short-duration electrical impulses applied across a cell membrane exceed a critical threshold, that membrane will become transiently, but highly permeable.[2-8] For short-duration pulses (1 – 10 μsec) this threshold

[1] U. Zimmermann and H. B. Urnovitz, this series, Vol. 151, p. 194.
[2] U. Zimmermann and G. Pilwat, Z. Naturforsch. 29c, 304 (1974).
[3] H. Weber, W. Forster, H.-E. Jacob, and H. Berg, in "Current Developments in Yeast Research, Advances in Biotechnology" (G. G. Stewart and I. Russel, eds.), p. 219. Pergamon, New York, 1981.
[4] T. Y. Tsong and E. Kingsley, J. Biol. Chem. 250, 786 (1975).
[5] U. Zimmermann, P. Scheurich, G. Pilwat, and R. Benz, Angew. Chem., Int. Ed. Engl. 20, 325 (1981).
[6] U. Zimmermann, Biochim. Biophys. Acta 694, 227 (1982).

is generally 1 V for a single membrane and 2 V for two membranes arranged in series at room temperature, provided that the pulse duration is not less than 1–5 μsec.[5–7,9] Membrane resealing occurs rapidly and spontaneously after cessation of the breakdown pulse.[8,9] If excessively strong or long pulses are applied, membrane breakdown is irreversible and cell death ensues.[10]

The field strength, E_c, required to achieve membrane breakdown for spherical cells (such as lymphocytes and myeloma/hybridoma fusion partners) can be calculated using the integrated Laplace equation [Eq. (1)]:

$$V_c = 1.5aE_c \cos \alpha \qquad (1)$$

where a is the cell radius and α, the angle between a certain membrane site in relation to the field direction.[6,11,12]

It is evident from Eq. (1) that smaller cells require application of a higher voltage to achieve membrane breakdown. Only a portion of the cell membrane actually breaks down with voltage application as a result of the angular and radial dependence of the electrical field[6,12] (Fig. 1). Field strength, and consequently transmembrane voltage, is maximal along the long axis of the electrical field. Thus, application of supercritical field strengths will result in a larger area of membrane breakdown, but a lethal threshold may be exceeded at the membrane sites oriented in the field direction. Optimal conditions for breakdown must, therefore, balance these factors. It is also evident that, wherever possible, fusions should be performed with the cells aligned along the long axis of the applied field to ensure uniform breakdown (see below).

Equation (1) yields a breakdown voltage for membranes in steady state with respect to the membrane potential. This state is attained after field application for a duration of approximately 5 times the relaxation time of the exponential membrane charging process.[12–14] This may be calculated using Eq. (2).

$$\frac{1}{\tau} = \frac{1}{R_m C_m} + \frac{1}{(\rho_i + 0.5\rho_e)} \qquad (2)$$

[7] R. Benz and U. Zimmermann, *Biochim. Biophys. Acta* **597**, 637 (1980).
[8] U. Zimmermann, F. Riemann, and G. Pilwat, U. S. Patent 4,154,668 (1974).
[9] A. E. Sowers, *Biochim. Biophys. Acta* **985**, 334 (1989).
[10] A. J. H. Sale and W. A. Hamilton, *Biochim. Biophys. Acta* **163**, 37 (1968).
[11] D. Gross, L. M. Loew, and W. W. Webb, *Biophys. J.* **50**, 339 (1986).
[12] U. Zimmermann, *Rev. Physiol. Biochem. Pharmacol.* **105**, 175 (1986).
[13] W. M. Arnold and U. Zimmermann, *in* "Biological Membranes" (D. Chapman, ed.), p. 389. Academic Press, New York, 1984.
[14] G. Pilwat and U. Zimmermann, *Biochim. Biophys. Acta* **820**, 305 (1985).

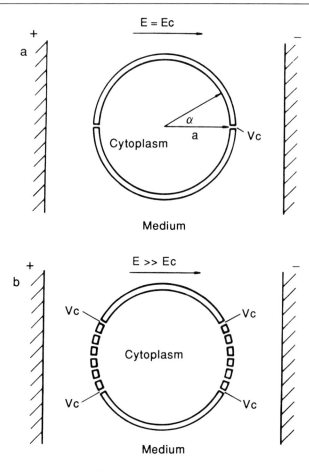

$$Vc = 1.5 \cdot a \cdot E \cdot \cos \alpha$$

FIG. 1. Schematic diagram of a cell exposed to high electrical field strengths. The membrane potential V_c, which is built up across the membrane in response to the external field, is at its highest value at membrane sites oriented parallel to the field direction (poles) and progressively decreases toward the equator (perpendicular to the field direction). For membrane sites in perpendicular orientation to the field, the potential is zero. The breakdown voltage is, therefore, first reached at sites parallel to the field direction. It is only reached in membrane sites oriented at a certain angle α to the field direction if supercritical field strengths ($E \gg E_c$) are applied. Breakdown of the membrane is indicated by the formation of transmembrane pores (b).

It will be seen that the relaxation time is proportional to both intrinsic passive membrane electrical properties (R_m, specific membrane resistance; C_m, specific membrane capacitance) and the conductivity of the cytosol (ρ_i) and extracellular solution (ρ_e). The membrane resistance of most of the cells is normally very high. Consequently, the relaxation time, τ, depends on only the second "conductivity" term of Eq. (2).

Electrofusion is normally performed in weakly conductive solutions (resistance $10^4 - 10^5$ Ω cm^{-1}) with τ of the order 500 nsec to 5 μsec for most relevant cells. Thus, we would expect that pulse durations of $10-20$ μsec should allow the steady state to be achieved. Pulse durations of this length have, indeed, proved to be optimal in our hands, as well as those of other authors,[15-19] although variation among cell types exists. It will be appreciated that the relaxation time is proportional to the cell radius. Hence, smaller cells will require shorter duration pulses to achieve the steady state. It is obvious from the foregoing that minor changes in cell size and electrofusion medium composition can have profound effects on membrane breakdown, requiring adjustment of field strength and pulse duration.

It must be noted that the intrinsic electrical membrane field results in an asymmetry of net field strength across the membrane during application of the electrical field pulse. This is diagrammed in Fig. 2 and has been demonstrated experimentally (Fig. 3).[20-22] Although this difference is relatively small under some circumstances, it may be of critical importance in the fusion process. This is particularly important for fusion of small cells such as lymphocytes to larger myeloma fusion partners.[22]

Cells apposed during electrical membrane breakdown will fuse under appropriate circumstances.[1,5,6,11,12,22-25] It is thought that membrane "intermixing" of closely applied cells occurs spontaneously during the resealing phase, resulting in membrane continuity of the involved cells. The fused syncytium will then "round up," forming a single multinucleated cell

[15] T. Hibi, H. Kano, M. Sugiura, T. Kazami, and S. Kimura, *Plant Cell Rep.* **7**, 153 (1988).

[16] H.-G. Broda, R. Schnettler, and U. Zimmermann, *Biochim. Biophys. Acta* **899**, 25 (1987).

[17] J. J. Schmitt, U. Zimmermann, and G. A. Neil, *Hybridoma* **8**, 107 (1988).

[18] G. Van Duijn, J. P. M. Langedijk, M. De Boer, and J. M. Tager, *Exp. Cell Res.* **183**, 463 (1989).

[19] D. A. Stenger, R. T. Kubiniec, W. J. Purucker, H. Liang, and S. W. Hui, *Hybridima* **7**, 505 (1988).

[20] D. L. Farkas, R. Korenstein, and S. Malkin, *FEBS Letts.* **120**, 236 (1980).

[21] W. Mehrle, U. Zimmermann, and R. Hampp, *FEBS Letts.* **185**, 89 (1985).

[22] W. Mehrle, R. Hampe, and U. Zimmermann, *Biochim. Biophys. Acta* **978**, 267 (1989).

[23] U. Zimmermann and J. Vienken, *J. Membr. Biol.* **67**, 165 (1982).

[24] J. Teissie, V. P. Knutson, T. Y. Tsong, and M. D. Lane, *Science* **216**, 537 (1982).

[25] A. E. Sowers, *J. Cell Biol.* **99**, 1989 (1984).

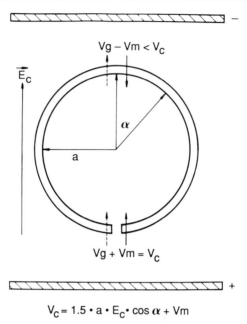

$$V_c = 1.5 \cdot a \cdot E_c \cdot \cos \alpha + V_m$$

FIG. 2. Schematic diagram of a cell exposed to a uniform electrical field (E_c) between two large electrodes. The intrinsic membrane potential is taken into account. a, radius of the cell; V_m, resting transmembrane potential; V_g, generated (superimposed) membrane potential; α, angle between a given membrane site and the field direction. The critical voltage required for membrane breakdown may be calculated using the equation given.

(Fig. 4). We have observed that this process requires 5–30 min in order for complete cell fusion to occur. If the field strengths of the fusion pulses is very high, the fusion and rounding up process can be considerably accelerated. Under these conditions, however, no viable hybrids are obtained, owing largely to cell death (which may not be immediately obvious microscopically after the fusion is completed). Mixing or shaking during "rounding up" may sometimes disrupt the fusion partners during this critical phase.

A variety of methods may be used to appose cells to be fused. These include high suspension density,[26] micromanipulation,[27,28] lectins,[29] mag-

[26] U. Zimmermann and G. Pilwat, *in* "Abstracts of the Sixth International Biophysics Congress," p. 140. Kyoto, Japan, 1978.

[27] M. Senda, J. Takeda, S. Abe, and T. Nakamura, *Plant Cell Physiol.* **20**, 1441 (1979).

[28] H. Berg, *Bioelectrochem. Bioenerg.* **9**, 223 (1982).

[29] M. Chapel, J. Teissie, and G. Alibert, *FEBS Lett.* **173**, 331 (1984).

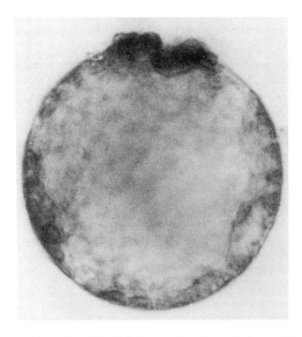

FIG. 3. Asymmetric uptake of the vital dye neutral red into a leaf mesophyll protoplast of *Nicotiana tabacum* exposed to a field pulse of 420 V cm^{-1} and 50 μsec duration. The accumulation of the dye was first restricted to the cytoplasm in the "pole" region of that hemisphere which had been oriented toward the anode (see Fig. 2).

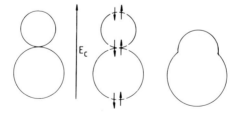

FIG. 4. Diagrammatic representation of electrofusion of two cells adhering to each other in the direction of the external electrical field by dielectrophoresis or other physical forces. Electrical breakdown occurs in the membrane area in the zones of contact which are oriented vertically with respect to the field lines (see arrows). Under the given condition, electrical breakdown not only induces exchange of substances between the cells and their medium, but also intercellular exchange. As a result, fusion occurs.

netic fields,[30] specific linking agents,[31,32] and sonic fields.[33] The most popular means of achieving cell contact is electrical alignment or "dielectrophoresis." [1,5,6,11,12,22,34,35] This method offers the advantages of simplicity, reproducibility, general applicability, and nuclear fusion (see below). Modern electrofusion apparatuses have this capability built in, and we thus recommend dielectrophoretic alignment for all but certain specialized applications.

Dielectrophoretic alignment occurs when cells are exposed to a slightly nonuniform alternating electrical current (ac) of relatively low field strength and long duration (5–30 sec). This process is illustrated in Fig. 5. An alternating dipole is generated by ac-induced charge separation within the cell. Cell movement toward one of the two electrodes results from asymmetry of the field and proceeds despite alternation of cathode and anode. Pearl-chaining of the cells ensues because the intracellular dipoles become the focus of attraction during cell migration. Furthermore, the attractive forces arising from the intracellular dipoles are much higher than the repulsive forces of net surface charge. Alternating fields of 10 kHz to 4 MHz are optimal for aligning most cell types under usual circumstances. Frequencies lower than 1 kHz result in electrolysis, whereas those in excess of 4 MHz are associated with deleterious effects on cellular components because of pronounced current flow through the cell interior. Above this frequency the membrane capacitance is short-circuited. The application of ac fields in high conductivity media results in appreciable heating and the potential for interruption of membrane contact because of turbulence; hence, caution must be exercised if such conditions are employed. Low conductivity media will favor higher hybrid yields owing to the decrease in membrane resistance following application of the breakdown pulse. For the production of murine and human hybridomas, the optimum alignment frequency is 1–2 MHz.

Another important aspect of the alignment phase is the probability of attaining two-cell fusion (versus polyploid heterokaryotic cells). Cell density in the fusion chamber is a critical parameter in determining the

[30] I. Kramer, K. Vienken, J. Vienken, and U. Zimmermann, *Biochim. Biophys. Acta* **772**, 407 (1984).

[31] M. Lo, T. Y. Tsong, M. K. Conrad, S. M. Strittmatter, L. D. Hester, and S. H. Snyder, *Nature (London)* **310**, 792 (1984).

[32] D. M. Wojchowski and A. J. Sytkowski, *J. Immunol. Methods* **90**, 173 (1986).

[33] J. Vienken, U. Zimmermann, H. P. Zenner, W. T. Coakley, and R. K. Gould, *Biochim. Biophys. Acta* **820**, 259 (1985).

[34] U. Zimmermann, J. Vienken, and P. Scheurich, *in* "Biophysics of Structure and Mechanism" (K. Gersonde, ed.), Vol. 6, p. 86. Springer-Verlag, New York, 1980.

[35] U. Bertsche, A. Mader, and U. Zimmermann, *Biochim. Biophys. Acta* **939**, 509 (1988).

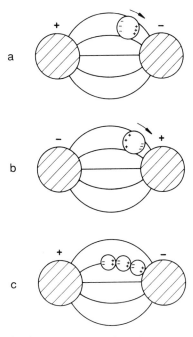

FIG. 5. Dielectrophoresis of cells. A cross section of the electrode arrangement is shown. The field is slightly inhomogeneous. Owing to the external field, the charges are separated within the cell (ultimately leading to the buildup of a dipole and, in turn, a membrane potential; see Fig. 1). The induced positive and negative charges are equal. However, the field strength on the two sides of the cell are different, thus giving rise to a net force which pulls the cell into the region of the highest field intensity [i.e., toward the electrodes (a)]. If the field direction is reversed (b), the charge separation within the cell is opposite to the conditions described above; however, the force still acts toward the region of highest field intensity. In other words, the field-induced movement of the cell is independent of the polarity of the field and therefore also occurs in the presence of an alternating field. The effect is known as dielectrophoresis. If two or three cells come close together, they will attract each other and make intimate membrane contact (c). The attractive forces of the dipoles are much larger than the repulsive forces arising from the net negative surface charge of the outer membrane surface (not shown) and from the Brownian motion. An alternating field with a frequency of 10 kHz to 4 MHz is usually used for dielectrophoresis and subsequent fusion.

number of cells fused to produce a single hybrid. Even though it is common for dielectrophoretically aligned "cell chains" to contain 4–15 members (depending on the distance between the electrodes), analysis of fusion products shows a preponderance of two nuclei per hybridoma.[35] Because membrane breakdown occurs only in the polar regions of "pearl-chained" cells, whereas the probability of fusion (F_n) between any two adjacent

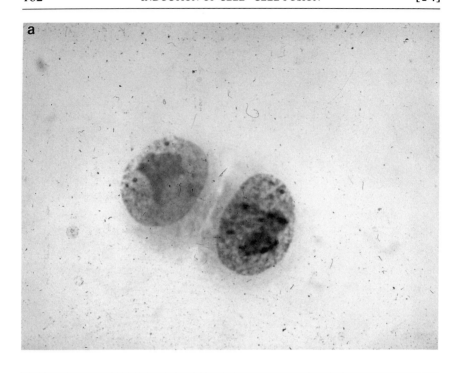

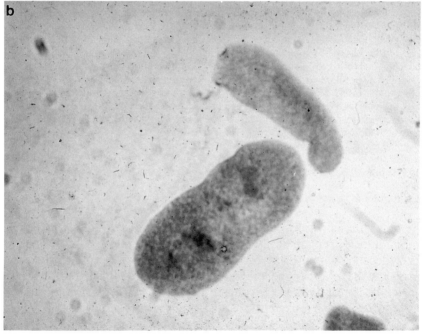

members (P) is relatively constant, the fusion index of n cells aligned in long chains may be estimated by Eq. (3):

$$F_n = P^{n-1} \tag{3}$$

Equation (3) is in reasonable agreement with experimental observations for a number of cell types, for cells in suspension densities of up to 10^7 per ml under isosmotic conditions. As higher suspension densities, adjacent chains come into lateral contact, and Eq. (3) is no longer valid. Under these conditions, multinucleated "giant cell" hybrids will result. When lower cell densities are used, particularly in the hypoosmotic fusion protocol given below, much smaller pearl chains will result. Two-cell heterokaryons are, of course, optimal for the production of antibody-secreting hybridomas.

Another potential benefit of dielectrophoretic alignment is the facilitation of nuclear membrane fusion. In conductive fusion medium and with alignment frequencies of 1–4 MHz, that is, at frequencies at which plasmalemma capacitance no longer shields the cell interior completely, intracellular nuclear dielectrophoresis and nuclear fusion are frequently observed.[35] This is illustrated in Fig. 6. This phenomena may be exploited for particular applications requiring nuclear fusion, although satisfactory hybrid yields may be obtained with more conventional protocols (see below).

After satisfactory alignment has been achieved, the breakdown pulse is applied and fusion occurs. These steps may be followed microscopically using transparent chambers (see below). The cells are then allowed to reseal and cultured under appropriate conditions. The hybrids are selected using standard methodology, for example, hypoxanthine, aminopterin, and thymidine (HAT)-supplemented media when HGPRT-deficient fusion partners are used.[36]

Biological Considerations

We have repeatedly observed that some subclones of standard cell lines used for electrofusion are superior to others with respect to yields of viable

[36] M. Shulman, C. D. Wilde, and G. Koehler, *Nature (London)* **276**, 269 (1978).

FIG. 6. Successive phases of nuclear membrane and nuclear fusion of Ehrlich ascites tumor cells subjected to electrofusion. An alternating field of strength 220 V cm⁻¹ and frequency 1.7 MHz was applied for 30 sec, followed by the application of three pulses of strength 5 kV cm⁻¹ and 15 μsec duration. After fusion, the cells were fixed in Carnoy's fixative and stained with Giemsa to better demonstrate the nuclei. (a) Orientation of nuclei to the plasmalemma contact zone of two dielectrophoretically aligned cells (cells fixed just after application of the alignment field). (b) Nuclear fusion (cells fixed 10 min after application of the fusion pulses).

hybridomas, antibody production, and long-term stability. This is not surprising in view of the complexity of these cells and variability of cell size, membrane characteristics, and growth kinetics. Spontaneous variants continuously arise, probably by somatic mutation, and the investigator must be constantly vigilant to maintain lines with appropriate characteristics. It is prudent to maintain frozen stock of suitable cells against the chance of contamination or emergence of deleterious mutants. We have also found that it is possible to select subclones optimal for electrofusion using a number of strategies including viability after prolonged exposure to fusion medium, dielectrophoresis, or repeated application of breakdown pulses or adherence to plastic culture flasks. Unfortunately, this methodology remains entirely empirical in that the biological characteristics which distinguish "good fusers" have not been elucidated and await systematic study.

It will be obvious to the experienced investigator that fastidious maintenance of tissue culture practices and reagents is also of paramount importance. There is experimental evidence to suggest that hybrid yields are increased by fusing cells in log growth phase.[35,37] This is particularly true for cells in G_1 phase. We have observed that a decrease in intracellular ATP is associated with a decrease in fusion efficiency among mammalian cells in plateau growth phase and that the addition of ATP may be beneficial in these circumstances.[38] It has been suggested that membrane fluidity and lateral diffusion of protein molecules within the membrane are inhibited by low concentrations of ATP.[38] Other factors including cell diameter, facilitation of nuclear dielectrophoresis, and associated membrane/intracellular changes may also be important, although these have not, as yet, been well characterized.

Although the basic configuration of the cell membrane is believed to be a lipid bilayer, membrane proteins are also a major constituent. By modifying cell–cell contact, these proteins may play either a facilitatory or inhibitory role in the fusion process. It is known that treatment of erythrocytes with various proteases alone may induce fusion under some circumstances[39] or enhance polyethylene glycol (PEG)-induced fusion.[40] Furthermore, our group and others have shown that pretreatment of cells with

[37] P. N. Rao, B. Wilson, and T. T. Puck, *J. Cell. Physiol.* **91**, 131 (1986).
[38] B. Verhoek-Koehler, R. Hampp, H. Ziegler, and U. Zimmerman, *Planta* **159**, 199 (1983).
[39] Q. F. Ahkong, A. M. J. Blow, G. M. Botham, J. M. Launder, S. J. Quirk, and J. A. Lucy, *FEBS Lett.* **95**, 147 (1978).
[40] J. X. Hartmann, J. D. Galla, D. A. Emma, K. N. Kao, and O. L. Gamborg, *Can. J. Genet. Cytol.* **18**, 503 (1976).

pronase and/or dispase may enhance the efficiency of electrofusion.[41-45] Although some of these effects may be attributed to the addition of contaminating Ca^{2+} ions with the enzymes, or the nonspecific membrane-stabilizing effects of proteins, it appears that, under some circumstances, stripping protein from the cell membrane may be beneficial to fusion. This might be explained by removal of repelling surface charge or changes in membrane fluidity, although the mechanism remains poorly understood. We have found the results of this approach to be inconsistent and rarely of substantial benefit. We therefore do not routinely employ it. A number of published protocols may be consulted for the methodology.[41-45]

We have observed that prefusion hypoosmotic shock treatment of cells (75 mOsm) results in dramatic increases in cell fusion and hybrid yield.[46-49] This process is illustrated in Figs. 7 and 8. Selected data from fusion experiments performed under hypoosmotic conditions are presented in Fig. 9. Lucy and Ahkong[50-52] have proposed that osmotic pressure gradients are largely responsible for fusion of the adherent but transiently disrupted membranes following electrical breakdown. These authors have proposed that the osmotic pressure gradients established by water uptake after chemical or electrical membrane disruption might force the lipid bilayers together at the contact zone, facilitating fusion. However, in contrast to this argument, we found that increased fusion and hybrid yield result even if the osmotically shocked cells are returned to isosmotic conditions prior to fusion (allowing their volumes to normalize[46]). We therefore believe that the increased fusion efficiency is more likely due to membrane/cytoskeletal changes and consequent increased fluidity and permeability of the membrane induced by hypoosmotic shock.

[41] U. Zimmermann, G. Pilwat, and H.-P. Richter, *Naturwissenschaften* **68**, 577 (1981).
[42] U. Zimmerman, G. Pilwat, and H. A. Pohl, *J. Biol. Phys.* **10**, 43 (1982).
[43] T. Ohno-Shosaku and Y. Okada, *Biochim. Biophys. Res. Commun.* **120**, 138 (1984).
[44] K. Ohnishi, J. Chiba, Y. Goto, and T. Tokunaga, *J. Immunol. Methods* **100**, 181 (1987).
[45] R. W. Glaser, H.-D. Bolk, Ch. Liebenthal, S. Jahn, and R. Grunow, *Hum. Antibodies Hybridomas* **1**, 111 (1990).
[46] J. J. Schmitt, U. Zimmermann, and P. Gessner, *Naturwissenschaften* **76**, 122 (1989).
[47] J. J. Schmitt and U. Zimmermann, *Biochim. Biophys. Acta* **983**, 42 (1989).
[48] U. Zimmermann, P. Gessner, M. Wander, and S. K. H. Foung, *in* "Electromanipulation in Hybridoma Technology" (C. A. K. Borrebaeck and I. Hagen, eds.), p. 1. Stockton, New York, 1989.
[49] U. Zimmerman, P. Gessner, S. Perkins, and S. K. H. Foung, *J. Immunol. Methods* (in press).
[50] J. A. Lucy and Q. F. Ahkong, *FEBS Lett.* **199**, 1 (1986).
[51] Q. F. Ahkong and J. A. Lucy, *Biochim. Biophys. Acta* **858**, 206 (1986).
[52] Q. F. Ahkong and J. A. Lucy, *Biochim. Soc. Trans.* **14**, 1129 (1986).

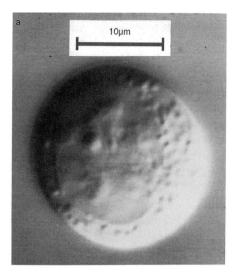

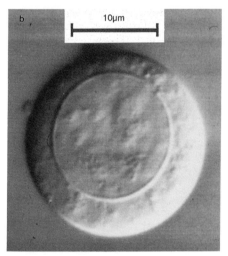

FIG. 7. Laser scan micrograph of a cell of the mouse–human heteromyeloma cell line H73CII incubated in isosmotic (a) and hypoosmotic (b) fusion medium. Note the swelling of the cell and the nucleus after transfer to hypoosmotic solutions. The nucleus changes from an irregular to a spherical shape. Magnification: ×2000. (Courtesy of Dr. Spring, DKFZ, Heidelberg, Germany.)

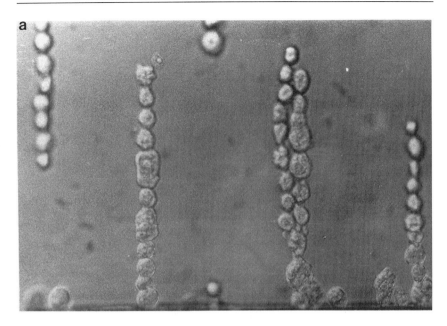

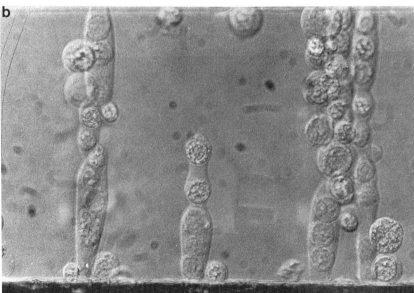

FIG. 8. Photomicrograph of alignment and fusion of the mouse–human heteromyeloma cell line H73CII in an open chamber under isosmotic (a) and hypoosmotic (b) conditions.

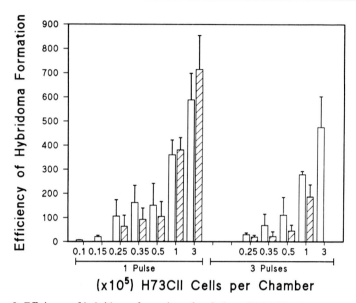

FIG. 9. Efficiency of hybridoma formation after fusion of H73CII cells to G8 cells in 75 mOsm fusion medium as a function of the number of H73CII cells in helical fusion chambers. The data shown on the left-hand side were derived from fusions performed using a single fusion pulse of 15 μsec duration. The efficiency of hybridoma formation was measured by scoring the number of separate colonies after 5–7 days postfusion in HAT selection medium. The field strength of the fusion pulse was 1.25 kV cm^{-1} (open columns) or 1.5 kV cm^{-1} (hatched columns). The data represent a total of 4–11 experiments.

Interestingly, and perhaps surprisingly, no significant reduction in cell viability results from this treatment. Its simplicity and high degree of reproducibility with several different mammalian and human fusion partners suggest that fusion in hypoosmotic medium is the current method of choice for many electrofusion applications.

Fusion Partners

Successful hybridoma production is highly dependent on selection of suitable cells for fusion. Many highly efficient nonsecreting hybridoma fusion partners have been described, but most have been selected for use in chemical (i.e., PEG) fusion protocols. A variety of as yet undetermined biological characteristics may influence the behavior of cells in electrical fields. When we subcloned the murine fusion partner SP2/0,[36] we found considerable variation in the electrofusibility of various lines (G. A. Neil and U. Zimmermann, unpublished). We recommend that this procedure be followed if inadequate fusion efficiencies are encountered with otherwise suitable fusion partners. Of the lines we have tested, SP2/0 has proved most suitable for both isomotic and hypoosmotic fusion to murine B cells, whereas H73CII[49] is equally suitable as a human fusion partner.

B Cell Activation

Successful hybridoma production requires adequate numbers of suitable activated B lymphocytes. The fusion of nonactivated B cells does not generally result in the formation of antibody-secreting hybridomas. Moreover, such activated cells are much larger than their nonactivated counterparts, and are thus more closely approximate the size of the myeloma fusion partners. This is an important factor in alignment and electrofusion. This phenomenon may be partly responsible for the higher likelihood of hybridomas produced by electrofusion to secrete specific antibody than hybridomas produced by other means.[17]

For murine hybridoma production, *in vivo* activation by parenteral immunization using any of a number of protocols is usually satisfactory. Large numbers of antigen-specific B cells may be derived from the spleens of immunized animals and require no additional preparation. Enrichment of the B cell preparation by red blood cell- and complement-mediated T cell lysis enhances the fusion frequency, perhaps by removing cells which might otherwise compete with B cells in the alignment phase. We have found that it is possible to further enrich for antigen-specific B cells by stimulating *ex vivo* B cells isolated from the spleens of immunized animals *in vitro* with lipopolysaccharide (LPS) and dextran sulfate (DXS).[53] This method greatly enriches for activated B cells, most of which appear to secrete antibody with specificity for the immunogen. One disadvantage is the observation that most of these hybridomas secrete IgM antibodies (although this may be desirable for some applications).

Activation of human B cells is usually accomplished by *in vitro* immunization and/or immortalization with Epstein-Barr virus prior to fusion.[54] Propagation of Epstein-Barr virus-immortalized cells similarly enriches for activated B cells and thus enhances the fusion process. Finally, some authors have found that methods designed to chemically "couple" antigen-specific B cells to myeloma fusion partners prior to fusion enhance the hybridoma yield.[31,32] Such procedures might be beneficial in given circumstances, but our experience with such techniques is limited.

[53] D. W. Sammons, U. Zimmermann, N. Klinman, and G. A. Neil, *Adv. Space Res.* **12**, 363 (1992).
[54] S. K. Foung and S. Perkins, *J. Immunol. Methods* **116**, 117 (1989).

Fusion Medium

The electrofusion medium is of critical importance for cell viability and successful hybridoma production. To minimize potentially deleterious manipulations and shock to the cells, we prefer a single medium for alignment, fusion, and resealing. Low-conductivity solutions are optimal for most applications. Water of the highest purity and reagent grade chemicals should always be used to avoid contamination with heavy meals and other toxins. Neutral sugars including sucrose, inositol, or sorbitol are used to adjust the osmolality of the solution. Mannitol is not suitable for this purpose. Solutions of 280 mM are used for isoosmolar media, and 75 mM solutions for hypoosmolar fusion media.

We and others have found that addition of the divalent cations Mg^{2+} and Ca^{2+} is essential[12,16,17,19,43,44] to achieve optimal results. This is, perhaps, to stabilize critical membrane proteins, to facilitate adhesion, or to permit satisfactory function of cell metabolism. We prefer the acetate salts at concentrations of 0.1–0.3 mM for Ca^{2+} and 0.3–0.5 mM for Mg^{2+}. We have found that the addition of 1–2 mg/ml of high-grade bovine serum albumin also improves the ultimate viability of cells and hybrid yields.

Where necessary, the pH of the solution should be adjusted from 7.0 to 7.4. Although pH is relatively unimportant to the electrical field parameters, it is obviously of importance in maintaining cell viability. The pH is best adjusted with small amounts of phosphate buffer or histidine. Phenol pH indicator dyes should be omitted as their uptake by permeabilized cells results in decreased cell viability.

For the production of murine and human hybridomas we found that the following medium is optimum: 0.1 mM Ca^{2+} acetate, 0.5 mM Mg^{2+} acetate, 1 mg/ml bovine serum albumin (BSA, Serva, Heidelberg, Germany, 11930), and sorbitol (Merck, Darmstadt, Germany) sufficient to adjust the osmolality to 75 or 300 mOsm as determined with an osmometer (Osmomat 030, Gonotec, Berlin, Germany). After fusion it is very critical that the cells are gently flushed from the chamber in complete growth medium.

Hardware

Power Supply

To support the full range of electrofusion protocols, suitable power supplies should be capable of producing an alternating field of frequencies ranging from 100 kHz to 10 MHz and a maximum peak amplitude of 15 V for sustained periods of time for both pre- and postfusion alignment (at least 30 sec is required pre- and postfusion). If fusion is performed in isosmotic solutions, several consecutive pulses must be generally applied in order to initiate the fusion process between two aligned cells. Therefore, the power supply must also be capable of delivering 1–9 breakdown pulses of 1–400 V for 1 to 100 μsec. The time interval between consecutive pulses should be programmable between 0.1 and 1000 sec. Fusion under hypoosmotic conditions normally requires only one breakdown pulse (Fig. 9). Application of two or more pulses under hypoosmotic conditions generally results in reduced hybrid viability. A programmable alignment off time (AOT) of 10 to 999 msec should also be available. This permits interruption of the ac alignment current during pulse application.

Temperature Control

It should be possible to maintain the temperature of the fusion chamber between 20° and 30°. Higher temperatures must be avoided because the resealing process of the individual membranes accelerates such that intermingling of the attached membranes cannot occur. Consequently, the yield of hybridomas decreases considerably. Allowances for heat dissipation to compensate for heat generated by the alignment current should, therefore, be made.

Fusion Chambers

A wide variety of suitable fusion chamber designs are available.[23] The simplest chamber consists of two parallel wires affixed to a microscopic slide or other suitable platform. The wires are connected to the power supply, and the cell suspension is introduced by direct application. A coverslip may be added to facilitate viewing. This chamber offers the advantage of simplicity and the possibility of visual control of the fusion process. Fused cells are difficult to recover from this chamber, however. Moreover, fusion must be performed in a laminar flow hood to maintain sterility when this chamber is used.

The so-called "helical chamber" (Fig. 10) has proved to be the most versatile design for average numbers of cells. This chamber consists of a central cylindrical core wrapped with two or four parallel platinum wires in a helical fashion. The wires are affixed to a pair of plugs by which the chamber is connected to the power supply. The electrode fits into a hollow chamber (200–350 μl capacity) into which the cell suspension is introduced.

High fusion efficiency and hybrid yield may be achieved, but it is difficult to obtain viable fusion products when cell numbers are limiting. In the case of small input numbers of parental cells, the helical chamber can

Fig. 10. Helical chamber designed for large-scale hybridoma production. The chamber consists of a hollow Perspex tube around which are wound two cylindrical platinum electrodes 200 μm apart, forming the electrode. A total of 1 m of electrode wire is used. The wrapped electrode is introduced into a cylindrical Perspex "jacket" (left) to which the cell suspension has been added. As the electrode is introduced into the tight-fitting jacket, the cells rise up between the wires, eliminating nearly all dead space in the chamber.

be used only when the volume is considerably reduced. This can be achieved by filling part of the chamber with a solution of higher density and viscosity (normally fusion media with 20 mg/ml of albumin). The cells (suspended in fusion medium) are layered on top of this medium. Newly developed chambers, for example, the "adjustable plate microchamber" (see below) are superior for handling very small cell numbers.[55]

The "adjustable plate microchamber" (Fig. 11) allows the fusion of as few as 5000 cells with efficiencies comparable to those attained in the helical chamber (Fig. 12). This chamber consists of two parallel electrical plates separated by a variable distance which is adjusted by a screw micrometer. The distance between the plates is calibrated such that the electrical field strength may be accurately calculated. Although the electrical field generated in such a chamber is theoretically homogeneous, such a field becomes inhomogeneous when a cell suspension is added because of the distortion of the field lines by the suspended cells.

Fusion Protocol

As discussed below, the fusion protocol must be adapted and optimized for specific applications. The most common application is the production of hybridomas using activated B lymphocytes and myeloma/hybridoma

[55] U. Zimmermann, P. Gessner, D. W. Sammons, and G. A. Neil, *Biochim. Biophys. Acta* (in press).

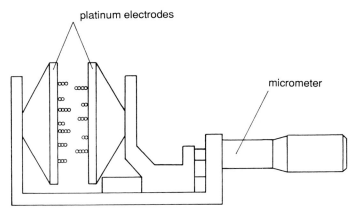

FIG. 11. Adjustable plate microchamber designed for the fusion of limiting numbers of cells. The distance between the platinum plate electrodes is adjusted with a micrometer to accommodate small volume cell suspensions (as little as 10 μl). Alignment and fusion field strengths are calculated based on the set distance between the electrodes.

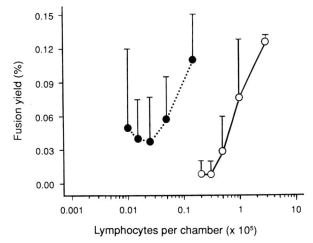

FIG. 12. Comparison of fusion yields using (●) the adjustable plate microchamber (see Fig. 11) and (○) the helical chamber (Fig. 10). The mouse myeloma fusion partner SP2/0 was mixed in a ratio of 1:1 with murine B cells stimulated *in vitro* with lipopolysaccharide and dextran sulfate, washed in 75 mOs*m* fusion medium, and added to either chamber. The cells were aligned in an alternating field of strength 250 V cm^{-1} for 30 sec. Fusions were performed using a single fusion pulse of 1.25 kV cm^{-1} and 15 μsec duration. The efficiency of hybridoma formation was measured by scoring the number of separate colonies visible 5–7 days postfusion in HAT selection medium. Note the relatively high fusion efficiency obtained with even low input lymphocyte numbers using the adjustable plate microchamber. Error bars depict standard errors calculated from five separate experiments.

fusion partners such as SP2/0. The following protocols are based on the foregoing principles as well as empirical observations. These have proved successful in the production of murine and human hybridomas, but may require adjustment for other cell lines and applications.

Isosmolar Fusion Protocol

We have developed a simplified and standardized protocol based on the previously described principles of electrofusion. This protocol, which has proved suitable for use with open, helical, and adjustable plate micro-chambers without substantial modification, is as follows.

1. Count and mix the cells together in the appropriate ratio. For standard mouse hybridoma production, a ratio of 1–5 lymphocytes per myeloma/hybridoma fusion partner is optimal (and can be adjusted according to individual needs).

2. Wash the cells twice at room temperature in two to three changes of 300 mOsm fusion medium in order to replace the conductive growth medium by the weakly conductive fusion medium. (*Note:* Sufficient washes to replace the conductive growth medium is critical to successful fusion.) Care should be taken to prevent the loss of appreciable numbers of cells during the wash steps. Because the final cell number and ratio are also critical, repeat counting and adjustment of cell numbers after washing should be performed. We have developed a small dialysis chamber that allows rapid replacement of the growth medium by fusion medium without cell loss (not shown).

3. Resuspend the cells at 3×10^5/ml in fusion medium. Carefully pipette the appropriate amount of the cell suspension into the fusion chamber (300 μl for the standard spiral chamber, 5–10 μl for the adjustable plate microchamber).

4. Connect the chamber to the power supply. Apply a constant alignment field of 250–300 V cm^{-1} at a frequency 1.5–2.0 MHz for 30 sec.

5. Apply three square wave fusion ("breakdown") pulses of 2.0–2.25 kV cm^{-1} strength and 15 μsec duration.

6. After the fusion pulses, apply the alternating current at 250–300 V cm^{-1} for 30 sec to keep the cells in close contact (thus facilitating fusion).

7. Allow the cells in the chamber to reseal undisturbed for 30 min at room temperature. (*Note:* This step is critical.)

8. Gently flush the cells from the chamber with complete growth medium (without phenol red indicator, which is toxic when admitted into the cell), wash twice and resuspend in appropriate selection medium, for example, complete growth medium supplemented with hypoxanthine,

aminopterin, and thymidine (or other appropriate selection medium). The hybridomas are then cultured according to standard protocols.[17,36]

Hypoosmolar Fusion Protocol

As noted above, the use of a hypoosmotic fusion medium both enhances and reduces the variability of fusion frequency.[47] The protocol is identical to the isosmolar protocol, except that a lower concentration of sorbitol is used and a single breakdown pulse is applied. We have found that adjustment of the osmolality to 75–100 mOsm is optimal for the standard murine fusion partner, SP2/0, and the human fusion partner, H7C3II. The optimal osmolality may vary for other fusion partners or for B lymphocytes of species other than mouse or man. Note that under hypoosmotic conditions much lower suspension densities must be used (compared with isosmotic conditions). Because of the enhanced fusion frequency, multinucleated "giant cells" tend to result when the cell suspension density exceeds about 10^6 cells. We have found that the optimum cell density for the production of human and murine hybridomas is 3×10^4 cells/ml. When fusing small numbers of cells (e.g., in the adjustable plate microchamber), the volume of the electrode gap must be appropriately changed in order to achieve this final suspension density. Note also that the field strength of the fusion pulse must be lowered because of the increase in the cell radius that results from hypoosmotic shock (to about 1.25 kV cm^{-1}).

Optimization of Fusion Conditions

Although we have found that the methods outlined in this chapter are highly reproducible, biological variability of cell lines and differences in culture techniques will sometimes require adjustment of fusion parameters in order to achieve optimal results. Because it is often excessively time-consuming, tedious, and expensive to attempt to optimize fusion conditions by generating hybridomas using a range of conditions, a more rapid method is needed. One such method suitable for the production of murine or human hybridomas is fusion of standard HGPRT hybridoma lines such as SP2/0[36] or H7C3II[49] to thymidine kinase-deficient variants such as G8.[56] Hybrid cells (but neither parent cell line) grow in HAT-supplemented medium, with visible hybrids evident as early as 2–3 days postfusion. We have found that fusion parameters optimized using these cells are

[56] R. Schnettler, P. Gessner, U. Zimmermann, G. A. Neil, H. B. Urnovitz, and D. W. Sammons, *Appl. Micrograv. Tech.* **2**, 3 (1989).

readily transferrable to fusion of *in vitro* or *in vivo* immunized lymphocytes without additional modification.

Conclusion

Electrofusion provides a controlled, simple, and efficient means of generating stable antibody-forming hybridomas. Application of the principles and methods outlined here should allow even neophytes to achieve satisfactory results. Many alternative methodologies are currently available, any number of which might provide comparable results, although these techniques have proved most satisfactory for us. As with other technologies, careful attention must be paid to the physical and biological properties of the system. The investigator who needs to generate large numbers of hybridomas rapidly and efficiently, or who needs to produce hybridomas from very small numbers of parental cells, will find the investment in the necessary equipment and time amply rewarded.

Acknowledgments

Research was supported by grants from the Deutsche Forschungsgemeinschaft (SFB 176, B5) and the Federal Ministry for Research and Technology (DARA 50WB9212-6) to U.Z., and by a grant from the National Aeronautics and Space Administration (NAG 7-816) to G.A.N.

[15] Membrane Electrofusion: A Paradigm for Study of Membrane Fusion Mechanisms

By Arthur E. Sowers

Introduction

Electrofusion is an interdisciplinary method which is simple in overall concept, has novel characteristics, and has been used successfully in many laboratories for various purposes and applications. Electrofusion depends on the use of a specific electric field treatment as a fusogen (Fig. 1). However, electrofusion may be difficult or impossible to use in some potential projects, owing to the fact that the mechanism of electrofusion is still very poorly understood. Enough progress has been made in terms of both the electrochemical and biological aspects of electrofusion to suggest that additional efforts in mechanistic studies will not only improve our understanding of electrofusion, but also lead to fundamental new informa-

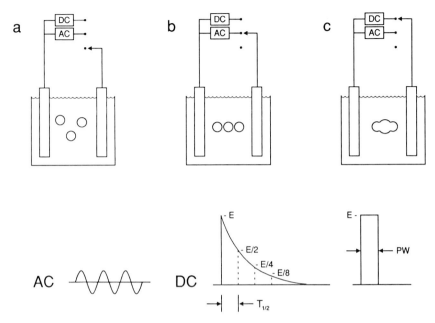

FIG. 1. Normal electrofusion protocol and waveforms of electric field treatment. (a) Spherically shaped membranes in random position in suspension. (b) Alignment of membranes into close membrane–membrane contact by pearl chain formation through application of a low-strength sine wave (lower left) alternating electric field. (c) Fusion of membranes in pearl chains on application of a single exponentially decaying (lower middle) or square wave (lower right) pulse. The exponentially decaying pulse has a decay half-time, $t_{1/2}$, whereas the square wave pulse has a pulse width, PW. Both pulses have a voltage (V) specification which, on application through the separation distance (s) between the electrodes, produces a field strength E ($= V/s$).

tion about properties of membranes, improved methodological protocols, and possible new insights into biological membrane fusion. This chapter describes the status of the electrofusion paradigm as it applies to the study of membrane fusion mechanisms, explains the requirements for obtaining electrofusion in erythrocyte membranes (and its applicability to other membrane systems), and guides the reader through specific problems, as they are presently understood, including both methodology and idiosyncrasies.

Additional information and perspectives can be found in several literature sources.[1-6] Other review papers have been primarily directed at the

[1] H. A. Pohl, K. Pollock, and H. Rivera, *Int. J. Quant. Chem.: Quant. Biol. Symp.* **11,** 327 (1984).

mechanism of electrofusion.[7-9] Some investigations have involved specific applications such as novel approaches to the measurement of lateral diffusion coefficients[10] and to the creation of membrane[11] and cell–tissue constructs.[12] The terms electrically induced fusion, electro-cell fusion, and pulse-induced fusion are sometimes also used more or less interchangeably with electrofusion. When membrane fusion mechanisms are to be studied, it is essential to have an understanding of the fundamental problems of membrane fusion, including rigorous criteria for fusion, fusion assays, and the interpretation of data. These problems are extensively and authoritatively discussed elsewhere.[13-17]

Advantages of Electrofusion

Electrofusion has characteristics which offer unique experimental access to the fusion event for mechanistic studies. For practical applications, electrofusion can provide much higher fusion yields than chemical fusogens, and it may allow fusion to occur at low yields when chemical fusogens completely fail or have other undesirable characteristics. The four features of electrofusion which may confer unique and advantageous char-

[2] G. A. Hofmann and G. A. Evans, *IEEE Eng. Med. Biol. Mag.* **5,** 6 (1986).

[3] T. Y. Tsong, *Biosci. Rep.* **3,** 487 (1983).

[4] E. Neumann, A. E. Sowers, and C. Jordan (eds.), "Electroporation and Electrofusion in Cell Biology." Plenum, New York, 1989.

[5] C. A. K. Borrebaeck and I. Hagen (eds.), "Electromanipulation in Hybridoma Technology: A Laboratory Manual." Stockton, New York, 1989.

[6] D. C. Chang, B. M. Chassy, J. A. Saunders, and A. E. Sowers (eds.), "Handbook of Electroporation and Electrofusion," Academic Press, San Diego (1992).

[7] A. E. Sowers, *in* "Electroporation and Electrofusion in Cell Biology" (E. Neumann, A. E. Sowers, and C. Jordan, eds.), p. 229. Plenum, New York, 1989.

[8] A. E. Sowers, *in* "Charge and Field Effects in Biosystems 2" (M. J. Allen, S. F. Cleary, and F. M. Hawkridge, eds.), p. 315. Plenum, New York, 1989.

[9] U. Zimmermann, *Biochim. Biophys. Acta* **694,** 227 (1982).

[10] A. E. Sowers, *Biophys. J.* **47,** 519 (1985).

[11] D. M. Miles and R. M. Hochmuth, *in* "Cell Fusion" (A. E. Sowers, ed.), p. 441. Plenum, New York, 1987.

[12] R. J. Grasso, R. Heller, J. C. Cooley, and E. M. Haller, *Biochim. Biophys. Acta* **980,** 9 (1989).

[13] G. Poste and G. L. Nicolson (eds.), "Membrane Fusion." Elsevier/North Holland, Amsterdam, 1978.

[14] A. E. Sowers (ed.), "Cell Fusion." Plenum, New York, 1987.

[15] S. Ohki (ed.), "Molecular Mechanisms of Membrane Fusion." Plenum, New York, 1988.

[16] F. Bronner and N. Düzgüneş (eds.), "Membrane Fusion in Fertilization, Cellular Transport, and Viral Infection" (*Curr. Top. Membr. Transport* **32**). Academic Press, New York, 1988.

[17] H. Plattner, *Int. Rev. Cytol.* **119,** 197 (1980).

acteristics to a mechanistic-oriented experiment are as follows: (1) it is nonchemical; (2) the fusogenic membrane alteration can be applied simultaneously to all membranes; (3) fusions can be induced in high yields; and (4) electrofusion is broadly applicable to a wide variety of cell and membrane types, including vesicular membranes.

The fact that electrofusion involves an electrical connection with the aqueous medium, however, permits a second physical phenomenon called "dielectrophoresis" [18] to be used to "align" membranes into the so-called pearl chain formation. Dielectrophoresis occurs when a relatively low-strength, but continuous, alternating electric field is passed through the medium. This causes the positions of the membranes to change from random positions into long parallel rows. Complete alignment can take as little as a few seconds or as long as 1–2 min, depending on other experimental conditions. It is a fortunate accident of nature that the points of contact during dielectrophoresis coincide with the locations on the membrane where the pulse causes fusion (Fig. 1).

When dielectrophorseis is used to align membranes into contact, the following three additional advantageous characteristics are realized: membrane–membrane contact is reversible; the phenomenon is essentially independent of the chemistry of the medium, as long as the conductivity is low (see below); and the phenomenon is mild. Membranes can be brought into close contact by a wide variety of other techniques, including unit gravity sedimentation into a monolayer,[9] cell growth to confluence in a monolayer,[19] and alignment with micromanipulators.[20]

As a general approach to the study of membrane fusion mechanisms, the dielectrophoresis plus electrofusion approach permits the experimental separation of the fusogen, the induction of membrane–membrane contact, and the chemistry of the medium from one step to another. Because the fusogen can be instantly applied to the suspension through an electrical signal, it is possible to use the same electrical signal to trigger other experimental data-recording instrumentation for time-resolved studies.

Shortcomings of Electrofusion

Disadvantages

The same strong electric field pulse which *may* induce membrane fusion, *will* also induce a transient increase in membrane permeabilization through the induction of membrane pores called electropores.[3,4,6] This is

[18] H. A. Pohl, "Dielectrophoresis." Cambridge Univ. Press, London, 1978.
[19] J. Teissie, V. P. Knutson, T. Y. Tsong, and M. D. Lane, *Science* 216, 537 (1982).
[20] M. Senda, J. Takeda, S. Abe, and T. Nakamura, *Plant Cell Physiol.* 20, 1441 (1979).

important because (1) this induced increase in permeability may be deleterious to the survival of viable cells and must be taken into consideration in
fusion protocols where cell survival is essential; (2) much early speculation
about the mechanism of electrofusion included the hypothetical involvement of electropores as an intermediate stage between the substrate and the
fusion product[21]; and (3) electropore induction may itself be a primary goal
of an investigation, and much of what follows will also be relevant. A
number of strategies can be used to design experimental protocols that will
diminish the decrease in cell viability caused by electroporation. Erythrocyte ghosts[22] can be used to avoid the viability problem altogether. In work
which requires that viability be maintained, the general strategy is to use
osmoticants to reduce colloid osmotic swelling.[23] Additional discussion on
the viability problem can be found in other reviews.[2,4,7]

There may be irreproducibility in fusion yield data[23] owing to some
experimental detail, or a biological characteristic, which is simply not
obvious. Insufficient knowledge of the mechanism may also be a limitation. For unknown reasons, some cells or membrane systems may be
resistant to electrofusion.[23] It remains to be determined whether this is due
to a fundamental biological detail or to a methodological parameter which
must be adjusted outside of the usual range. Experimental reproducibility
is quite acceptable for a visual scoring technique.[8] Also, fusion yield variability is higher for human erythrocyte ghosts[24,25] than in ghost membranes
of erythrocytes from laboratory rabbits, possibly because of uncontrollable
factors in human diets compared to the controlled diet of a laboratory
animal.

There is good evidence that the phenomenon of electroosmosis takes
place in the electropores that are induced along with electrofusion.[26,27] This
phenomenon may prevent water-soluble molecules from being used as
traditional contents-mixing indictors,[28,29] since electroosmosis may generate an artifactually high fusion yield if appropriate controls are not used.[26]

[21] G. Pilwat, H.-P. Richter, and U. Zimmermann, *FEBS Lett.* **133**, 169 (1981).
[22] J. T. Dodge, C. Mitchell, and D. J. Hanahan, *Arch. Biochem. Biophys.* **100**, 119 (1963).
[23] G. Bates, J. Saunders, and A. E. Sowers, *in* "Cell Fusion" (A. E. Sowers, ed.), p. 367. Plenum, New York, 1987.
[24] A. E. Sowers, *Biochim. Biophys. Acta* **985**, 334 (1989).
[25] D. C. Chang, J. B. Hunt, and P.-Q. Gao, *Cell Biophys.* **14**, 231 (1989).
[26] A. E. Sowers, *Biophys. J.* **54**, 619 (1988).
[27] D. S. Dimitrov and A. E. Sowers, *Biochim. Biophys. Acta* **1022**, 381 (1990).
[28] N. Düzgüneş and J. Bentz, *in* "Spectroscopic Membrane Probes" (L. M. Loew, ed.), Vol. 1, p. 117. CRC Press, Boca Raton, Florida, 1988.
[29] S. J. Morris, D. Bradley, C. C. Gibson, P. D. Smith, and R. Blumenthal, *in* "Spectroscopic Membrane Probes" (L. M. Loew, ed.), Vol. 1, p. 161. CRC Press, Boca Raton, Florida, 1988.

It is possible that short pulse width square wave pulses may overcome this problem. In contrast, these molecules may still be useful as "markers."[26]

Finally, electrofusion is truly an interdisciplinary biophysical technique. Therefore, success with it requires either an appreciation and familiarity with physics and electronics that is not common among biologists or an approach which duplicates the necessary setup, whether it be a homemade system or a commerically available system. The minimum level of knowledge needed would approximately correspond to that needed at the graduate level in electrophysiology. In some cases, the electrical signal generators may have to be custom built. The required chambers, electrical signal generators, and the biology of the experiment require appropriate interfacing and matching to one another. Although some commercial equipment is specifically available for the electrofusion and electroporation market, it is often very expensive and inflexible in design.

If commercially available pulse generators are to be used for electrofusion or electroporation research, those produced for the industrial or engineering market may be more likely to have the most reliable technical specifications. However, the disadvantage of these devices is that there is usually no provision for dielectrophoresis. The capability for inducing dielectrophoresis can be arranged, however, through an appropriate relay or a simple external passive network.[7] Up to now the preference has been to custom build devices, because access to interior circuits for modification and adjustment is often necessary in research applications or repair. Additional information on construction of these devices has been published.[7]

Limitations

There is a minimum effective diameter needed before membranes can be aligned into pearl chains. It has been predicted[2,12] that dielectrophoresis of small-diameter ($< 1 \mu$m) vesicles should not occur. However, alignment (and fusion) with vesicles about $0.5-1.0 \mu$m diameter of mitochondrial inner membranes has been observed.[30] Attempts to electrofuse $0.5-2.0 \mu$m diameter azolectin lipid vesicles have been successful but not satisfactory because (1) the vesicles are not stable; (2) the diameters are too heterogeneous for stable and uniform pearl chain formation (visual observation by phase-contrast light optics shows that Brownian motion causes significant fluctuations in position and membrane–membrane distance); (3) the dielectrophoresis is weak; (4) the vesicles cannot be prepared in a high enough concentration for large numbers of long pearl chains to be formed; and (5) the fusion yield is extremely low.

[30] A. E. Sowers, *Biochim. Biophys. Acta* **735**, 426 (1983).

The use of dielectrophoresis to generate the force which brings membranes into pearl chains decreases as the electrical conductivity of the buffer–medium increases. Alignment of erythrocyte ghost membranes in sodium phosphate buffer is generally straightforward between 0 and 40 mM buffer strength, but much more difficult at 40 mM and above.[7,8,27,31,32] The main issues are whether the membranes come into contact, whether the bulk field strengths needed for alignment are high enough to become comparable to the strong dc pulse, and whether the aqueous suspension heats up to change conditions. Further discussion of dielectrophoresis can be found in a review[33]; earlier information, dealing largely with inorganic particles in organic fluids, can be found in a treatise.[18]

Membranes with greatly dissimilar diameters may not be good fusion substrates. Electrofusion of two spherically shaped cells may be difficult if their diameters are very different. This is due to (1) the general experimental observation that cells with small effective diameters require pulse field strengths above a higher threshold than cells with larger effective diameters and (2) the theoretical prediction[33,34] that the external field-induced membrane potential will be directly proportional to the radius of the cell (assuming spherical geometry). The induced membrane potential in the smaller cell may be too low to induce the fusogenic state in the membrane of that cell, whereas the induced membrane potential in the larger cell may cause extensive membrane damage, if not fragmentation. The quantitative aspects of this problem are far from resolved, let alone defined, but additional discussions may be found elsewhere.[34-37]

Membrane suspensions must have concentrations which are within a certain range in order to be effective (see below), because the rate of pearl chain formation is directly related to concentrations of membranes in solution. For example, dielectrophoresis of human erythrocyte ghosts in 20 mM sodium phosphate buffer (pH 8.5) requires 0.1 to 0.5 min if a pellet of membranes is resuspended in 10 to 15 volumes of buffer. If concentrations are too high, some pearl chains which are parallel and close to each other will come into contact and possibly cause interference with

[31] A. E. Sowers, *Biochim. Biophys. Acta* **985**, 339 (1989).

[32] D. S. Dimitrov and A. E. Sowers, *Biochim. Biophys. Acta* **1023**, 389 (1990).

[33] H. P. Schwann, *in* "Electroporation and Electrofusion in Cell Biology" (E. Neumann, A. E. Sowers, and C. A. Jordan, eds.), p. 1. Plenum, New York, 1989.

[34] D. C. Chang, *Biophys. J.* **56**, 641 (1989).

[35] D. L. Farkas, S. Malkin, and R. Korenstein, *Biochim. Biophys. Acta* **767**, 507 (1984).

[36] D. Gross, L. M. Loew, and W. Webb, *Biophys. J.* **50**, 339 (1986).

[37] K. Kinosita, Jr., I. Ashikawa, N. Saita, H. Yoshimura, H. Itoh, K. Nagayama, and A. Ikegami, *Biophys. J.* **53**, 1015 (1988).

the electric fields in the vicinity of the membranes, or lead to confusion of the optical pathways needed for unambiguous observation of selected membranes and fusion events.

Misconceptions

A number of misconceptions concerning electrofusion are prevalent.[7,8] These include the stated or implied absolute and general requirements[12] that, if dielectrophoresis is to be used as the means of inducing close membrane – membrane contact, (1) the alternating electric fields must have a high frequency (i.e., 1 – 10 MHz); (2) the buffer strength must be very low (2 – 10 mM), or zero in ionic strength; and (3) the field geometry in the chamber must have a bulk inhomogeneous strength and orientation. High frequency may be necessary to minimize gas bubble formation and/ or electrolysis at electrodes[38] when these electrodes are so close to each other (100 – 200 μm) as to be visible in the field of view of the light microscope. High frequency may not be necessary when the electrodes are far from the viewing part of long-path (2 mm) chambers[7] and when bubble formation is unimportant. The presence of membranes in a suspension may cause *local* inhomogeneities, but a bulk inhomogeneous field does not appear to be necessary. The chamber used in our studies is symmetrical and produces a uniform electric field (Fig. 2).

General Requirements

A particularly important factor in observing fusion is a rigorous definition of membrane fusion (Fig. 3). Often what is incorrectly recognized as fusion is the cell rounding step, which takes place after true fusion.[39] Nucleation indexes are thus a count of nuclei within a common cytoplasm rather than a measure of fusion yield. It is actually possible for fusion to take place without cell rounding,[40] or for cell rounding to take place after a delay.[39] In either case, true fusion yield may be seriously underestimated. In fact, membrane fusion can be observed in erythrocyte ghosts under conditions where the rounding is not obvious.[41] The simplest approach to obtaining evidence for fusion is to use a fluorescent lipid analog and examine the cells in contact for evidence of lateral diffusion of the fluores-

[38] B. Rosenberg, L. Van Camp, and T. Krigas, *Nature (London)* **205**, 698 (1965).
[39] S. Knutton and C. A. Pasternak, *Trends Biochem. Sci.* **4**, 220 (1979).
[40] J. W. Wojcieszyn, R. A. Schlegel, K. Lumley-Sapanski, and K. A. Jacobson, *J. Cell Biol.* **96**, 151 (1983).
[41] L. V. Chernomordik and A. E. Sowers, *Biophys. J.* **60**, 1026 (1991).

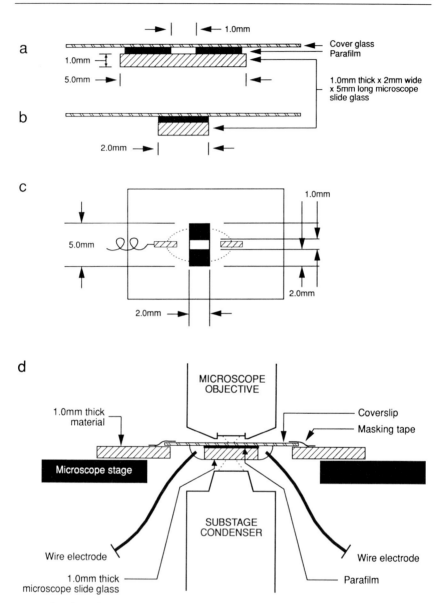

FIG. 2. Microslide chamber for observation of electrofusion by light microscopy. (a) End view showing "tunnel" (1.0 mm wide × 0.075 mm high) cross section. (b) Side view showing tunnel length of 2.0 mm. (c) Top view showing location for "pools" of buffer (dotted lines) and placement of electrodes (hatched). (d) Inverted position of the chamber when carried by a chamber carrier (hatched) on an upright light microscope stage, including path of wires and optical path.

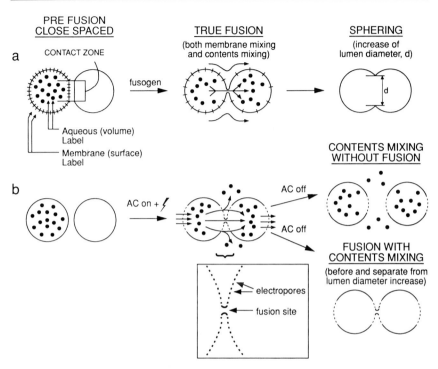

FIG. 3. Rigorous criteria for membrane fusion. (a) Evidence must be obtained for movement of either aqueous (volume) or membrane (surface) labels, preferably both from the originally labeled to the originally unlabeled membrane (the "sphering" step, which leads to an increase in lumen diameter, d, until a perfect sphere is reached, is a phenomenon separate from fusion). (b) Electrofusion (alignment by dielectrophoresis followed by the pulse) is accompanied by electropore induction (center and inset) which can show artifactual evidence for contents mixing,[26,27] unless it is supplemented with a test for physical connection, following the removal of the dielectrophoretic force after the pulse.[26] *Note:* Fusion with contents mixing is shown without aqueous label detail (dots).

cence from the originally labeled membrane to the unlabeled membrane. It is essential that rigorous criteria be applied to scoring fusion events.[28,29,40]

Fusion is usually observed with a light microscope with phase optics and epiillumination fluorescence optics. Phase optics can be used to observe morphological changes, whereas fluorescent lipid analogs can be used to detect membrane mixing. Both the purpose of the electrofusion and the nature of the cell membrane will determine all other aspects of electrofusion. In all cases, both a signal generator and a chamber are needed. The signal generator must be electrically matched to both the experiment and the chamber.

Besides the methodology used to arrange close membrane-membrane contact, the two most critical aspects of the fusogenic electric pulse are the field strength, E, as generated in the medium (the voltage applied to the chamber divided by the gap distance, V/mm) and the duration of the pulse. Also important, but less apparent, are two other factors: the waveform of the pulse (Fig. 1) and the geometry of the chamber. The former will determine how the pulse duration will be specified. Square wave pulses (Fig. 1) will have a true pulse "length" or "duration" whereas exponentially decaying pulses can be correctly specified only in terms of a decay half-time $t_{1/2}$.

Chamber Design

The chamber is designed for (1) observing electrofusion at a relatively large distance (at least 1.0 mm) from both electrodes, (2) duplicating the standard optical microscope light path, (3) providing a simple means for connecting an electrical current to the membrane suspension (Fig. 3), and (4) being easily and inexpensively assembled. It provides a 2 mm long, 1 mm wide, and 0.075 mm deep "tunnel" (formed by a coverslip on top, a 1 mm thick glass microscope slide on the bottom, and Parafilm walls on each side). The electrodes are made of solder "tinned" hookup wires which "hang" in pools of buffer on each end of the tunnel. Electrolytic bubble formation and other electrochemical reactions may take place at least 1–2 mm from the center of the chamber, and it can be calculated that diffusion of the electrochemical products to the center of the chamber can take up to 1–2 min.[30] Usually all observations and measurements can be made substantially before these products arrive at the site in the chamber where they may interfere with the processes being measured.

The chamber can be constructed easily, quickly, and inexpensively, with commonly available materials, and either discarded after single use or disassembled, washed, and reassembled. The disadvantage of the chamber design is that it requires a pulse with a relatively high voltage (1000–2000 V), but at relatively low currents. Larger volume chambers can be easily constructed using parallel sheets of, for example, stainless steel,[42] and they can similarly be disassembled for cleaning or reassembled for reuse. Alternating currents may have to be in the high frequency (1 MHz) range instead of the ac-utility (60 Hz) frequency to avoid or reduce sparking, arcing, or other electrochemical processes.

Many chamber descriptions can be found in the literature.[2–6] For visual observation of experiments a simple, inexpensive, throwaway chamber

[42] J. F. Speyer, *BioTechniques* **8**, 28 (1990).

(Fig. 2) is the most practical. It generates a bulk homogeneous field (although the presence of membranes may distort this homogeneity locally), while both permitting excellent compatibility with standard optical paths and keeping electrodes at an optimum distance to minimize electrochemical effects. Determining the field strength of the pulse is a matter of knowing the voltage, V (in volts), applied across the chamber and dividing by the length, L (in mm or cm), and obtaining the field strength, E (in volts/mm or volts/cm).

The chamber geometry and the conductivity of the solution will determine how much power will be needed in the pulse to achieve fusion. It may be sufficient for electrofusion to be induced in very small (microliter) volumes of an aqueous medium, because the experiments involve monitoring the result in real time by optical microscopy. It is usually relatively straightforward to construct homemade pulse generators for such volumes. However, other protocols may call for the preparation of volumes of 0.1 to 1.0 ml. Resourcefulness and imagination will make chamber construction, even in this case, quite simple and feasible.[42] However, the conductivity of the media can affect significantly the power needed from the pulse generator. Such high-volume protocols require considerable planning. Another constraint is compatibility between chamber, specimen, and pulse generator. Some chambers will not work with certain pulse generators. and vice versa.

The fact that complex electrolytic and electrochemical reactions will almost always take place at the electrode–solution interface should be a consideration in the planning of experiments. In the chamber design used for these experiments, the electrodes are placed at a considerable distance from the point at which observations of the specimen take place. This minimizes the effects of electrochemical reactions, especially if the observations can be made before electrochemical products diffuse from the electrode to the point of observation.[30]

The generation of pulses with a square shape (Fig. 1) requires circuitry which is relatively complex. Commercial pulse generators are used because of the desire for the square waveform, or to duplicate a previously existing setup. In contrast, pulses with an exponentially decaying waveform (Fig. 1) can be easily obtained from the discharge of a charged capacitor.[7] Such circuitry is relatively simple to design, construct, and operate. Some investigators use a mechanical switch, or spark gap, as a switch to connect the charged capacitor to the chamber. Indeed, some significant genetic studies have utilized such a crude switch.[43] Because mechanical switches can undergo considerable mechanical vibration, contact bounce, and contact

[43] M. Fromm, L. P. Taylor, and V. Walbot, *Proc. Natl. Acad. Sci. U.S.A.* **82**, 5824 (1990).

pitting, they may lead to both positional uncertainty and waveform irreproducibility. An appropriate alternative is a solid-state device (thyristor or silicon-controlled rectifier) or glass tube (vacuum or thyratron) used as a switch element,[2,7] as there is no mechanical component to undergo wear. Furthermore, when mechanical switches are used in a gas atmosphere, there is usually an arc or spark, which means that some of the energy is wasted in making the spark and in pitting the contacts. Electronic switches completely avoid this problem and have outstanding lifetimes and essentially perfect reproducibility. A storage-screen oscilloscope is also needed for monitoring the pulse parameters, on either a continuous or regular basis, to demonstrate waveform fidelity.

Adjustment of Parameters

The protocol will generally involve the application of a low-strength ac electric field, until dielectrophoresis causes alignment to come to completion. At any time after this condition is met, at least one strong dc pulse is applied. Electrofusion will most likely occur with a pulse field strength (E) between 100 and 1000 V/mm and a decay half-time ($t_{1/2}$) of about 0.2–1.0 msec if an exponentially decaying waveform is used, or a pulse width of 10–200 μsec if a square wave pulse is used.[44]

The diversity in fusion response of various cells to a given pulse is so high that use of membrane systems for which no previous experimental experience exists will require that the lower E threshold for fusion yield as well as the upper threshold E for fragmentation be determined by surveying the entire range of combinations of pulse field strength and duration. As either or both the field strength or the duration of the pulse is increased, the fusion yield will increase. At some upper limit, the pulse may cause membrane fragmentation. There is some reciprocity between pulse field strength and pulse duration, namely, a pulse of shorter length needs to have a higher field strength to produce the same fusion yield. Although human erythrocyte ghosts show a reciprocal relationship between pulse field strength and pulse duration for a given fusion yield (Fig. 4), rabbit erythrocyte ghosts have a qualitatively nonreciprocal electrofusion response to changes in pulse strengths and durations (Fig. 4). This clearly demonstrates the need to thoroughly explore the entire range of both pulse duration and strength to determine the fusion response to an electric field pulse.

It is also known that buffer strength can have a large effect on fusion yield. Over the range 0 to 60 mM in sodium phosphate, fusion yield in

---

[44] U. Zimmermann, P. Scheurich, G. Pilwat, and R. Benz, *Angew. Chem., Int. Ed. Engl.* **20,** 325 (1981).

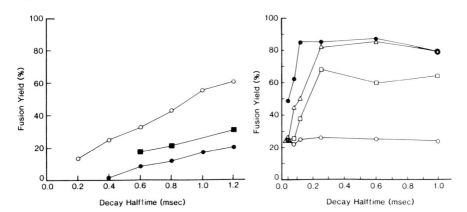

FIG. 4. Fusion yield as a function of pulse strength (E) and decay half-time (msec) for human erythrocyte ghosts (left) and rabbit erythrocyte ghosts (right). *Left:* ●, 400 V/mm; ■, 600 V/mm; ○, 800 V/mm. *Right:* ○, 300 V/mm; □, 400 V/mm; △, 500 V/mm; ●, 600 V/mm. (Reproduced with permission from Ref. 31.)

erythrocyte ghosts is zero below a buffer strength of 2.5–5 mM and reaches a peak in the 20–30 mM range.[31] In sodium phosphate buffer (pH 8.5) of 60 mM, very long duration (> 1.5–2 msec) decay half-time pulses can cause fragmentation of human erythrocyte ghosts instead of fusion. In nucleated cells there is controversy about the effect of buffer strength.[45,46]

Miscellaneous Considerations

Pulse generators produce either an exponentially decaying pulse or a square wave pulse. In those cases where neither an exponentially decaying nor square wave pulse is used, the pulses do not have simple waveform descriptions.[34,47] Although the advantages of one waveform over another have been debated, and some research has been published,[48] definitive studies, in which both pulse strength and duration for both the exponentially decaying waveform and the square waveform have been varied, have not been reported.

The erythrocyte ghost membrane shows a profound shape change when treated with fusogenic electric pulses.[7] If the ghost is a nonspherical shape before the pulse, then one or more pulses cause the membrane to become

[45] S. I. Sukharev, *Bioelectrochem. Bioenerg.* **21,** 179 (1989).

[46] C. Blangero and J. Teissie, *J. Membr. Biol.* **86,** 247 (1985).

[47] J. Teissie, M. P. Rols, and C. Blangero, *in* "Electroporation and Electrofusion in Cell Biology" (E. Neumann, A. E. Sowers, and C. A. Jordon, eds.), p. 203. Plenum, New York, 1989.

[48] J. A. Saunders, C. R. Smith, and J. M. Kaper, *BioTechniques* **7,** 1124 (1989).

spherical. Conditions can be found which cause the ghost membrane to become nonspherical if it is spherical before the pulse. These conditions are not the same as conditions needed to achieve fusion. A hypothesis for the origin of this shape change has been presented.[7]

Most of the electrofusion literature describes electrofusion protocols involving the induction of membrane–membrane contact before the application of the fusogenic pulse. Electrofusion may take place with the reverse of this ordination[49,50]; fusion may occur when the pulse treatment is applied before close membrane–membrane contact is induced. This observation has been confirmed and recognized as a fundamentally distinct phenomenon in the electrofusion protocol.[51–53] Thus, the fusogenic alteration induced by the electric pulse has a long-lived component. There are no definitive data on whether this long-lived component is also responsible for fusion when the normal protocol is used or for fusion obtained by the normal protocol that proceeds through a different pulse-induced membrane alteration. This is of fundamental significance because commonly used fusion protocols involve a chemical change in the medium which not only induces fusion, but also aggregation. Using erythrocyte ghost membranes in this modified protocol, fusion is obtained in more moderate yields, but requires multiple pulses. Mechanistic studies require additional controls to prevent the inclusion of artifactual fusion events which are caused not by the long-lived fusogenic state, but by a pulse-induced attraction between cells or membranes which are not in contact but nearby before the pulse.

Studies suggest that there is a small but finite interval between the application of a single electric pulse and the earliest appearance of evidence for fusion, when the lateral diffusion of a fluorescent lipid analog from a labeled membrane to an adjacent unlabeled membrane is used as the fusion indicator.[54] This interval can be as short as 10–20 msec, but also as long as 2–3 sec, and it is strongly dependent on pulse parameters and environmental properties (buffer strength, viscosity, and temperature). Whether this interval represents a fusion intermediate, a prefusion conformational change in the membrane, the effect of a postfusion temporary restriction in lateral mobility, or the effect of a variable number of fusion sites is yet to be determined. The interval appears to be analogous to a lag

[49] A. E. Sowers, *J. Cell Biol.* **102**, 1358 (1986).

[50] A. E. Sowers, *Biophys. J.* **52**, 1015 (1987).

[51] J. Teissie and M.-P. Rols, *Biochem. Biophys. Res. Commun.* **140**, 258 (1986).

[52] M. H. Montane, E. Dupille, G. Alibert, and J. Teissie, *Biochim. Biophys. Acta* **1024**, 203 (1990).

[53] I. Tsoneva, I. Panova, P. Doinov, D. S. Dimitrov, and D. Stahilov, *Stud. Biophys.* **125**, 31 (1988).

[54] D. S. Dimitrov and A. E. Sowers, *Biochemistry* **29**, 8337 (1990).

phase between the instant of the creation of fusogenic conditions and the observation of fusion of some viruses with cell membranes.[55]

Fusions may be attempted between membranes of dissimilar size or origin. Size is important because, in terms of a spherical membrane geometry, there is an inverse relationship between the transmembrane voltage induced by a bulk electric field and the diameter of the sphere.[33] This relationship is not well understood, but in general it means that smaller diameter spheres are likely to require pulses with larger field strengths. If membranes are dissimilar in origin, it may be expected that fusions between A + B partners will be about halfway between fusion yields for A + A and B + B partners. There is evidence that this can happen, but in other cases the fusion yield will be dominated by the presence of one of the two membranes. The use of very short square wave pulses may overcome the problem of electroosmosis. Also, because electroosmosis causes movement of the contents-mixing indicator toward the negative electrode only, it may be possible to observe true contents-mixing events by looking only for fluorescence movement toward the positive electrode.[26,27]

Simple calculations based on total conversion of electrical energy to heat without dissipation indicate that temperature increases of $1-10°$ can occur in the medium with each pulse. In large chambers this may be a problem if multiple pulses are needed. In smaller chambers, use of low-melting waxes suggests that significant thermal buffering may be possible.[49] Definitive experiments have not yet been reported.

Ultrastructural studies using freeze–fracture electron microscopy have shown several features. First, electrofusion in mitochondrial inner membranes is accompanied by vesiculation toward the matrix space, and the vesicles have an inside-out orientation.[30] Electrofusion in erythrocytes is accompanied by the formation of microscopic structures with distinct lifetimes.[56,57] Application of a fusogenic pulse to erythrocyte ghosts held in pearl chains normally leads to a fusion zone which is actually a septum with perforations (fusion sites).[41] Treatments known to disrupt the spectrin network (heating, low ionic strength, and low pH) release these constraints, which allows the linear chains to go on to form giant spherical fusion products.[41]

Acknowledgment

This work was supported by Office of Naval Research Grant N00014-89-J-1715.

[55] M. J. Clague, C. Schoh, L. Zech, and R. Blumental, *Biochemistry* **29**, 1303 (1990).
[56] D. A. Stenger and S. W. Hui, *J. Membr. Biol.* **93**, 43 (1986).
[57] D. C. Chang and T. S. Reese, *Biophys. J.* **58**, 1 (1990).

[16] Electrofusion of Cells: Hybridoma Production by Electrofusion and Polyethylene Glycol

By Sek Wen Hui and David A. Stenger

Introduction

A number of physical, chemical, and biological techniques have been used to induce fusion between cells. The best known chemical technique employs polyethylene glycol (PEG). Viruses and viral proteins are commonly used as biological fusogens. Although these fusion methods have been widely used in genetics and immunology studies, their molecular mechanisms are still not well understood. As a result, improvement of these methods is achieved mainly on an empirical basis. A physical method, namely, electrofusion, has been developed that rivals other fusion methods in efficiency. Electrofusion has been reported to produce hybridomas with a relatively high yield, using a small number of cells. With increasing knowledge about the electrofusion mechanism, the myth and suspicion usually associated with newer techniques, especially those involving high-voltage application in cell biology, should soon be resolved.

Electrofusion has several distinct advantages over other fusion methods in producing cell hybrids. (1) The physical nature of the method guarantees it to be free of chemical or biological contaminants, eliminating the necessity of repeated washing of cells after fusion. (2) The fusion parameters can be easily and accurately controlled by electronics. (3) Electrofusion can be widely applied because the dielectric properties of most cells are within a narrow range. There are no known cells resistant to electrofusion, and properly prepared liposomes also undergo electrofusion. (4) Very few cells (minimum of 2) are needed per run. (5) The entire process can be observed using a microscope. (6) At present, the efficiency, without reaching its limit, is already greater than PEG-induced fusion in many instances.

The electrofusion method, combining the use of dielectrophoresis alignment and reversible membrane breakdown for fusion, was first reported by Zimmermann et al.[1] An overall description of the method was published in this series in 1987.[2] New research has since led to a better understanding of the theory. The method has emerged from a trial-and-error basis to a more analytical one, which enables rational design of new protocols and hardware, thus promising a bright future for new technical

---

[1] U. Zimmermann, J. Vienken, and G. Pilwat, *Bioelectrochem. Bioenerg.* **7**, 533 (1980).

[2] U. Zimmermann and H. B. Urnovitz, this series, Vol. 151, p. 194.

development and innovation. This chapter deals with the rationale for the protocols, describes common protocols, and compares electrofusion with other fusion methods, especially PEG-induced cell fusion.

Forces Involved in Electrofusion

When an external electric field E_0 is imposed on cells suspended in medium (usually less conductive than the cytoplasm), cells become polarized and behave as electric dipoles in the medium. Dipoles are aligned in an electric field and are driven toward or away from areas of high field gradient by the dielectrophoresis force, depending on the electric permittivities of the cell and of its environment. Quantitatively, the strength of the dielectrophoretic force is determined by the effective polarizability B of the cell in its environment:

$$F = B \nabla E_0^2 \tag{1}$$

B is determined by the frequency-dependent permittivities ε_i of the spherical cell, of radius a, and ε_e of the medium:

$$B = [2\pi a^3 (\varepsilon_i - \varepsilon_e) \varepsilon_e/(\varepsilon_i + 2\varepsilon_e)] \tag{2}$$

For nonspherical cells, different shape factors may apply.[3]

The expression for the polarizability indicates that ε_i must exceed ε_e for uncharged cells to approach each other and concentrate along the convergent field near the electrodes. Because the imaginary part of the permittivity is the conductivity/frequency ratio, the external medium should always be less conductive than the cell interior at that particular frequency. The dependencies of F on conductivity and frequency arise from the fact that the effective dipole moment is reduced by Debye shielding of the counterions of the external medium. The values of F, hence B, have been modeled in detail and are experimentally measurable.[3,4]

This dielectrophoretic force is responsible for forming cell–cell "contact" along the field lines. The cell membranes are not really in contact, for an aqueous gap of 15–25 nm still exists between cells according to electron microscopy studies.[5] This gap in between the membrane sandwich cannot be closed to less than 5 nm even if the cells are pretreated with proteolytic enzymes (pronase or dispase) to remove most of the cell surface glycoproteins.

[3] D. A. Stenger, K. V. I. S. Kaler, and S. W. Hui, *Biophys. J.* **59**, 1074 (1991).
[4] K. V. I. S. Kaler and T. B. Jones, *Biophys. J.* **57**, 173 (1990).
[5] D. A. Stenger and S. W. Hui, *J. Membr. Biol.* **93**, 43 (1986).

a

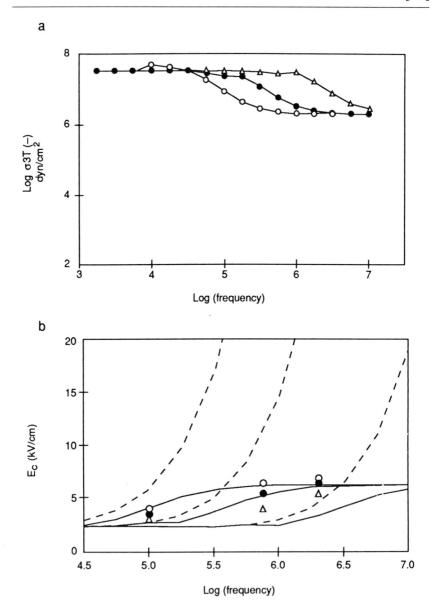

FIG. 1. (a) Calculated combined pressure σ_{3T} on the contact regions of erythrocyte membranes from induced membrane potential and intercellular dipole interaction for members of a three-cell chain in electrode contact. The external media conductivities are 15 (O), 150 (●), and 650 μS/cm (△). The pressure is calculated in response to a 4 kV/cm sinusoidal field at the indicated frequencies. Membrane potential pressure dominates the low-frequency range, whereas the dipole pressure determines the total pressure at high frequency. The

The applied electric field also gives rise to a potential difference, in addition to the existing membrane potential, across the less conductive membranes or the membrane sandwich between contacting cells. This potential exerts a capacitive pressure against the resilient membranes or membrane sandwich, leading to their mechanical breakdown. To ensure the pressure is sufficient to break through the membranes, a high-voltage pulse is used. The pressure across an individual membrane of thickness h is

$$\sigma_m = 0.5\varepsilon_0\varepsilon_e[1.5aE/h\,(1 + \omega^2\tau^2)^{1/2}] \qquad (3)$$

where

$$E = E_0[1 + B/(8\pi\varepsilon_0\varepsilon_e a^3)] \qquad (4)$$

and B is given by Eq. (2). E represents the effective field strength across the membrane in the presence of other dipoles; $\tau = aC(\rho_i + \rho_e/2)$ is the charging time of the membrane of capacitance C, in a cell and medium of resistivities ρ_i and ρ_e, respectively. The pressure σ_m is responsible for the breakdown of individual membranes.

A model has been proposed which states that the total pressure σ_t is, to a first-order approximation, a linear sum of σ_m and the pressure σ_D across the membrane sandwich, due to the dipole–dipole attraction.[3] The dipole pressure is given by

$$\sigma_D = (BE_0)^2/8a^2\varepsilon_0\varepsilon_e \qquad (5)$$

Unlike σ_m, the pressure σ_D may contribute to the breakdown of both the membrane and the intermembrane aqueous barrier. These pressures are complicated functions of frequency-dependent permittivities of the cell and the medium and the number of cells in the pearl chains.[3] B may be measured experimentally by cell dielectrophoretic migration or by cell levitation. The pressures may then be deduced from the above relations. An example is given in Fig. 1 for pronase-treated human erythrocytes, where an ellipsoidal geometry was assumed.[3] Fusion occurs only when the total pressure exceeds that which the membrane and/or junction can withstand.

---

changeover frequency depends on the conductivity of the medium. (b) Calculated and measured critical field strength (E_c) for the membrane breakdown and fusion of erythrocytes by a 20-μsec pulse of sinusoidal waves at the indicated frequencies. Solid and dashed curves represent, respectively, theoretical values with and without the contribution of the dipole pressure. Solid curves (top to bottom) and dashed curves (left to right) correspond in each case to media conductivities of 15, 150, and 650 μS/cm, respectively. Symbols are experimental values for different media conductivities, following the convention in (a).

The above analysis points to two important but not well appreciated concepts. First, the magnitudes of both the dielectrophoretic force and the breakdown pressure vary with the conductivities of the cell and the media, and only the latter is experimentally controllable. Second, the conductivities also determine the frequency range in which the pressure is effective.

Criteria for Successful Fusion

Cell Counting

Because electrofusion can be observed easily under a microscope, many earlier works reporting electrofusion efficiency were based on the number of observed fused cells per number of cells used. The method of counting included manual counting of multinuclear cells, counting by image analysis, and cell sorting, using size change as the criterion for fusion. The efficiency reported in this way can be in the 70–80% range.

A variation of the counting method makes use of fluorescent dyes. A percentage of cells can be stained with fluorescent probes on the surface membrane at a self-quenching concentration. Fusion is detected as the degree of dequenching due to the dilution of membrane probes from labeled to unlabeled cells. If cells can be loaded with fluorescent probes and/or their quenchers, the quenching or dequenching may also serve as indicators of fusion.

Viability

A slightly more stringent criterion is the quantitation of fusion survivors. Exclusion of vital dyes, such as trypan blue or erythrocin B, has been applied to distinguish survivors of electrofusion. However, this method does not distinguish reproducing survivors from nonreproducing ones. Furthermore, a large portion of the cells become leaky immediately after electrofusion; some remain leaky for as long as 1 hr but eventually recover.[6] Thus the vital dye exclusion test is at most a rough estimate of successful fusion.

Selective Cloning

Quantitation of vital hybrids by cloning is a crucial test of electrofusion success if the process is used for hybrid cell production. Standard selection in HAT medium (containing hypoxanthine, aminopterin, and thymidine)

---

[6] M. L. Escande-Geraud, M. P. Rols, M. A. Dupont, N. Gas, and J. Teissie, *Biochim. Biophys. Acta* **939,** 247 (1988).

is usually applied as the first step to measure the success rate of hybridoma formation. The success rate is measured as a ratio of the number of surviving clones to the number of lymphocytes used. This has been reported to be in the range of 10^{-3} to 10^{-6}. In fact most of the fusion pairs observable under the microscope do not survive the selective cloning, and high-yield experiments seldom produce immediately observable fusion pairs. Therefore, optimization of procedures using the microscope criterion often fails to satisfy the vital hybrid criterion.

Antibody Secretion

Only a portion of the hybridomas formed from nonpreselected splenocytes produce the desired immunoglobulin. To increase the specific hybridoma percentage, a preselection of lymphocytes is desirable. This step reduces drastically the number of lymphocytes available for fusion. It is for the fusion of a small number of cells, such as in the production of human hybridomas,[7] that electrofusion shows its superior potential over other fusion methods. Specific hybridomas produced by electrofusion of preselected lymphocytes may constitute more than half of the viable clones after HAT medium selection.[8] The quantitation of specific antibody-producing clones constitutes the only meaningful criterion for a successful fusion for hybridoma production.

Electrical Factors Contributing to Successful Hybrid Formation

The basic protocol for electrofusion is to apply an ac field of just sufficient dielectrophoretic strength to form cell–cell "contacts" for a few seconds. One or more higher voltage dc or ac pulses, of microsecond durations each, are applied to initiate fusion. The dielectrophoretic ac field is then restored for several seconds to ensure postfusion cell contact. The cells are then treated for recovery. Detailed electrical parameters vary from hardware to hardware, but the general parameters as discussed above are ubiquitous.

Frequency

Although it is not necessary to construct a pressure–frequency–conductivity curve for each cell system to be fused, one should bear in mind its importance in designing a protocol. For instance, at a media conductivity of 150 μS/cm ($\rho_e = 6.67 \times 10^3 \ \Omega$ cm), B has a finite value

[7] M. Glassy, Nature *(London)* **333**, 579 (1988).
[8] M. Pratt, A. Mikhalev, and M. Glassy, *Hybridoma* **6**, 469 (1987).

only in the range of approximately 100 kHz to 10 MHz and, for the present purpose, is negligible elsewhere. Therefore, the dipole pressure σ_D contribution exists only during the rise and fall of a rectangular high-field pulse when frequencies within the polarizability bandwidth may contribute. However, the membrane capacitive pressure σ_m, derived mainly from the low-frequency components of the pulse, contributes for most of the duration of the pulse.

Because additional high-frequency pressure is obtained from the dipole effect [Eq. (5)], a high-frequency ac burst is often as effective as a rectangular dc pulse.[3] By the same token, the frequent rise and fall of a train of dc pulses may contribute more pressure than a single one with the same sum of duration. Perhaps for this reason, a series of several pulses are sometimes more effective than a long, single pulse in inducing fusion.[9,10]

Medium Conductivity

In general, the lower the medium conductivity, the more effective is the dielectrophoresis and the dipole pressure σ_D. The low medium conductivity also reduces the Joule heating which damages the cells. Typically, applying a 280 V/cm, 1.5 MHz ac field for 15 sec across a 100 μm electrode gap of a 60 μS/cm medium results in less than 2° increase in temperature.[11] However, the pressure on the membrane σ_m is correspondingly reduced [Eq. (3)], and the crossover frequency from the range dominated by membrane pressure to that by dipole pressure is also lowered. Thus a medium of higher conductivity may reduce the threshold field strength for membrane breakdown[3] (Fig. 2). However, because there is a vast difference in the size and, hence, in the critical breakdown fields for lymphocytes and myelomas, there is an upper limit to the conductivity of the fusion medium used. Most fusion media are solutions of 250–300 mM inositol, mannitol, or sorbitol for maintaining osmolality, and they have conductivities in the 50 μS/cm range. Higher conductivity may lead to reduced yield owing to myeloma lysis.[11] An additional limiting factor is the requirement of certain ions in the medium for cell recovery (to be discussed later). Taking these aspects into consideration, a base medium may be formulated to serve as a starting point for experimental optimization.

Pulse Field Strength

Myeloma lysis imposes a limit on the pulse field strength. The onset field strength of decreasing yield coincides with the onset for increasing

[9] K. Ohnishi, J. Chiba, Y. Goto, and T. Tokunaga, *J. Immunol. Methods* **100**, 181 (1987).
[10] J. J. Schmitt and U. Zimmermann, *Biochim. Biophys. Acta* **983**, 42 (1989).
[11] D. A. Stenger, R. T. Kubiniec, W. J. Purucker, H. Liang, and S. W. Hui, *Hybridoma* **7**, 505 (1988).

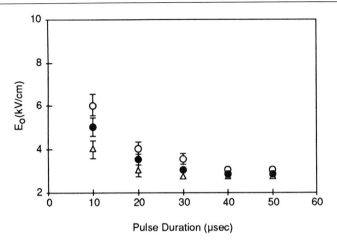

FIG. 2. Threshold electric field amplitude necessary for fusion of erythrocytes as a function of external conductivity (O, 15; ●, 150; △, 650 μS/cm) and pulse duration for a pulse of sinusoidal waves at 100 kHz.

observable myeloma lysis[11] (Fig. 3). The optimal "high-voltage" pulse field is usually in the range of 2–3 kV/cm. It should be noted that a "high-voltage" pulse that produces a 3 kV/cm field across a 100 μm interelectrode space is only 30 V; therefore, electrical hazard is sometimes exaggerated, although caution should nevertheless be used at all times.

Nonelectrical Factors Contributing to Successful Hybrids

Osmotic Pressure

Osmotic pressure is an important factor in the successful formation of hybridomas. Increasing osmolality in the fusion media reduces the yield of hybridomas.[11] On the other hand, fusion carried out in hypotonic media, as well as using cells treated in hypotonic media and returned to isotonic media for fusion, results in significant increases in hybridoma yield.[10] This is believed to be a consequence of stress damage to the membrane, leading to a higher susceptibility to the fusion pressure. The importance of osmotic force has long been recognized in PEG-induced[12] and in electrically induced[11,13] fusion. The exact pressure/stress relationship has been verified for liposome membranes prepared from erythrocyte lipids.[14] Because of differential cellular susceptibility to osmotic swelling, the degree of appli-

[12] S. W. Hui, T. Isac, L. T. Boni, and A. Sea, *J. Membr. Biol.* **84**, 137 (1985).
[13] Q. F. Ahkong and J. A. Lucy, *Biochim. Biophys. Acta* **858**, 206 (1986).
[14] D. Needham and R. M. Hochmuth, *Biophys. J.* **55**, 1001 (1989).

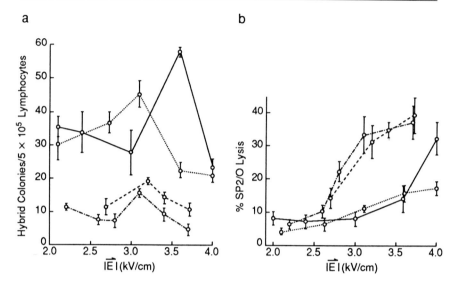

FIG. 3. (A) Hybrid production efficiency as a function of electric field strength, pulse width, and dispase pretreatment of SP2/0 myelomas. Dashed and dot–dash lines represent the efficiency curves using a 20-μsec pulse with and without dispase pretreatment, respectively. Use of a 15-μsec pulse resulted in higher efficiency curves with (solid line) and without (dotted line) dispase pretreatment. (B) SP2/0 myeloma lysis that occurred within 30 sec following pulse application during the experiments described in (A). The 20-μsec pulses with (dashed line) and without (dot–dash line) dispase pretreatment produced considerably more lysis than 15-μsec pulses with (solid line) and without (dotted line) dispase pretreatment.

cable hypotonicity varies from cell system to cell system.[10] For those cells which swell significantly in hypotonic media, compensatory adjustments of the pulse field should be made according to Eqs. (3) and (5).

Divalent Cations

The presence of calcium and/or magnesium ions are apparently essential for the viable recovery of fused cells, rather than for the osmolality or conductivity they provide. Without these ions, especially calcium, fusion yield drops precipitously,[15] and hybridoma formation is suppressed.[9,11] Other divalent cations can partially compensate for the need of calcium for fusion[15] but not for hybridoma formation. Because calcium is commonly required for membrane resealing, and because a similar calcium dependence is found in the fusion of various secretory vesicles,[15a] its involvement in the electrofusion process is to be expected.

[15] T. Ohno-Shosaku and Y. Okada, *J. Membr. Biol.* **85**, 269 (1985).
[15a] N. Düzgüneş, *Subcell. Biochem.* **11**, 195 (1985).

Protease Treatment

The need for protease treatment of the cell surface to attain close cell contact prior to fusion has been appreciated since the early days of electrofusion. It is now clear that the removal of cell surface glycoproteins by protease treatment, which reduces intercellular spacing from 25 to 15 nm in the case of human erythrocytes[5] during dielectrophoresis, also facilitates the breakdown of the intercellular aqueous boundary by the dipolar force of the high field pulse, further reducing it to below 5 nm.[5] Such close contact may bring the lipid bilayers of apposed membranes together, thereby facilitating their fusion.[15] A detailed account of various effects of proteases is given by Ohno-Shosaku and Okada.[16] The protease treatments not only increase fusion yield, but also improve cell survival. Neuraminidase increases the yield of protease-treated cells but cannot substitute for the role of the latter.[16]

Recovery Period

Immediately after the fusion pulses, fused cells are very fragile and may still be leaky.[6] For the first 10–30 min after pulsing, it is more advantageous not to disturb them mechanically (e.g., by pipetting or centrifuging). The period and temperature of recovery vary from cell line to cell line. Returning cells to the normal culture medium while the cells are not resealed may lead to high mortality, perhaps owing to the sodium/potassium imbalance. After the resealing period, cells may be returned to the normal culture medium, centrifuged, and plated.

Typical Protocol for Electrofusion

Usually 10^5 cells are sufficient for a typical electrofusion experiment. The ratio of lymphocytes to myelomas may vary from 1:1 to 10:1 depending on the source of lymphocytes and the strain of myeloma used. To increase yield, cells are treated for 5 min with pronase (0.05–0.1 mg/ml) or dispase (0.01 mg/ml). Treated cells are washed two or three times in the fusion medium, which contains 250–300 mM of inositol, mannitol, or sorbitol, 0.1 mM Ca^{2+}, 0.5 mM Mg^{2+}, and 1 mg/ml bovine serum albumin. Lowering the osmolality to 100 mOs/kg and below, with appropriate lowering of the pulse field to account for the cell swelling according to Eq. (3), is said to increase the yield significantly.[10] The washed cells, suspended at a density of $2–7 \times 10^7$ cells/ml, are then loaded into the fusion chamber.

[16] T. Ohno-Shosaku and Y. Okada, *in* "Electroporation and Electrofusion in Cell Biology" (E. Neumann, A. E. Sowers, and C. A. Jordan, eds.), p. 193. Plenum, New York, 1989.

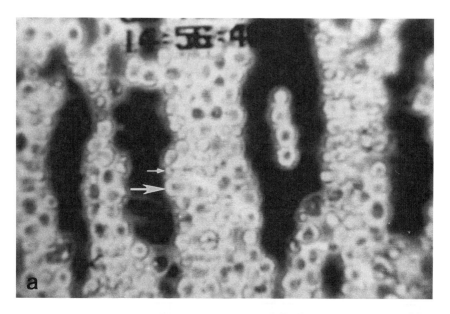

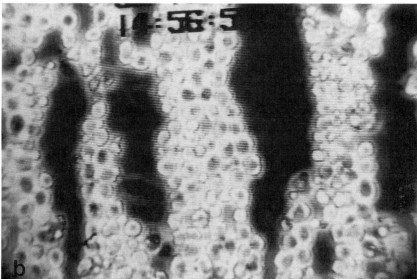

FIG. 4. Video phase-contrast micrographs of the sequence of SP2/0 myeloma and murine splenocyte fusion. (a) Eight seconds after an ac field of 280 V/cm, 1.5 MHz was applied. Most myeloma cells (large arrow) were pearl-chained, with smaller splenocytes (small arrow) on the side or in between. The free-standing chain of four myeloma cells later attached to the main column before the pulse. (b) Immediately after a 3 kV/cm, 15 μsec rectangular pulse was

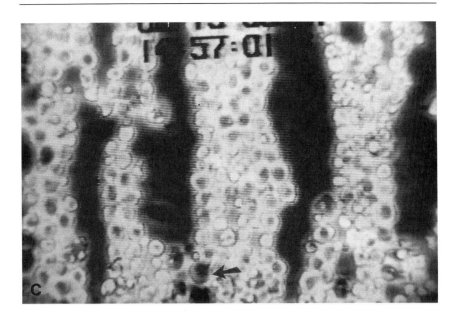

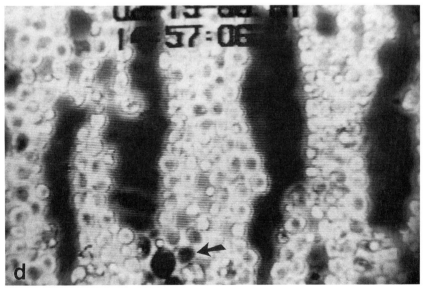

applied. Little configurational change was observed after the mild jolt. (c) Three seconds after the pulse. A few myeloma cells were lysed (black arrow). (d) Eight seconds after the pulse. Myeloma lysis became more prominent. This run resulted in only 137 HAT-selected clones from 6×10^7 splenocytes because of myeloma lysis.

A number of fusion chambers, either commercial or custom-made, are available. The designs range from a pair of parallel platinum wires to complicated interdigitated patterns evaporated on an insulating substrate.[2,11] To keep the voltage low, the electrodes are usually kept 100–500 μm apart. The cavity for the cell suspension is sufficiently small to keep most cells within the applied field and to provide adequate heat dissipation during the short process.

An ac field of 100–300 V/cm, at 1–3 MHz, is applied to form pearl chains (Fig. 4a). Unfortunately, lymphocytes and myelomas are often aligned in segregation because of their different sizes, but there is usually enough intermixing to ensure heterocytotic fusion. Dielectrophoresis for 10–15 sec is sufficient to bring most cells in contact with others (Fig. 4b). A 15–30 μS rectangular dc pulse, or a burst of sinusoidal waves of 100 kHz to 2.5 MHz, at a peak field strength of 2–4 kV/cm, is applied. In some experiments, a train of two to five pulses is also used.[2,9,17] The response to the electric pulse is barely noticeable by microscopic observation, and less noticeable in repeating pulses, probably owing to the increase in the local conductivity of the medium from the leakage of ions from cells. A visible disturbance resulting in the disruption of pearl chains indicates that the applied pulse field is too intense. In high-yield runs, there are no apparent fused pairs immediately after the pulse (Fig. 4b), in contrast to expectation. The dielectrophoretic ac field is resumed for 15 sec after the pulse. During this time, some myeloma cell lysis can be seen (Fig. 4c,d). The electronics providing the pulse sequence is not complicated, especially since the highest voltage required normally does not exceed 200 V, even for a 500 μm electrode separation. A typical setup is shown in Fig. 5.

After remaining in the fusion medium for 10–30 min for recovery, the cells may be removed for further resealing in phosphate-buffered saline (PBS) supplemented with 10% fetal bovine serum (FBS), or in culture media, for 15 min at 37°. After centrifugation, the cells are suspended in HAT medium for initial plating.

Comparison between Electrofusion and Chemical Fusion Methods

Polyethylene glycol causes cell fusion through the colloidal–osmotic force which leads to the aggregation of cells and the aggregation of membrane proteins; the latter creates large patches of membranes devoid of intramembranous particles to facilitate bilayer–bilayer contact.[18] The lipid

[17] D. Chang, in "Electroporation and Electrofusion in Cell Biology" (E. Neumann, A. E. Sowers, and C. A. Jordan, eds.), p. 215. Plenum, New York, 1989.
[18] S. W. Hui and L. T. Boni, in "Membrane Fusion" (J. Wilschut and D. Hoekstra, eds.), p. 231. Dekker, New York, 1991.

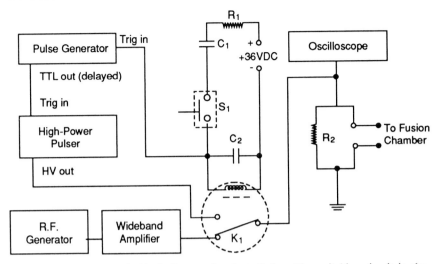

FIG. 5. Basic electronic block diagram for electrofusion. The switching circuit in the middle disconnects temporarily the RF source through K while a delayed pulse is applied. The switching time interval is controlled by R_1C_1.

bilayers of the apposing cell membranes are driven to within 0.5 nm by the colloidal–osmotic force. Structural defects in the bilayer, either endogenous or induced by PEG, promote fusion between the bilayers. Cell lines which do not form bare membrane patches in response to PEG are not susceptible to fusion.[19]

To create sufficient colloidal–osmotic force, over 35% of PEG (M_r 6000–8000) is needed. At this concentration, all water molecules in the bulk are structured by the PEG molecules. In response to the extreme osmotic pressure, intracellular water is also depleted, and the cell volume collapses. Collapsed cells aggregate in clumps, with their membranes tightly in contact. Lipid vesicles would have fused at this stage, yet cells apparently remain unfused, judging by the mixing of cytoplasmic or membrane components between contacting cells. Detectable fusion through lumens begins with the dilution or removal of PEG.[20] Membrane and cytoskeletal proteins may have kept the cells unfused until this stage, when the osmotic swelling further stretches and weakens the membranes in contact. Although PEG is nontoxic to most cells, many cells perish during

---

[19] D. Roos, in "Molecular Mechanisms of Membrane Fusion" (S. Ohki, D. Doyle, T. D. Flanagan, S. W. Hui, and E. Mayhew, eds.), p. 273. Plenum, New York, 1987.
[20] S. K. Huang, M. Cheng, and S. W. Hui, *Biophys. J.* **58**, 1119 (1990).

TABLE I

Electrofusion and Polyethylene Glycol-Induced Fusion Efficiencies

Fusion partners	Yield[a]		Ratio[b]	Criterion[c]	Ref.
	PEG	EF			
Murine spleen × SP2/0	2.2	116	52.6	Hybrid yield	11
Murine spleen × SP2/0	10.3%	20.4%	1.98	Ab production	9
TAPC301 × KR4	9.3	96.8	10.40	Hybrid yield	9
TAPC301 × KR12	7.0	25.4	3.63	Hybrid yield	9
Murine spleen × SP2/0	4	1500	375	Hybrid yield	23
	0	120	—	Ab production	23
Murine lymph node × SP2/0	5	700	140	Hybrid yield	23
	0.4	230	575	Ab production	23
Murine spleen × SP2/0	1.4	11.5	14.3[d]	Hybrid yield	24
Murine spleen × F0	5.5	22	4	Hybrid yield	25
	39%	25%	0.64	Ab production	25
Murine spleen × X63	6	32	5.3	Hybrid yield	25
	0.1	4	40	Ab production	25

[a] Yield in units of 10^{-6}. Values quoted as percentages indicate the fractions of antibody-secreting clones per total surviving clones after HAT selection.
[b] Ratio of (EF yield)/(PEG yield).
[c] Fusion was measured by the number of clones after selection in HAT medium (hybrid yield) or by the number of specific antibody-producing clones (Ab production).
[d] Value given by authors.

the PEG treatment and the subsequent dilution step owing to osmotic shock.

Procedures for using PEG to produce hybrid cells have been given elsewhere.[21,22] Typically, 10^7 cells are used per run because of the ease of handling during the PEG removal steps. Between 2 : 1 and 5 : 1 mixtures of lymphocytes to myelomas are collected in a pellet by centrifugation, and 1 ml of 35–50% PEG (M_r 1000–8000) in original medium is added. PEG is diluted by adding the original medium dropwise to the pellet, after 1–5 min of incubation. The cells are then collected by centrifugation and plated. Hybridoma survival rate is typically 10^{-6}.

Several direct comparisons between the PEG and electrofusion (EF)

[21] G. Kohler and C. Milstein, Eur. J. Immunol. 6, 511 (1976).
[22] R. L. Davidson and P. S. Gerald, in "Methods in Cell Biology," p. 325. Academic Press, New York, 1977.

methods have been made, using the same fusion partners.[23-25] Several parameters of such comparison experiments are given in Table I. From the data gathered so far, the efficiency of electrofusion varies from equivalent to two orders of magnitude better than that of PEG-induced fusion. The results are not surprising, in view of the many possibilities of fine tuning the electrofusion technique. For investigations in which high yield is not an important consideration, and the supply of lymphocytes is not restricted, PEG remains a simple and reliable method.

Conclusion

As the nature of electrofusion becomes better understood, improvements in the technique will continue to be made. Several modifications of the original protocol have been reported, leading to improvements in fusion yield and selectivity.[26] Electrofusion not only produces the highest hybridoma yield on record, its true strength lies in the capability to deal with a small number of cells, too small to be handled by other fusion methods. This opens the door to the production of human–human hybridomas, which must be made with a limited supply of lymphocytes from biopsy.[7] The method is also suitable for the production of hybridomas from a small number of preselected lymphocytes.

Electrofusion circumvents the difficulty of working with PEG-resistant cell lines, and it can also be applied to liposomes.[27] The precision timing of this fusion technique enables a number of biophysical experiments on fusion to be realized.[28] Although it will not entirely replace other fusion methods, electrofusion has already contributed much to the knowledge and application of cell fusion, and use of the method will certainly continue to grow.

Acknowledgments

This work was supported by Grant GM30969 from the National Institutes of Health. The assistance of R. Kubiniec is appreciated.

[23] G. van Duijn, J. P. M. Langedijk, M. de Boer, and J. M. Tager, *Exp. Cell Res.* **183,** 463 (1989).
[24] B. Seidel and H. Fiebig, *Stud. Biophys.* **130,** 197 (1989).
[25] U. Katenkamp, I. Schulmann, and I. Wolf, *Stud. Biophys.* **130,** 205 (1989).
[26] M. S. Lo and T. Y. Tsong, *in* "Electroporation and Electrofusion in Cell Biology" (E. Neumann, A. E. Sowers, and C. A. Jordan, eds.), p. 259. Plenum, New York, 1989.
[27] T. Teissie, M. P. Rols, and C. Blangero, *in* "Electroporation and Electrofusion in Cell Biology" (E. Neumann, A. E. Sowers, and C. A. Jordan, eds.), p. 203. Plenum, New York, 1989.
[28] A. Sowers, *Biophys. J.* **47,** 519 (1985).

[17] Polyethylene Glycol and Electric Field-Mediated Cell Fusion for Formation of Hybridomas

By Uwe Karsten, Peter Stolley, and Bertolt Seidel

Introduction

Since its first successful applications, electrofusion has been used increasingly for the formation of hybridomas.[1,2] Whereas the replacement of Sendai virus by polyethylene glycol (PEG) as a fusion-inducing agent was a straightforward process, it is not so obvious when using electric field impulses instead of PEG, since special technical equipment is required. This chapter compares the two techniques and discusses areas where electrofusion may be advantageously employed. In addition, it may serve as a practical guide to our relatively simple, broadly applicable variant of electrofusion, which has been used by two laboratories in experiments with comparable results.

Materials and Methods

The principles of both PEG-mediated and electric field-induced cell fusion have already been described in this series.[3,4] Many other sources of information are also available.[5-7] Because of differences in particular steps in the techniques which may quantitatively influence the outcome, the procedures are described in detail. Types of antigens, immunization protocols, and details of antibody assays, however, are not outlined.

Composition of Stock Solutions

2-Mercaptoethanol, 5 mM (100× stock solution) in water, stored at 20° and stable for 1 year

[1] M. M. S. Lo, T. Y. Tsong, M. K. Conrad, S. M. Strittmatter, L. D. Hester, and S. H. Snyder, *Nature (London)* **310**, 792 (1984).

[2] U. Karsten, G. Papsdorf, R. Roloff, P. Stolley, H. Abel, I. Walther, and H. Weiss, *Eur. J. Cancer Clin. Oncol.* **21**, 733 (1985).

[3] G. Galfrè and C. Milstein, this series, Vol. 73, p. 3.

[4] U. Zimmermann and H. B. Urnovitz, this series, Vol. 151, p. 194.

[5] J. H. Peters, H. Baumgarten, and M. Schulze, "Monoklonale Antikörper: Herstellung und Charakterisierung." Springer-Verlag, Berlin, 1985.

[6] A. H. Bartal and Y. Hirshaut, "Methods of Hybridoma Formation." Humana Press, Clifton, New Jersey, 1987.

[7] H. Friemel, "Immunologische Arbeitsmethoden," 4th Ed., Fischer, Jena, 1991.

L-Glutamine, 200 mM (100× stock solution) in water, stored at $-20°$ and stable for 1 year

Hypoxanthine, 5 mM (50× stock solution) in isotonic NaCl, dissolved at around 50°; stored in the dark at room temperature and stable for 6 months

Azaserine, 50 μg/ml (50× stock solution) in isotonic NaCl, stored at $-20°$ and stable for 1 year

Composition of Media

RPMI 1640 medium: Dry powder without HEPES is dissolved in water of tissue culture quality; after the addition of 2 g/liter NaHCO$_3$, the pH is adjusted to 7.2 with CO$_2$, and the medium is sterilized by membrane filtration. *Note:* If RPMI 1640 medium is stored for more than 1 week, L-glutamine is resupplemented from 100× stock solution.

Growth medium:

RPMI 1640	to 100%
Fetal calf serum (FCS), heat-inactivated	10%
2-Mercaptoethanol stock solution	1%
L-Glutamine stock solution	1%
(see note above)	

Preselection medium:

RPMI 1640 (without phenol red)	to 100%
FCS, heat-inactivated	20%
2-Mercaptoethanol stock solution	1%
L-Glutamine stock solution	1%
(see note above)	
Hypoxanthine stock solution	2%
Gentamicin (optional)	100 μg/ml

Selection medium: Same as preselection medium but with the addition of azaserine (2% of azaserine stock solution). *Note:* Instead of the preselection and selection media given, the better known HAT (hypoxanthine, aminopterin, and thymidine) selection system can be employed with similar results.[3]

Low ionic strength medium for electrofusion (BUW-8401 fusion medium[6]):

Inositol	280 mM
Calcium acetate	0.1 mM
Magnesium acetate	0.5 mM
Phosphate buffer, pH 7.0	1 mM

Hypo-osmolar fusion medium:[13]
Sorbitol	70 mM
Calcium acetate	0.1 mM
Magnesium acetate	0.5 mM
Bovine serum albumin	2 mg/ml

Postfusion medium:
NaCl	132 mM
KCl	8 mM
Calcium acetate	0.1 mM
Magnesium acetate	0.5 mM
Phosphate buffer (pH 7.0)	10 mM

Cell Cultivation: General Remarks

Good cell culture practice must include regular tests for the absence of mycoplasmas by DNA staining.[8] We recommend the use of pipetting devices which (1) can be autoclaved, and (2) are used exclusively for one single cell line or one fusion experiment at a time. Cell lines contaminated with mycoplasmas should be eliminated from the laboratory. CO_2 incubators are run at 36.5° and, in the case of RPMI 1640 medium without HEPES, 8% CO_2. The humidity should guarantee less than 5% loss by evaporation between medium changes. Feeding of cells should be done with prewarmed media. Cell cultures, especially myeloma cells, should be maintained for logarithmic growth. Most hybridomas require support by feeder cells for some time after fusion.

Preparation of Feeder Cells

1. An adult BALB/c mouse is sacrificed by cervical dislocation, placed on a board, and wiped with 70% (v/v) ethanol.
2. A sufficient area of abdominal skin is removed, leaving the peritoneum intact.
3. The peritoneal cavity is rinsed with 1–2 ml cold RPMI 1640 medium 3 times using a loosely fitting syringe.
4. Cells are collected in a siliconized glass tube on ice, spun down (150 g, 5 min, 2°), and resuspended in 5 ml of growth medium containing 100 μg/ml gentamicin. While still on ice, cells are counted in a hemocytometer. Mean yield is 6×10^6 peritoneal cells per mouse.
5. The cell suspension is diluted to a final density of 2×10^5 cells/ml and immediately distributed to tissue culture microtiter plates (100

---

[8] T. R. Chen, *Exp. Cell Res.* **104**, 255 (1977).

μl per well). *Note:* Microtiter plates with feeder cells can be prepared in advance and kept for 2–3 days in a CO_2 incubator. In this case, however, an additional 100 μl of growth medium should be added.

Preparation of Parental Cells

1. An immunized BALB/c mouse is sacrificed by cervical dislocation, placed on a board, and wiped with 70% ethanol.
2. The spleen is aseptically prepared and put into 5 ml of RPMI 1640 medium in a petri dish, where remaining fat or connective tissue can be removed. After changing to another dish with 10 ml of RPMI 1640 medium, the spleen is disrupted with a pair of special forks.[9] Alternatively, the spleen cell suspension may be prepared by a few strokes in a loosely fitting Dounce-type homogenizer.
3. The cell suspension is pipetted and filtered through nylon gauze to remove tissue fragments, and spun down. The supernatant is discarded and the cell pellet is suspended in 10 ml of RPMI 1640 medium. During a brief undisturbed period of 1–2 min aggregates form, which consist mainly of dead cells, and settle to the bottom of the tube. The supernatant containing the majority of living cells is carefully taken and counted. Average yield is 10^8 splenocytes. The spleen cell suspension may be kept at room temperature for at least 30 min.
4. Sufficient X63-Ag8.653 myeloma cells held in growth medium in the logarithmic growth phase and refed the day before are collected, spun down (150 g, 5 min, 2°), resuspended in RPMI 1640 medium, and counted.[10] The suspension should contain less than 10% dead cells.

Polyethylene Glycol-Induced Fusion

1. Preweighed portions of PEG (M_r 1500, Serva, Heidelberg, Germany) in 50-ml Erlenmeyer flasks (2.10 and 1.25 g, respectively, closed with tight cotton plugs and aluminum foil caps) are autoclaved at 100 kPa for 15 min. While still warm, 2.90 and 3.75 ml of RPMI 1640 medium, respectively, are added and thoroughly mixed to give 42 and 25% (w/v) solutions of PEG.
2. Aliquots containing 2×10^7 spleen cells and 2×10^6 myeloma cells are mixed in a 25-ml round-bottomed tube, spun down (150 g, 5 min, 2°), resuspended in about 10 ml of RPMI 1640 medium, and

[9] T. Kaspi, *Experientia* **31**, 386 (1975).
[10] J. F. Kearney, A. Radbruch, B. Liesegang, and K. Rajewsky, *J. Immunol.* **123**, 1547 (1979).

centrifuged again. About 9 ml of the supernatant is removed, and the tube is placed in a water-jacketed heating device, held at 37°.

3. After temperature equilibration, the remaining supernatant is thoroughly and carefully withdrawn from the cell pellet. Then 0.5 ml of 42% PEG is added in drops to the pellet, with gentle agitation during a 1-min period, followed by the addition of 0.5 ml of 25% PEG during a 2-min period and a 4 ml of RPMI 1640 medium within a further 3-min period. These steps are performed while the reaction tube is still in the heating device.

4. The fused cells are carefully transferred to 17 ml of prewarmed preselection medium and kept in the CO_2 incubator for 4 h.

5. Azaserine (0.4 ml of stock solution) is added, carefully mixed, and the cell suspension plated in two microtiter plates (100 μl per well) together with previously prepared peritoneal feeder cells. The number of spleen cells per well amounts to approximately 0.9×10^5. *Note:* In a number of fusions, a slightly different fusion protocol was followed, using only one PEG solution of 50% (w/v, M_r 4000, gas chromatography quality, Merck, Darmstadt, Germany).

Electrofusion Equipment

Several instruments suitable for electrofusion are available commercially. The apparatus developed and built at the institute was designed to provide (1) high frequency voltage for dielectrophoresis, adjustable in both frequency (0.5, 1.0, or 2.0 MHz) and amplitude (0 to 40 V), and (2) single square pulses suited for membrane breakdown, variable in height (up to 300 V) and duration (1–15 μsec). This powerful instrument allows fusion chambers with relatively wide electrode distances (0.5 mm or more) to be operated. The simple construction of the fusion chamber (Fig. 1),[11] with all parts (except the insulating ring) made of stainless steel, provides for easy handling, cleaning, and sterilization. All electrofusion experiments mentioned in this chapter are performed with a chamber 28 mm in diameter and 0.5 mm in height. Other types also have been used successfully.

Electrofusion Procedure

1. Erythrocytes are removed from the parental spleen cell suspension by Ficoll–Visotrast (or equivalent) density gradient centrifugation. Four milliliters of spleen cell suspension (containing about $5–10 \times 10^6$ splenocytes/ml) are carefully layered on top of 4 ml of Ficoll–

[11] U. Karsten, P. Stolley, I. Walther, G. Papsdorf, S. Weber, K. Conrad, L. Pasternak, and J. Kopp, *Hybridoma* 7, 627 (1988).

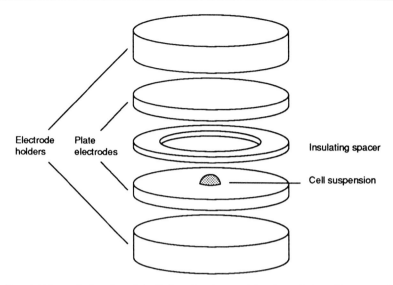

FIG. 1. Schematic drawing of the fusion chamber used for electrofusion. The proportions are as follows: outer diameter, 50 mm; inner diameter of the insulating spacer, 28 mm; thickness of the spacer, 0.5 mm. For a visual impression of the device, see Karsten et al.[11]

Visotrast solution with a specific density of 1.083 in a 10-ml centrifuge tube. The gradient is centrifuged at 400 g for 30 min at room temperature.

2. The layer of white cells is carefully removed with a Pasteur pipette, washed twice with 5 ml phosphate-buffered saline (PBS), resuspended in RPMI 1640 medium, and counted.

3. Samples of myeloma parental cells containing 10^6 cells are spun down in conical centrifuge tubes, treated for 30 sec with pronase (0.05 mg/ml in PBS), immediately spun down (300 g, 45 sec), and then resuspended in 300 μl of low ionic strength medium. *Note:* Results show that step 3 can be omitted.[12]

4. In a conical centrifuge tube 10^6 myeloma cells are mixed with 2×10^6 splenocytes, spun down, and washed 3 times with 300 μl each of low ionic strength medium. Between washings, cells are spun at 300 g for 45 sec, and the supernatant is removed very carefully. During these manipulations, an estimated 30% of the cells may be lost.

5. Finally, the cells are resuspended in 20 μl of low ionic strength medium and transferred to the fusion chamber as a small droplet at

[12] J. J. Schmitt, U. Zimmermann, and G. A. Neil, *Hybridoma* **8**, 107 (1989).

the center of the lower electrode (see Fig. 1) with the insulating ring already in place. Then the upper electrode is put in place, and the remaining parts are assembled.

6. Electric fields are applied as follows:
   a. Dielectrophoresis for cell alignment is applied for 10 sec (alternating field, 2 MHz, 500 V/cm).
   b. While dielectrophoresis is interrupted, a single square pulse of 5000 V/cm, 4 μsec, is applied.
   c. Dielectrophoresis is continued as above for 5 sec.
7. After a 1-min recovery period, the fusion chamber is opened. Approximately 200 μl of preselection medium is added to the fusion droplet, and the mixture is transferred into 1 ml of preselection medium. It is advisable to examine a small aliquot of the cell suspension under the microscope immediately after this transfer to determine overall viability and the presence of fusion products. Because the whole procedure takes only about 12 min, it is possible to perform a series of electrofusions using a single parental cell source. Only the electrode disks need to be replaced by a new pair of sterile disks.
8. The cells are kept in a CO_2 incubator for 3–4 hr.
9. Then 9 ml of prewarmed selection medium is added, and the suspension is distributed to 96 wells (100 μl each) of a microtiter plate with feeder cells. The number of spleen cells per well is about 1.5×10^4.

It has been shown that the use of a hypertonic fusion medium leads to even higher hybridoma yields.[13] An additional advantage is that red cells are lysed in this medium, and therefore the density gradient serving this purpose (steps 1–3) can be omitted. We recommend the following procedure: (1) Start with step 4, using hypo-osmolar medium instead of low ionic strength medium (steps 4 and 5). (2) Reduce the field strength by 40% by applying 300 V/cm and 3000 V/cm, respectively, in step 6. (3) In step 7 (after fusion), add 100 μl of postfusion medium (instead of 200 μl of preselection medium), and keep the mixture in the CO_2 incubator for 30 min. (4) Thereafter, add 1 ml of preselection medium and follow steps 8–9 as described.

Further Treatment

Further steps are identical in both types of fusion. On days 1, 2, 3, 6, and 8, 100 μl of medium per well is withdrawn and replaced by fresh

---

[13] J. J. Schmitt and U. Zimmermann, *Biochim. Biophys. Acta* **983,** 42 (1989).

TABLE I
ELECTRIC FIELD-MEDIATED AND POLYETHYLENE GLYCOL-MEDIATED CELL FUSIONS

Parameter	Electrofusion	PEG fusion
Number of experiments	36	36
Hybridoma yield[a]		
Overall range	0.1–13.3	0.01–1.4
Mean[a,b]	2.46 ± 2.55[c]	0.29 ± 0.37[c]
Mean ratio[d]	15.7 ± 12.4[c]	

[a] Number of hybridoma colonies per 10^5 spleen cells.
[b] Difference is statistically highly significant ($P < 1\%$).
[c] Mean ± standard deviation.
[d] Mean of individually calculated ratios of hybridoma yields (electrofusion/PEG fusion).

selection medium. Thereafter, medium is changed 3 times a week; during the first week azaserine is omitted, and thereafter both azaserine and hypoxanthine are omitted. Hybridoma growth is evaluated and colonies are counted from day 6 onward. Hybridoma yield is calculated as the number of colonies growing 10–14 days after fusion from 10^5 spleen cells. Antibody tests are performed as soon as possible using different methods according to the type of antigen and the purpose of the fusion experiment, keeping in mind the assay specificity of monoclonal antibodies. In the experiments summarized in Table I, enzyme-linked immunosorbent assays (ELISA), radioimmunoassays (RIA), immunocytochemistry, flow cytometry [fluorescence-activated cell sorting (FACS) analysis], and hemagglutination assays have been employed. Further manipulations of hybridomas including cloning, freezing, adaptation to growth without feeder cells, adaptation to medium with lower serum content, and scaling up are presented elsewhere.[3,5–7]

Results and Conclusions

In contrast to other modes of electrofusion, the batch-type variant is relatively simple. It is applicable to soluble as well as to cellular antigens, and it allows comparison with PEG-induced fusions starting from the same parental cell suspensions. Table I and Fig. 2 summarize the results of 36 experiments, in which both fusion techniques were compared. The accumulated data confirm that electrofusion results in higher hybridoma yield as compared to PEG-induced fusion.[2,11] On the average, 5 to 10 times more hybridomas per unit number of spleen cells arise from electrofusion. This holds true for a great variety of experimental settings, such as type of

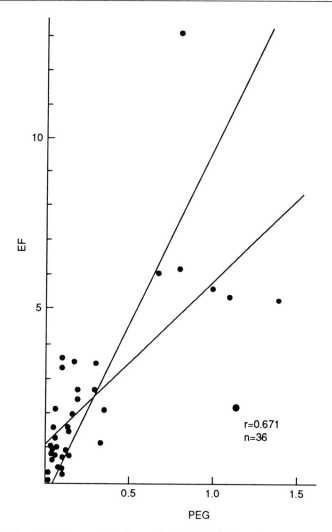

FIG. 2. Correlation between hybridoma yields of PEG-induced fusion versus electrofusion. Each point represents one independent comparative experiment. Both regression lines (y versus x and x versus y) and the correlation coefficient (r) were calculated; r is statistically highly significant ($p = 0.05$). n, Number of experiments.

antigen (tumor cells, leukocytes, immunoglobulins, haptens, enzymes, hormones, etc.); species (mouse or rat); type of serum (fetal calf or horse); source of PEG; type of selection medium (azaserine versus HAT); pretreatment of myeloma cells with pronase; or (limited) modification in the ratio of spleen cells to myeloma cells.[11] Figure 2 demonstrates the correla-

TABLE II
ADVANTAGES AND DISADVANTAGES OF ELECTRIC FIELD-MEDIATED AND POLYETHYLENE
GLYCOL-MEDIATED FUSION TECHNIQUES[a]

Technique	Advantages	Disadvantages
PEG fusion	Inexpensive Easy to perform	Only selected batches of PEG suitable Toxic if not properly handled Conditions poorly defined
Electrofusion	Hybridoma yield 5–10 times higher Small numbers of lymphocytes can be fused Hybridomas grow more vigorously immediately after fusion Defined fusion parameters allow further improvement	Special equipment necessary Special fusion medium necessary

[a] The conclusions summarized here are in agreement with those of other groups.[14–17]

tion between both techniques with respect to the hybridoma yield of individual experiments. There are "good" and "bad" experiments independent of the fusion technique applied. It follows that overall variability is mainly due to factors such as experimental design, type of antigen, as well as the immune response of the animal.

Another advantage of electrofusion is a more vigorous growth of hybridomas immediately after fusion. This is visually very obvious, although the reason for it is not known. All other parameters examined (e.g., isotype distribution, antibody production, and transplantability) do not reveal any difference between hybridomas or their products generated by either method. The percentage of hybridomas secreting specific antibodies is no lower in electrofusions than in PEG-induced fusions.[11] This means that electrofusion leads to an increase in hydridoma yield in real terms.

Electrofusion is applicable to small numbers of lymphocytes, which may be essential in production of human monoclonal antibodies or in fusions with selected (e.g., antigen-specific) lymphocytes. Finally, fusion parameters are better defined in electrofusion and can therefore be further optimized or adapted for specific purposes if necessary (see Table II).

[14] K. Ohnishi, J. Chiba, Y. Goto, and T. Tokunaga, J. Imm. Meth. 100, 181 (1987).
[15] G. van Duijn, J. P. M. Langedijk, M. de Boer, and J. M. Tager, Exp. Cell Res. 183, 463 (1989).
[16] S. K. H. Foung and S. Perkins, J. Imm. Meth. 116, 117 (1989).
[17] M. Tomita and T. Y. Tsong, Biochim. Biophys. Acta 1055, 199 (1990).

Acknowledgments

We appreciate the excellent technical assistance of M. Haase, M. Kiefer, and I. Walther, and the essential contributions provided by K. Degenhardt and A. Kaarz in the construction of the apparatus.

[18] Selective B Lymphocyte – Myeloma Cell Fusion

By TIAN YOW TSONG and MASAHIRO TOMITA

Introduction

Although electrically induced cell fusion was observed in the work of H. Pohl in the early 1970s, its potential applications were not clearly recognized. Later, Zimmermann and Pilwat reported fusion of cells with intense electric pulses.[1] This early work stimulated the systematic investigation of the electrofusion of protoplasts, *Dictyostelium discoideum,* and cultured fibroblasts.[2-4] Electrofusion has since been applied to agricultural, molecular biological, and immunological research.[5,6]

Most experiments utilize the dielectrophoresis method[5-7] to congregate cells before fusion-inducing electric pulses are applied. However, the formation of cell aggregates by dielectrophoresis depends on cell polarization and the physical characteristics of the cell membranes. Selection of cells according to their biological ability is not possible by this method. This chapter presents some designs for fusing myeloma cells to lymphocytes that are destined to secrete antibodies against a specific antigen.[8-11] The

[1] U. Zimmermann and G. Pilwat, *Sixth Int. Biophys. Congr., Kyoto, Abstr.* **VI-19**(H), 140 (1978).

[2] M. Senda, J. Takeda, A. Shunnosuke, and T. Nakamura, *Plant Cell Physiol.* **20,** 1441 (1979).

[3] E. Neumann, G. Gerisch, and K. Opatz, *Naturwissenschaften* **67,** 414 (1980).

[4] J. Teissie, V. P. Knutson, T. Y. Tsong, and M. D. Lane, *Science* **216,** 537 (1981).

[5] E. Neumann, A. E. Sowers, and C. A. Jordan, "Electroporation and Electrofusion in Cell Biology." Plenum, New York, 1989.

[6] A. E. Sowers, "Cell Fusion." Plenum, New York, 1987.

[7] H. A. Pohl, "Dielectrophoresis." Cambridge Univ. Press, Cambridge, 1978.

[8] M. M. S. Lo, T. Y. Tsong, M. K. Conrad, S. M. Strittmatter, L. D. Hester, and S. Snyder, *Nature (London)* **310,** 792 (1984).

[9] T. Y. Tsong, M. Tomita, and M. M. S. Lo, *in* "Molecular Mechanisms of Membrane Fusion" (S. Ohki, D. Doyle, T. D. Flanagan, S. W. Hui, and E. Mayhew, eds.), p. 223. Plenum, New York, 1988.

selection of a particular cell population for electrofusion is achieved by membrane receptor–ligand recognition. Although the design is for hybridoma production, the concept introduced here would be equally applicable to other areas where targeting of cells or tissues for chemical treatment is desired. The general procedures for electrofusion are discussed elsewhere in this volume.[12,13] The reader should refer to other sources for hybridoma technology and the preparation of monoclonal antibodies using conventional nonselective cell fusion procedures, for example, the PEG (polyethylene glycol) method[14-16] or electrofusion using the dielectrophoresis method.[12,13]

Avidin–Biotin Bridging Method

Approximately 10^8 lymphocytes in the immune system can produce as many as 10^{10} antibodies having different specificities. The presence of an antigen stimulates B cells to proliferate.[17] Only these B cells have the ability to secrete high-affinity monoclonal antibodies against the antigen. The hybridomas resulting from the fusion of such B cells and myeloma cells continue to divide and make monoclonal antibodies. The avidin–biotin bridging method preselects these B cells from the general population to fuse with myeloma cells using the electrofusion method.[8-11] The method purposely avoids nonproductive fusion by preventing the fusion of irrelevant B cells. Although the efficiency of fusion may appear to be low, when the experiments are performed properly, each hybridoma harvested should secrete monoclonal antibody. Thus, the need for hybridoma screening, which is required when using the conventional PEG fusion method or electrofusion using dielectrophoresis, is eliminated. The high selectivity of the method has been demonstrated in several cases.[8] In some experiments, however, the selectivity has been compromised by the nonspecific binding of avidin to the cell surface. Further improvement of the method is required. However, the avidin–biotin selection method and its variations provide alternatives and a greater chance of success where conventional approaches fail.

[10] M. K. Conrad, M. M. S. Lo, T. Y. Tsong, and S. Snyder, in "Cell Fusion" (A. E. Sowers, ed), Plenum, New York, 1987.
[11] M. M. S. Lo and T. Y. Tsong, in "Electroporation and Electrofusion in Cell Biology" (E. Neumann, A. E. Sowers, and C. A. Jordan, eds.), p. 259. Plenum, New York, 1989.
[12] G. A. Neil and U. Zimmermann, this volume [14].
[13] A. E. Sowers, this volume [15].
[14] G. Galfre and C. Milstein, this series, Vol. 73, p. 1.
[15] S. W. Hui and D. A. Stenger, this volume [16].
[16] U. Karsten, P. Stolley, and B. Seidel, this volume [17].
[17] E. B. Golub, Cell (Cambridge, Mass.) 48, 723 (1987).

Basic Design

Antigen is used to recognize the desired B cells, which are then selectively conjugated to myeloma cells for electrofusion. Five steps are involved in the original design: (1) chemically cross-linking avidin to the antigen; (2) mixing the avidin–antigen complex with B cells (the avidin–antigen complex will preferentially bind to the desired B cells through antigen–receptor interaction); (3) biotinylation of myeloma cells; (4) mixing the B cell preparation with the biotinylated myeloma (only B cells with membrane-bound avidin–antigen complexes will pair with the myeloma cells through avidin–biotin interactions); and (5) electrofusion of the mixed cells (only B cell–myeloma pairs established via avidin–biotin complexing have a high probability of membrane fusion). The hybridomas formed are expected to be antibody-producing cells. Figure 1 schematically summarizes these procedures.

Chemical Cross-Linking of Avidin to Antigen

Each avidin molecule has four biotin binding sites. When using a chemical to cross-link avidin to an antigen, these binding sites may be modified and lose their affinity for biotin. To achieve a high specificity, therefore, at least one of the four sites needs to be preserved. The cross-linking reaction is performed in an iminobiotin–sepharose column (Fig. 1). One micromole of avidin is mixed with 0.2 ml of iminobiotin–sepharose gel containing 10 μmol iminobiotin/ml gel. After 2 hr of incubation, a 100-fold excess of 1,5-difluoro-2,4-dinitrobenzene (DFDNB) is added to react with the immobilized avidin, at pH 10.5, for 10 min, wherein the bound avidin becomes chemically activated. One-tenth milliliter of this mixture is packed into a glass column, and the resin is thoroughly washed with 10 ml 0.1 M NaCl, 50 mM borate buffer, pH 8.5. One microgram of antigen in the same buffer is added to the column and reacts with DFDNB. After 10 min the column is washed with either glycine or lysine to block unreacted DFDNB sites. The antigen is then cross-linked to the immobilized avidin. The column is washed with the borate buffer at pH 8.5 until no protein is detected in the wash. The antigen–avidin complex is then washed off the column using a buffer of pH 3.5. At this pH, the affinity between avidin and iminobiotin is greatly reduced, and the antigen–avidin complex, which conserves at least one biotin binding site, is collected.

Biotinylation of Myeloma Cells

Myeloma cells, either P3x63Ag8.653 or FOX-NY (10^7–10^8 cells), are collected and washed with phosphate-buffered saline (PBS) and gentami-

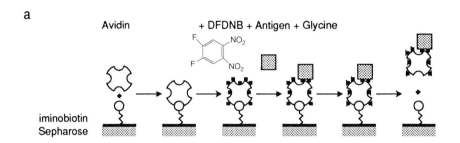

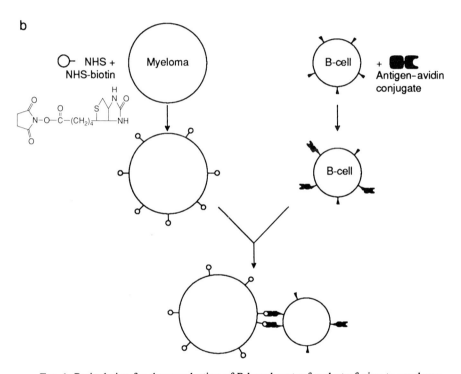

FIG. 1. Basic design for the preselection of B lymphocytes for electrofusion to myeloma cells using the avidin–biotin bridge. (a) An iminobiotin–Sepharose column is used to cross-link an antigen to avidin. (b) The surface of myeloma cells is coated with biotin, using NHS-biotin, and the surface of relevant B cells is coated with the avidin–antigen conjugate through binding of the antigen moiety to the membrane receptor. When the two populations of cells are mixed, only relevant B cells adhere to myelomas to form cell doublets. Subsequent electrofusion of the mixture will produce only hybridomas that are destined to secrete monoclonal antibodies against the antigen used to select the B cells. See text for details. [After M. M. S. Lo, T. Y. Tsong, M. K. Conrad, S. M. Strittmatter, L. D. Hester, and S. Snyder, *Nature (London)* **310**, 792 (1984).

cin (10–20 μg/ml). Twenty microliters of 0.1 M N-hydroxysuccinimide biotin (NHS-biotin) in dimethylformamide (DMF) is carefully added to 10 ml of a myeloma cell suspension. The mixture is kept at room temperature for 15–30 min. Then the myeloma–biotin conjugate is separated from unreacted NHS-biotin by overlaying the mixture on 10 ml fetal calf serum (FCS) and centrifuging at 400 g for 10 min at room temperature. The cells are washed once with 30–40 ml of Dulbecco's modified Eagle's medium (DMEM) containing 10 μg/ml of DNase I and gentamicin (10–20 μg/ml) before being used for electrofusion.

Fusion of B Cell–Myeloma Conjugates by Pulsed Electric Field Method

A suitable amount of antigen–avidin complex is mixed with B cells in DMEM containing gentamicin (10–20 μg/ml). After 20 min of incubation, the suspension is combined with biotinylated myeloma cells in the same medium. The mixture is spun at 400 g for 5–10 min and resuspended in 1 ml of the same medium. The suspension is kept at room temperature for 30 min with gentle stirring. Then the mixture is carefully overlaid on 10 ml of sterilized isotonic sucrose (320 mOsm) containing 2 mM sodium phosphate buffer, pH 7.2, and spun at 400 g for 5–10 min. The pellet, containing B cell–antigen–avidin–biotin–myeloma conjugates, is gently suspended in 1 ml of the isotonic sucrose buffer.

Electrofusion of the cell mixtures is carried out with a Cober 605P high-voltage generator (Cober Electronics, Stamford, CT). Model 605P can deliver up to 2.2 kV, with a maximum power output of 24 kW. This setup has been described in earlier publications.[18–20] There are many other commercial instruments with comparable ability, and any of them should work as well. Basically, the suspension of myeloma and B cells is placed between two platinum electrodes 2–3 mm apart which are enclosed in a cylindrical Plexiglas container. Typically 10^7 myeloma and 10^7 B cells are placed in a sample volume of 0.1–0.2 ml. The temperature of the chamber is controlled by circulating water. For electrofusion of myeloma and B cells, we use five square wave pulses of 2.5 kV/cm and 5 μsec duration, at 35°. Intervals of 15 sec are given between pulses to allow dissipation of Joule heat. The cells are then incubated at 37° for 15 min and plated at about 10^5 cells per well into plates seeded with about 10^4 murine peritoneal macrophages per well. The medium contains aminopterin to select for hybridomas. Wells with growing colonies are screened for antibody production by standard procedures.

[18] K. Kinosita and T. Y. Tsong, *Proc. Natl. Acad. Sci. U.S.A.* **74**, 1923 (1977).
[19] T. Y. Tsong, this series, Vol. 149, p. 248.
[20] T. Y. Tsong, this series, Vol. 157, p. 240.

Assessment of the Method

Preparations of monoclonal antibodies against angiotensin-converting enzyme, enkephalin convertase, bradykinin, mitochondrial F_0F_1-ATPase, and other protein antigens have been attempted using the above avidin–biotin bridging method, with mixed results. In the case of rat lung angiotensin-converting enzyme, two different immunizations resulted in 42 wells containing growing cells from a total of 216 wells. All of the 42 active wells produced monoclonal antibodies against angiotensin-converting enzyme. Subclass-specific antisera were used to determine the immunoglobulin subtypes of 31 monoclonal antibodies. Most of these antibodies belonged to the IgG_{2a} or IgG_{2b} class; exceptions found were two IgM, one IgG_1, and one IgG_3.

An experiment with enkephalin convertase as antigen was also satisfactory. Four wells containing growing cells were obtained out of 72 wells. All 4 wells had antibody activity against enkephalin convertase.

To raise antibodies against bradykinin, bradykinin was first conjugated to human albumin before mixing with Freund's complete adjuvant for immunization of mice. After electrofusion by this procedure, 7 wells containing growing cells were obtained from a total of 120 wells. All 7 wells demonstrated antibody activity against bradykinin as determined by radioimmunoassay using methionyllysylbradykinin labeled with [125]I.

Other protein antigens, such as F_0F_1-ATPase of beef heart mitochondria and cAMP receptor protein of *Dictyostelium discoideum*, were also tested but with less satisfactory results. In general, the frequency of active hybridoma produced by using the avidin–biotin bridging method was higher than that using the PEG method. In many experiments, however, hybridoma screening was still required, and the advantage of specific selection was thus compromised.

Variations of Avidin–Biotin Bridging Method

Improvements for Avidin–Biotin Method

There are some technical problems which make the avidin–biotin bridging method less than ideal. Avidin tends to bind nonspecifically to hydrophobic surfaces. We have found that this tendency of avidin can cause difficulties in at least two steps of the bridging method. First, the recovery of antigen–avidin conjugates from the iminobiotin–Sepharose column has not been consistent. The usually minute quantity of a conjugate is easily lost through nonspecific binding to Sepharose beads or to the glass wall of the column. Second, when an antigen–avidin conjugate is added to B cells, the conjugate does not always bind to the antigen receptor. Nonspecific interaction of the avidin moiety with cell membranes often reduces the efficiency of cell selection and forfeits the purpose of the design. These problems may be circumvented to a certain extent by replacing avidin with streptavidin. Streptavidin has an isoelectric point near neutral pH and does not bind nonspecifically to hydrophobic surfaces. We have also explored other variations of the avidin–biotin bridge, and some of them are discussed here.

Antigen–Biotin–Streptavidin–Biotin Bridge

The solid-phase cross-linking of antigen to avidin, namely, the step that uses the iminobiotin–Sepharose column, is a complicated procedure unsuitable for routine laboratory use. To avoid such a procedure, an antigen is cross-linked to biotin in solution, and the antigen–biotin conjugate is then used to form a bridge between a B cell and a myeloma cell. The complete chain is B cell–antigen–biotin–streptavidin–biotin–myeloma. An example using the F_0 subunit of beef heart mitochondrial ATPase as an antigen is discussed below.

Two hundred microliters of F_0 solubilized in 0.05% sodium cholate, 0.1 M NaHCO$_3$ (pH 8.2), at a concentration 0.4 mg/ml, is mixed with a 20-fold molar excess of NHS-biotin [8 μl of NHS-biotin in dimethyl sulfoxide (DMSO)] over protein and reacted at room temperature for 2 hr with stirring. The concentration of DMSO in the mixture is approximately 4% (v/v). Then an equal volume (8 μl) of 1 M NH$_4$Cl is added to the mixture, and the reaction is allowed to continue for another 10 min. At the end of the reaction, 200 μl of PBS plus 1% bovine serum albumin (BSA) is added to the mixture, and the mixture is dialyzed against PBS at 4° overnight. The antigen–biotin conjugate solution is stored frozen until use.

Preparation of the B cell–antigen–biotin–streptavidin conjugate is as follows. Twenty-five micrograms of F_0-biotin diluted with DMEM plus 1% BSA is carefully added to the spleen cell suspension in 5 ml of DMEM, containing 10 μg/ml DNase I and 10–20 μg/ml gentamicin, and gently mixed by stirring at 4° for 2 hr. The B cell–antigen–biotin conjugate is obtained by overlaying the cell suspension on an appropriate volume (5–10 ml) of a mixture of equal volumes of the above medium and isotonic sucrose. The suspension is centrifuged at 400 g for 10 min at room temperature. Streptavidin (62.5 μg) in DMEM plus 1% BSA is carefully added to the centrifuge tube containing 2.5 ml of the cell suspension, and the contents are mixed by topping the tube gently at 4° for 30 min. In this case, a 5- to 10-fold molar excess of streptavidin is used against antigen in order to avoid intermolecular nonspecific binding between B cell–antigen–biotin complexes. The final conjugate of B cell–antigen–biotin–

streptavidin is washed once with the above medium and kept at 37° for 20–30 min before electrofusion is carried out.

Use of Bifunctional Chemical Cross-Linkers

An antigen may be cross-linked to avidin using bifunctional chemical cross-linkers, such as *m*-maleimidobenzoyl-*N*-hydroxysuccinimide (MBS) and *N*-succinimidyl-3-(2-pyridyldithio)propionate (SPDP), which react with both NH$_2$ and SH groups. Avidin is reacted with a 5-fold excess of either SPDP or MBS in PBS for 1 hr at room temperature. The mixture is then chromatographed on a Sephadex G-25 column to remove unreacted cross-linker. The SPDP- or MBS-avidin is stored at −70° before use.

To cross-link an antigen to SPDP- or MBS-avidin, the antigen is first reduced with 10 m*M* dithiothreitol for 15 min. The reduced antigen is then added to SPDP- or MBS-avidin and reacted for 20 hr. Cysteine is added to block any remaining unreacted maleimide groups. The antigen–SPDP–avidin conjugate prepared in this way is used without further purification.

Myeloma–Avidin Conjugates

Avidin may be directly conjugated to myeloma cells. Myeloma cells (P3x63Ag8.653) grown in complete medium containing 100 μ*M* 2-mercaptoethanol are collected by centrifuging at 400 *g* for 10 min at room temperature, washed once with PBS plus 100 μ*M* 2-mercaptoethanol, and resuspended in 5 ml PBS (cell concentration 1–5 × 10^6/ml). SPDP-SH

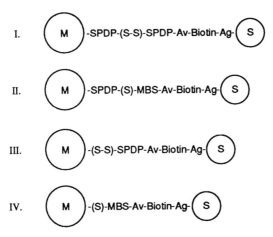

FIG. 2. Variations of avidin–biotin bridges discussed in the text. M designates a myeloma cell; S, spleen cell; Ag, antigen; Av, avidin molecule; SH, sulfhydryl group; -S-, one-half of a disulfide link; -S-S-, disulfide link; SPDP, *N*-succinimidyl-3-(2-pyridyldithio)propionate; MBS, *m*-maleimidobenzoyl-*N*-hydroxysuccinimide.

(reduced SPDP) is immediately added, to a final concentration of 125 μM, and the mixture is incubated at room temperature for 15–30 min. SPDP-SH is prepared by reacting SPDP in DMF (0.1 M) with an equal volume of 0.1 M dithiothreitol, also in DMF, for 5–10 min at room temperature. The mixture containing myeloma–SPDP-SH conjugate is collected by centrifuging at 400 g. The pellet containing myeloma–SPDP-SH is resuspended in 9 ml PBS for use. The procedure using MBS is similar. Four types of chemically modified myeloma cells are obtained: myeloma–SPDP–SPDP–avidin, myeloma–SPDP–MBS–avidin, myeloma–SPDP–avidin, and myeloma–MBS–avidin. Each would form a bridge with the B cell–antigen–biotin conjugate by a biotin–avidin linkage. The length of the bridges is different with each type, and the efficiency of electrofusion is also different. Several bridges based on the high affinity of avidin–biotin binding and the recognition of an antigen by relevant B cells are shown in Fig. 2.

Perspective

Cell surface-specific selections discussed here have certain advantages over nonselective hybridoma preparation methods. The most obvious is the elimination of the antibody screening step for hybridomas. Another advantage is the potential of preparing monoclonal antibodies of high affinity. By adjusting the concentration of antigen–avidin or antigen–biotin complexes and B cells in solution, it is possible to choose a range of affinities for antibodies to be prepared. In experiments with angiotensin-converting enzyme, monoclonal antibodies with affinities in the range of 10^8 to 10^{11} M^{-1} have been prepared. A third advantage is to shorten the immunization time. In the examples given above, the immunization time was much shorter than that generally required using the PEG method.

There are other designs to link a myeloma cell with a B cell through the specific binding of an antigen with its receptor on the B cell. Use of the Fab fragment of antibodies against myeloma cell surface antigens represents another approach. An antigen may be linked to such an Fab by various chemical cross-linkers. The Fab moiety would bind to the myeloma cell, and the antigen moiety would bind to the B cell. Systematic studies are needed to assess such a design or other designs.

Acknowledgment

We thank our collaborator, Dr. M. M. S. Lo, for contributions to this project.

Section III

Fusion of Viruses with Target Membranes

[19] Biochemical and Morphological Assays of Virus Entry

By MARK MARSH

Introduction

A detailed understanding of the cellular and molecular mechanisms involved in enveloped virus entry has emerged. The experiments have been conducted with several well-characterized enveloped viruses, namely, Semliki Forest virus (SFV, an alphavirus), influenza virus (an orthomyxovirus), and vesicular stomatitis virus (VSV, a rhabdovirus).[1] In this chapter, the methods that we have used to study the entry of SFV are described. Many of these methods, especially the ones pertaining to binding and endocytosis, may be adapted for use with other enveloped and nonenveloped viruses. Before discussing the techniques in detail, the pathway of SFV entry, as understood at the present time, is described briefly.

Pathway of Virus Entry

SFV is a relatively simple enveloped virus. Each virion contains a single strand of positive-sense RNA, four structural polypeptides, and a bilayer membrane. One of the polypeptides, the capsid protein, combines with the viral RNA to form the nucleocapsid. During assembly of new virions in infected cells, the nucleocapsid particles bud from the plasma membrane and in so doing acquire a lipid membrane, or envelope. In addition to lipid, the envelope contains multiple copies of the viral envelope or spike glycoprotein. These spike glycoproteins are assembled from the other three viral polypeptides; two of which are transmembrane glycoproteins (E1 and E2), while the third is an external peripheral protein (E3). Although SFV is structurally simpler than many other viruses, this basic organization of nucleocapsid, membrane, and spike protein complex is a common feature of the enveloped viruses.

During the entry of an enveloped virus into a cell, the nucleocapsid must be transferred from within the viral membrane to the cytosol of the target cell. Following binding to cell surface receptors, SFV virions are relocated in the plasma membrane to coated pits and are internalized in coated vesicles. These events are not virally induced, but occur as part of the constitutive cellular endocytic activity. Coated vesicles deliver the

---

[1] M. Marsh and A. Helenius, *Adv. Virus Res.* **36**, 107 (1989).

internalized virions to larger tubulovesicular structures termed endosomes. These organelles are acidic (pH ≤ 6) and are normally involved in endocytic membrane trafficking. Although the virions at this stage are within the confines of the cell, the viral genome remains within both the viral and endosomal membranes and outside the cytoplasm. However, the acidic environment inside endosomes triggers a conformational change in the spike glycoprotein that initiates the fusion reaction. As a consequence the spike glycoproteins interact with the membrane of the endosome and induce fusion between the viral and endosomal membranes. This fusion brings about penetration (i.e., the entry of the viral nucleocapsid into the cytoplasm) and the onset of viral replication. For SFV all of the reactions involved in replicating the viral RNA occur in the cytoplasm of the infected cells. The initial RNA synthesis can be detected within 1–2 hr of infection, protein synthesis occurs in 2–3 hr, and progeny virus is released 4–5 hr after penetration.

The fusion reaction for SFV is absolutely dependent on exposure of the virions to low pH. Under normal circumstances the extracellular medium will be near neutral pH. Thus, fusion cannot occur at the plasma membrane. Acid pH also provides a trigger for the fusion of a number of other viruses (e.g., influenza and VSV). However, low pH is not a universal viral fusion trigger, and it is now quite clear that paramyxoviruses, such as Sendai virus, and many of the retroviruses, including the human immunodeficiency viruses (HIV-1 and -2), do not require exposure to acid pH for fusion. Whether these pH-independent viruses require an alternative fusion trigger is unknown at present. However, the fact that fusion is not well understood for all enveloped viruses means that care should be taken when studying less well-characterized viruses. For example, fusion leading to productive infection may occur at the cell surface, and the assays described below may not distinguish between cell surface fusion and endocytosis (Fig. 1). Thus it is important to utilize different approaches to address each specific question, and to clearly establish what each assay measures. At least in part, the successful analysis of SFV entry derives from the fact that a combination of biochemistry, morphology, virology, and immunology has been used to address the problems.

Preparation of Virus

In general, purified virus is preferable to virus containing tissue culture supernatants. For SFV either radiolabeled ([^{35}S]methionine, [^{3}H]uridine, or ^{32}P) or unlabeled virus can be purified by differential and gradient

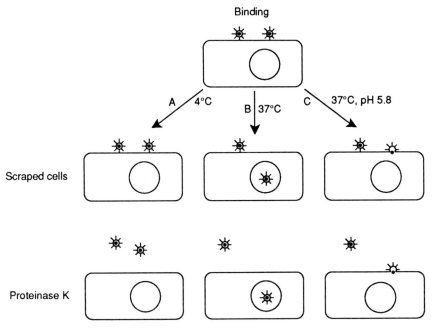

FIG. 1. Diagrammatic representation of the use of proteinase K to distinguish between cell surface-bound, endocytosed, and cell surface-fused virus. (A) When binding is carried out at 4°, all the virus remains on the cell surface and is removed by proteinase K at 4°. (B) After warming to 37° some of the cell surface-bound virus is internalized. The virus remaining on the plasma membrane is still accessible to proteinase K and is removed at 4°. The internalized virus is resistant to proteinase K elution. (C) Following incubation at 37° in mildly acid medium, some of the bound virus is fused at the cell surface. The fused virus is resistant to proteinase K treatment at 4° while the nonfused virus is removed by proteinase K.

centrifugation.[2] Following infection and virus production, virus-containing supernatants are centrifuged first at 5000 g for 20 min to remove cellular debris. The supernatant is then centrifuged at 25,000 rpm for 2.5 hr in an SW27 rotor (Beckman Instruments, Inc., Palo Alto, CA, or equivalent), or 40,000 rpm for 1 hr in an SW40 rotor, to pellet the virus. Pelleted virus will begin to resuspend as soon as the centrifuge has stopped. Thus, supernatants should be removed immediately and 250 μl of TN buffer (50 mM Tris, pH 7.4, 100 mM NaCl) added to each tube. The pellet can be resuspended either by leaving it overnight on ice or by gentle pipetting.

[2] L. Kääriäinen, K. Simons, and C. H. von Bonsdorff, *Ann. Med. Exp. Biol. Fenn.* **47**, 235 (1969).

The resuspended virus is centrifuged at 1500 g for 5 min to remove aggregates, and the supernatant is layered onto a velocity/density or density gradient. For radiolabelled virus, a velocity/density gradient (containing an 8 ml velocity gradient of 10–20% sucrose in TN buffer over an equilibrium gradient containing 5 ml of 25–50% sucrose in TN) is preformed in an SW40 tube. The gradient is centrifuged at 40,000 rpm for 3 hr and the viral band collected by fractionation. The band of radiolabeled virus is not always visible but should be present at about 35% sucrose. For large-scale virus preparations, 5–50% gradients of potassium tartrate in TN are poured in SW27 tubes, and centrifuged at 25,000 rpm for 2.5 hr. With large-scale preparations a prominent band of virus should be visible on the gradient. Following gradient purification, the sucrose or tartrate can be washed away by diluting the virus-containing fractions into TN, centrifuging as described above, and resuspending the pellet in TN. Other viruses are more fragile and require additional care while purifying. For example, VSV should not be pelleted, but it can be centrifuged onto a 60% sucrose cushion.[3]

Virus stocks should be stored carefully. The purified virus preparations should be split into appropriate aliquots, snap-frozen in liquid nitrogen, and stored at −70°. Stocks should not be repeatedly thawed and refrozen (the titer of SFV drops by about 10% for each round of freezing and thawing even when snap-frozen). Careful storage is especially important for fusion experiments. With fresh virus the core proteins are inaccessible to externally added proteases. After snap-freezing and thawing, however, the viral membrane may be damaged and no longer provide an effective barrier to the enzymes (i.e., it becomes leaky). With poorly prepared and stored virus, 50% or more of a virus preparation may be damaged.[4]

Virus Binding

Biochemical analysis of binding and internalization is usually carried out using tissue culture cell lines grown in plastic tissue culture plates or wells. Suspension cells can also be used, but we have found it easier to make media or temperature changes in plates where the media can be quickly removed and replaced. Most of the assays described here have been established using baby hamster kidney (BHK-21) cells, but they are also effective with a number of other cell lines.

[3] K. Matlin, D. F. Bainton, M. Pesonen, D. Louvard, N. Genty, and K. Simons, *J. Cell Biol.* **97**, 627 (1983).
[4] J. White, J. Kartenbeck, and A. Helenius, *EMBO J.* **1**, 217 (1982).

It is important to establish the growth conditions which generate cells best suited for binding or internalization of a specific virus. For example, we find that the expression of CD4, the receptor for HIV-1 and HIV-2, in HeLa cells transfected with CD4 cDNA reaches a maximum 3 days after subculturing.[5] Consequently, for these cells, 3-day-old cultures are used in all experiments. Once the optimal conditions have been established, they should be used routinely in all experiments, unless there is good reason to change.

Tissue culture plastic can provide an adherent surface for many viruses, and large areas of a plate without cells may confuse binding data. Consequently the cells should be cultured to form a monolayer.

The choice of dish or well will depend on the methods to be used for recovering the cells and bound virus. Methods using detergents or NaOH to remove cells from a well do not distinguish between virus bound to cells and virus bound to plastic. Consequently, if a portion of the input virus is adsorbed nonspecifically to the plate, it is likely to be extracted by detergent or NaOH and recorded as cell-associated. We have routinely removed cells into binding medium using a scraper of the type manufactured by Costar (Cambridge, MA, Cat. No. 3010) or, for 24-well plates, scrapers cut from Teflon-coated card and held in scissor clamps. If, by comparing scraped cells with detergent- or NaOH-extracted cells, it can be demonstrated that recovery is quantitative and that background binding is not a problem, then 0.2 M NaOH or 1% Triton X-100/0.5% sodium dodecyl sulfate (SDS) are effective means for harvesting cells and can be used in 96-well plates.[5]

The steps in virus entry can be synchronized to some extent by using conditions which inhibit the cell-dependent events in viral entry. Endocytosis is blocked at 4°, so incubating virus with cells on ice will allow binding but not endocytosis. Note that pH-independent viruses may still undergo fusion at low temperature and may fuse at the cell surface.

Quantitative determinations of virus binding to the cell surface should be performed with radiolabeled virus and analyzed by scintillation counting. With SFV, binding is carried out by diluting virus into medium [RPMI 1640 or Glasgow's Modified Eagle Medium (G-MEM) containing 0.2% bovine serum albumin (BSA) and 10 mM HEPES, pH 6.8]. In most cases binding will be carried out on ice outside a CO_2 incubator. In these conditions media without bicarbonate should be used to limit changes in pH during the binding reactions. Parameters such as pH have a significant

[5] A. Pelchen-Matthews, J. E. Armes, and M. Marsh, *EMBO J.* **8,** 3641 (1989).

effect on the efficiency of virus binding. SFV binds to BHK-21 cells optimally at pH 6.8.[6] With VSV, however, the optimum is around pH 7.5. The binding conditions should be determined empirically for each virus and cell line combination. The binding of SFV and VSV increases dramatically as the pH of the medium is lowered toward 6.0, the fusion pH. Although with these viruses it is comparatively straightforward to distinguish binding from fusion, this may not be the case with other viruses. The binding parameters established using radiolabeled virus can then be used with unlabeled virus.

With SFV, BSA is included in the binding medium to minimize nonspecific binding of virus to cells and plastic. However, BSA may not be the most appropriate "carrier" for all viruses; for example, with myxoviruses, which bind to sialic acid residues on glycoproteins and glycolipids, binding may be inhibited by the presence of sialoglycoproteins. For SFV, binding is carried out for 1 hr with the cells, on ice, rocked gently on a reciprocal shaker. The time for optimal binding should be determined for individual viruses and cell types.[6] Subsequently, the free virus should be removed by washing with 4–5 changes of fresh, 0° medium. Identical medium to that used to bind the virus should be used. The cells are then scraped into binding medium, transferred to 10-ml tubes, and centrifuged at 1500 rpm for 5 min in a refrigerated centrifuge precooled to 4°. Subsequently the supernatant is removed and the cell pellet recovered for analysis.

Endocytosis

At 20–37° most cells undergo constitutive endocytosis. This is believed to occur mainly through coated pits and coated vesicles.[7] Following the binding of virus to the cell surface and the removal of free virus, endocytosis is initiated by warming the cells to 37°. The onset of endocytosis can be initiated relatively synchronously in a number of wells by substituting the 4° medium for prewarmed medium and placing the cells in a 37° incubator. Following a period at 37°, endocytosis can be stopped by returning the cells to 4°; the 37° incubation medium is removed, transferred to appropriate tubes, and stored on ice, and the cells are washed with 4° medium.

To quantitate endocytosis, assays should be established which enable intracellular virus to be distinguished from cell surface virus (Fig. 1). With many ligands, such as antibodies, cell surface label can be removed by washing the cells in low pH media under conditions in which the integrity

[6] E. Fries and A. Helenius, *Eur. J. Biochem.* **97,** 213 (1979).
[7] M. Marsh and A. Helenius, *J. Mol. Biol.* **142,** 439 (1980).

of the plasma membrane in retained and internalized ligand remains cell-associated.[5] With viruses, especially pH-dependent viruses, low pH washes are likely to trigger the fusion activity and should therefore be avoided. As an alternative, proteolytic enzymes can be effective in removing cell surface-bound viruses. The enzymes should be active at 4° and, in control experiments where cells have been maintained at 4° following virus binding, should be capable of removing the majority of cell surface virions. For SFV, proteinase K (Boehringer Mannheim, Germany, 0.5 mg/ml) in phosphate-buffered saline (PBS) containing calcium and magnesium (PBS[++]) releases 90–95% of cell surface virus.

For the assay, virus is initially bound to duplicate plates of cells as described above. Following endocytosis, one plate of cells is washed and scraped in binding medium to give the total cell-associated activity, while the other is incubated in proteinase K. The incubation is carried out on ice with the cells on a rocking platform. After 45 min the cell suspension (proteinase K releases the cells from the plastic) is transferred to 10-ml tubes in PBS[--] containing 0.2% BSA (w/v), the proteinase K is inhibited by addition of 1 mM phenylmethylsulfonyl fluoride (PMSF; Sigma, St. Louis, MO), and the cells are centrifuged at 1500 rpm for 5 min at 4°. The supernatant is removed and the pellet assayed for cell-associated, proteinase K-resistant, virus.[7]

Although proteinase K has been used routinely for SFV, trypsin will also remove the virus. Depending on the virus, other enzymes may be more appropriate; for example, for viruses such as influenza that bind to sialic acid residues, neuraminidase can be used.[8]

One of the consequences of endocytosis is that many ligands, including viruses, are delivered to lysosomes and degraded. Often polypeptides are degraded to single amino acids, or small (di- and tri-) peptides that are released into the medium. With [^{35}S]methionine-labeled virus, degradation is readily detected by assaying the 37° incubation medium. To distinguish between intact virus, which may have been eluted from the cell surface, and degraded virus, the media collected from the 37° incubation are made 10% with trichloroacetic acid (TCA) and incubated on ice for 1 hr. Subsequently, the samples are centrifuged and the TCA-soluble (degraded) activity in the supernatant determined. The TCA-precipitable activity can be analyzed for the integrity of the viral proteins by sodium dodecyl sulfate–polyacrylamide gel electrophoresis (SDS-PAGE).

The data should be plotted to show the total cell-associated activity, the proteinase K-resistant activity, and the TCA-soluble activity. Endocytosis

[8] K. S. Matlin, H. Reggio, A. Helenius, and K. Simons, *J. Cell Biol.* **91**, 601 (1981).

can then be estimated by adding the proteinase K-resistant and TCA-soluble activities and comparing this figure to the initial cell-associated activity.

Fusion

For pH-dependent viruses endocytosis leads to the delivery of virions into acidic endosomes. The fusion reaction occurs within these organelles. Although it has been possible to observe fusion within endosomes mor-phologically,[9] it has been difficult to quantitate the fusion events. However, the properties of the fusion reaction can be studied using lipo-somes[4,10,10a,b,11] or by artificially inducing fusion at the cell surface.[11,11a,b] As fusion with liposomes does not directly involve interaction between a virus and its target cell, only fusion at the cell surface is discussed here. For methods using liposomes, see Refs. 4, 10, 10a, and 10b.

If fusion of a virus can be triggered by acid pH, then transient acidifica-tion of the medium should also trigger the fusion activity. Radiolabeled virus (either [³H]uridine or [³⁵S]methionine) is bound to the cell surface at 4° as described above and the free virus washed away. Fusion of this surface-bound virus can then be induced by rapidly exchanging the neutral pH medium with 37° medium buffered to pH 6.0 or below. After 1 min the acid medium is washed away and replaced with 4° neutral pH me-dium. Note that some virus may be endocytosed during the 1-min incuba-tion at 37°. However, this is likely to be minimal (< 10% of surface-bound virus), as coated vesicle-mediated endocytosis is inhibited when cells are acidified.[12] Moreover, the amount of endocytosis can be estimated and controlled for by quantitating the proteinase K-resistant activity in cells that have been warmed for 1 min in neutral pH media.

Fusion is quantitated using the proteinase K assay described above for endocytosis (Fig. 1). At 4°, proteinase K will release surface-bound virions, apparently by cleaving the cellular receptors for virus. The enzyme has little effect on the viral envelope proteins under these conditions. After fusion, the virus and cell membranes are continuous, and the viral proteins are resistant to cleavage by the enzyme. Following fusion and reneutraliza-

[9] A. Helenius, Biol. Cell **51**, 181 (1984).
[10] J. M. White and A. Helenius, Proc. Natl. Acad. Sci. U.S.A. **77**, 3273 (1980).
[10a] D. Hoekstra and K. Klappe, this volume [20].
[10b] S. Nir, this volume [28].
[11] J. White, J. Kartenbeck, and A. Helenius, J. Cell Biol. **87**, 264 (1980).
[11a] N. Düzgüneş, M. C. Pedroso de Lima, L. Stamatatos, D. Flasher, D. Alford, D. S. Friend, and S. Nir, J. Gen Virol. **73**, 27 (1992).
[11b] A. Puri, M. J. Clague, C. Schoch, and R. Blumenthal, this volume [21].
[12] J. Davoust, J. Gruenberg, and K. E. Howell, EMBO J **6**, 3601 (1987).

tion, the medium is removed and replaced with 0.5 mg/ml proteinase K in PBS^{++} at 4°. The cells are rocked on ice for 45 min and then transferred to 10-ml centrifuge tubes as described above. The cells are washed twice in 10 ml of 4° PBS^{--}, containing PMSF and 0.2% BSA, and the cell-associated radioactivity is determined. A fusion efficiency is calculated by comparing the proteinase K-resistant activity, minus the activity due to endocytosis, to the total activity initially bound to the cells.

Although endosome fusion is difficult to assess directly, it can be monitored indirectly. The effect of low pH on the viral spike glycoproteins is to induce a conformational change in the protein. This change is prerequisite for fusion and can be detected biochemically by a change in the susceptibility of the protein to proteolytic cleavage.[13-15] ^{35}S-labeled SFV is bound to cells and allowed to endocytose for given times. The conformational change will have occurred on those virions that have entered the acidic endosomes, and which may have undergone fusion. However, it will not have occurred on virions which have not entered acidic organelles.

To detect the conformational change, the cells are washed in cold PBS, lysed in PBS^{++} containing 1% Triton X-100, and incubated with 100 μg/ml L-1-tosylamide-2-phenylethylchloromethyl ketone (TPCK)–trypsin. The incubations can be carried out at 4° or 37°, but different digestion patterns are seen at the two temperatures. After 10 min the digestion is stopped by adding a 3-fold excess of soybean trypsin inhibitor, and, after a further 10 min, the samples are processed for SDS-PAGE.[13] With samples incubated at 0° the acid-treated E2 glycoprotein is sensitive to trypsin digestion, but the E1 glycoprotein remains largely resistant. In contrast, on digestion at 37°, the E2 protein is degraded whether treated at neutral or acid pH, but the E1 protein becomes trypsin resistant after acid treatment.[13] Similar assays can be applied to detect the acid-induced conformational changes in influenza hemagglutinin.[15]

Penetration and Uncoating

The fusion reaction, whether induced at the cell surface or occurring normally through endosomes, leads to delivery of the viral nucleocapsid to the cytosol (penetration) and to uncoating of the viral RNA. Penetration can be determined biochemically using virus labeled with [^{3}H]uridine.[16,17]

[13] M. Kielian and A. Helenius, *J. Cell Biol.* **101**, 2284 (1985).

[14] J. Edwards, E. Mann, and D. T. Brown, *J. Virol.* **45**, 1090 (1983).

[15] J. Skehel, P. Bayley, E. Brown, S. Martin, M. Waterfield, J. White, I. Wilson, and D. Wiley, *Proc. Natl. Acad. Sci. U.S.A.* **79**, 968 (1982).

[16] W. C. Koff and V. Knight, *J. Virol.* **31**, 261 (1979).

[17] A. Helenius, M. Marsh, and J. White, *J. Gen. Virol.* **58**, 47 (1982).

These assays are best carried out using confluent 35 or 50 mm plates of cells.

The virus is bound to the cell surface at 4° and allowed to internalize for given times as described above. Subsequently, the cells are washed in PBS⁻⁻ and removed from the dish by scraping. The cells are transferred to a 10-ml centrifuge tube, made up to 10 ml with PBS⁻⁻, and pelleted at 1500 rpm for 5 min at 0°. During incubation at 37° some of the surface-bound virus will have internalized, arrived in endosomes, and undergone fusion. The nucleocapsids of these viruses should now be uncoated in the cytoplasm, whereas virions which have not fused will remain within their own membranes. To distinguish between the fused and nonfused virus, the cells are broken open using conditions which rupture but do not solubilize membranes, and which rupture the plasma membrane but not the viral membrane. The viral genomes which have undergone penetration and are present in the cytosol can then be detected by their accessibility to added ribonuclease (RNase); the nonfused viral nucleocapsids will be protected from these enzymes by the viral membrane (see Fig. 1).

For lysis the cell pellets are resuspended into 200 μl PBS⁻⁻ and then washed with 5 ml of 4° lysis buffer [0.25 M sucrose, 10 mM triethanolamine, 10 mM acetic acid, 1 mM ethylenediaminetetraacetic acid (EDTA), pH 7.4–7.5]. Some cells become very fragile in lysis buffer, and this first wash step may not be required. To avoid premature cell lysis, care should be taken at this stage not to vortex, pipette, or otherwise subject the cells to shear forces. The cells are again collected by centrifugation (3000 rpm, 10 min, 4°) and the pellet resuspended into 1 ml lysis buffer at 4°.

The cells can be broken in this buffer using a number of different techniques, for example, pipetting, passage through a fine-gauge needle, or homogenization in a small-volume Dounce homogenizer. Any technique is appropriate as long as the cells are effectively broken and the nonfused virus left intact. Effective lysis can be measured by assaying the release of a cytosolic enzyme or by microscopic examination. The integrity of nonfused viruses can be determined by assaying the amount of viral RNA degraded in control samples maintained at 0° throughout, and therefore containing no uncoated virus. RNase digestion is carried out by adding 0.5 mg (37 Kunitz units) RNase A and 1 unit of micrococcal nuclease (both from Sigma) in 50 μl of lysis buffer to each sample, then incubating at 37° for 30 min. The digest is precipitated by adding an equal volume of 20% TCA at 4° and incubating on ice for 1 hr. The precipitates are removed by centrifugation (3000 rpm, 10 min, 4°). The uncoated viral RNA will be digested and is recovered as TCA-soluble activity in the supernatants. Total degradable RNA in the sample is estimated by including 1% Triton X-100 during the digestion.

Cell Fractionation

The cellular distribution of viruses during entry can be determined both morphologically (see below) and by cell fractionation. Extensive documentation of cell fractionation procedures has been published elsewhere[18,19] and is discussed only in brief here. Percoll gradient centrifugation and free flow electrophoresis have both been used, either independently or in combination, to separate endocytic organelles, and they can be used to characterize the intracellular distribution of internalized virions.

Essentially cells are prepared in much the same way as described in the uncoating assay above. Radiolabeled virus is bound in the cold and then internalized by warming the cells. Endocytosis is stopped subsequently by returning the cells to 4°. The cells are scraped in PBS⁻ and lysed using the same lysis buffer described above. Additional care should be taken in these experiments to ensure that the plasma membrane is broken but the organelles maintained intact. The release of lysosomal hydrolases (e.g., β-hexosaminidase) provides a useful marker for the integrity of lysosomes, and an internalized fluid phase marker, such as horseradish peroxidase, can be used as a content marker for endosomes. The lysate is then centrifuged at 3000 rpm for 10 min at 4° to remove nuclei, intact cells, and cell debris, and the postnuclear supernatant is subjected to Percoll density centrifugation or free flow electrophoresis as described.[18,19]

Morphology

Virus binding, endocytosis, and fusion can be followed morphologically. For light microscopy, specific antibodies can be used to detect viral proteins using conventional immunofluorescence procedures.[8] At the electron microscope level, the 75 nm diameter virus particles are readily identified in conventional Epon thin sections. Following binding and/or endocytosis as described above, the cells are fixed for 30 min with 4°, 2.5% glutaraldehyde in 50 mM cacodylate buffer, pH 7.2, and subsequently dehydrated, embedded, sectioned, and stained as described.[20]

To visualize fusion or steps in uncoating, it is essential to synchronize the reactions as much as possible. Fusion figures are very transient structures, and unsynchronized fusions or lengthy time intervals between fusion and fixation will result in either difficulty in finding the relevant structures or loss of the structures. Thus, for cell surface fusion experiments the cells

[18] M. Marsh, E. Bolzau, and A. Helenius, *Cell (Cambridge, Mass.)* **32,** 931 (1983).
[19] M. Marsh, S. Schmid, H. Kern, E. Harms, P. Male, I. Mellman, and A. Helenius, *J. Cell Biol.* **104,** 875 (1987).
[20] A. Helenius, E. Fries, and J. Kartenbeck, *J. Cell Biol.* **75,** 866 (1977).

are usually grown on glass coverslips or plastic coverslips that can be embedded and sectioned (Lux Thermanox, Nunc, Inc., Naperville, IL). Following virus binding, the fusion reaction is induced by dipping the coverslip into prewarmed pH 5.8 medium for 15 sec. The coverslip is quickly dipped into 4° medium at neutral pH and then fixed immediately as described above. For intracellular fusions, virus particles are collected in endosomes by internalizing virus at 20° (see below) under conditions where fusion is prevented by neutralizing endosomes using 20 mM NH_4Cl. Subsequently, the endosomes are allowed to acidify by replacing the NH_4Cl medium with NH_4Cl-free medium at 37°. After 30 sec the cells are fixed as described above.[9,21]

Viral antigens can be localized at the electron microscope level with antibodies specific for individual viral proteins using immunoperoxidase,[3] or by staining ultrathin cryosections of virus-treated cells with antiviral antibodies and protein A–gold.[21]

Inhibitors

Several specific steps in the entry pathway can be inhibited using inhibitors of different cellular functions. As described above, low temperature (4°) blocks endocytosis completely and restricts virus–cell interactions to binding. The effects of low temperature are readily reversed on warming the cells. Furthermore, endocytosis events and intracellular transport of internalized viruses can be stopped at any time by returning the cells to 4°. Low temperature also affects intracellular transport. At 20° (in BHK cells) endocytosis will proceed, but transfer of internalized virus from endosomes to lysosomes is inhibited.[18] Thus, incubations at 20° can be used to collect virus in endosomes. It should be noted that low temperature may not block viral fusion with cellular membranes. *In vitro* studies with SFV and liposomes indicate that 40–50% of the normal viral fusion activity is observed at 4°.[10]

Endosomal fusion and infection of acid-dependent viruses is blocked by reagents that inhibit endosome and lysosome acidification. These reagents fall into two classes. One group is the weak bases and includes NH_4Cl, methylamine, amantadine, and chloroquine. In the nonprotonated form, these reagents diffuse across membranes, but in acid environments they become protonated and cross membranes at considerably slower rates. As a consequence the reagents sequester into acid organelles and raise the pH. The second group is the carboxylic ionophores and includes monensin and nigericin. These drugs partition into membranes

[21] G. Griffiths, A. McDowell, R. Back, and J. Dubochet, *J. Ultrastruct. Res.* **89,** 65 (1984).

and mediate the exchange of protons for monovalent cations according to the electrochemical gradient. In cells, therefore, these agents remove protons from acidic organelles in exchange for Na^+ and K^+ ions and raise the pH within these compartments. Both weak bases and carboxylic ionophores effectively block SFV infection by raising the pH within endosomes above that required for fusion.[17,22] The action of the drugs is reversible, and the reagents do not have major effects on intracellular transport. However, their actions are not restricted to endosomes, and the fact that they also interfere with exocytotic transport should be considered when designing experiments to assay viral infectivity.

Acknowledgments

The author is supported by the Medical Research Council AIDS Directed Programme.

[22] M. Marsh, J. Wellsteed, H. Kern, E. Harms, and A. Helenius, *Proc. Natl. Acad. Sci. U.S.A.* **79**, 5297 (1982).

[20] Fluorescence Assays to Monitor Fusion of Enveloped Viruses

By DICK HOEKSTRA and KARIN KLAPPE

Introduction

Viral nucleocapsids contain all the necessary genetic information for replication. However, viruses lack the synthesizing machinery to do so, which is why these particles exploit host cells for this purpose. The introduction of the nucleocapsid into the cytoplasm of a host cell is therefore a prerequisite for viral replication. Enveloped viruses, that is, viruses whose nucleocapsids are surrounded by a membrane, deliver their capsids into cells by membrane fusion between their envelope and a cellular membrane, either the plasma membrane or the endosomal membrane (Fig. 1). Fusion with the former occurs at neutral pH, whereas merging with the latter takes place at mildly acidic pH, after internalization of the virus by receptor-mediated endocytosis. The route of entry depends on the family to which a virus belongs.[1-6]

[1] J. White, M. Kielian, and A. Helenius, *Q. Rev. Biophys.* **16**, 151 (1983).

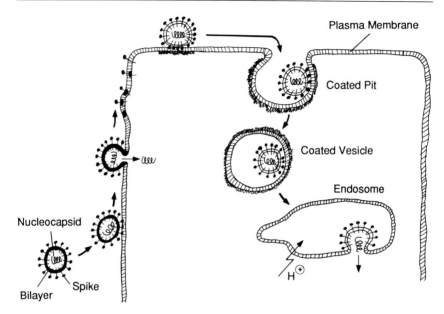

FIG. 1. Penetration of enveloped viruses into cells. As a result of fusion between a viral envelope and a cellular membrane, the viral nucleocapsid gains access to the cytoplasm of the cell. This step is crucial to productive replication of the virus. Some viruses fuse at neutral pH directly with the plasma membrane of the target cell. Others are first internalized by receptor-mediated endocytosis. In this case fusion occurs in the endosomal compartment (from within), triggered by a lowering of the pH to mildly acidic.

Many enveloped viruses have a relatively simple membrane composition, consisting of lipids derived from the host cell and some (which can be as few as one) virus-specific membrane glycoproteins. It has been known for many years that the ability of a virus to fuse is contained in these glycoproteins. That intracellular membrane fusion events, which are crucial for the biological life of a cell, are also mediated by proteins makes viral systems particularly attractive as a model for protein-induced fusion of biological membranes.

[2] R. W. Compans, A. Helenius, and M. B. A. Oldstone (eds.), "Cell Biology of Virus Entry, Replication, and Pathogenesis," UCLA Symp., Vol. 90. Alan R. Liss, New York, 1989.

[3] D. Hoekstra, and J. W. Kok, *Biosci. Rep.* **9**, 273 (1989).

[4] T. Stegmann, R. W. Doms, and A. Helenius, *Annu. Rev. Biophys. Biophys. Chem.* **18**, 187 (1989).

[5] M. Kielian and S. Jungerwirth, *Mol. Biol. Med.* **7**, 17 (1990).

[6] D. Hoekstra, *J. Bioenerg. Biomembr.* **22**, 121 (1990).

To examine and reveal the mechanisms underlying virus–membrane fusion, studies of fusion of pure phospholipid vesicles have paved the way toward developing the basic strategy for experimentation. This involves the application of assays that detect the fusion of viruses with target membranes in a sensitive and continuous fashion, thus providing kinetic and quantitative insight into the fusion process. In conjunction with these assays, by employing a mass action kinetic model[7] it has also been possible to dissect the overall virus fusion process further, and to obtain aggregation and fusion rate constants. A detailed view can thus be obtained as to molecular parameters that govern the binding and the actual fusion step of a virus interacting with a target membrane. Before discussing in detail the application of fluorescence assays in monitoring virus fusion, we briefly describe some procedures that may often serve as complimentary tools for analyzing virus membrane fusion.

Detection of Virus Fusion Activity

Electron microscopy provides a means to observe the interaction of a virus with a cell directly. The fusion event itself is usually difficult to discern, but lateral diffusion of viral antigens, a consequence of the fusion process, can be revealed by using gold- or ferritin-labeled antibodies against viral envelope proteins.[8] With similar techniques, the initial internalization of viruses by endocytosis can be shown.[9]

Cell–cell fusion is a typical manifestation of enveloped viruses when, after an incubation at 4° which causes virus-induced cell agglutination, the incubation temperature is raised to 37°. Without preagglutinating the cells at 4°, cell fusion is usually not observed, in spite of the fact that virions avidly fuse with the cells when added directly at 37°. Cell–cell fusion occurs at a relatively large exogenous dose of viruses. Typically, 10^6–10^7 cells are mixed in a small volume (~ 100–200 μl) of incubation buffer, containing 10–30 μg of virus (based on protein, or 200–400 hemagglutination units). After 5–15 min on ice, the sample is warmed to 37° for 5–10 min, followed by a 10-fold dilution with prewarmed buffer. After another 5–10 min, the fusion reaction can be stopped by adding cold buffer, and the cells can be resuspended in a smaller volume after centrifugation. Fusion is scored by estimating the fusion index, obtained from the ratio of the number of nuclei present in fused cells to the total number of nuclei present in a particular microscopic field.

[7] S. Nir, this volume [28].
[8] S. Knutton, *J. Cell Sci.* **43**, 103 (1980).
[9] A. Helenius, J. Kartenbeck, K. Simons, and E. Fries, *J. Cell Biol.* **84**, 404 (1980).

Human erythrocytes, which are frequently used as biological target membranes in virus fusion studies, are obviously not very convenient for quantifying virus-induced cell fusion, since these cells, in contrast to chicken erythrocytes, lack nuclei. In this case, however, another parameter can be monitored, namely, the release of hemoglobin. Hemolysis occurs after merging of the viral envelope and the red cell membrane, and it takes place both after a low temperature preincubation and when the virus is added directly at 37°. This implies that release does not occur as a result of cell–cell fusion. The release of hemoglobin depends on several factors: (1) the type of virus, namely, the susceptibility of the erythrocyte membrane for a particular virus, (2) the viral dose, and (3) the viral membrane structure. With respect to the latter, it should be noted that hemolysis is likely related to some "damage" induced in the viral membrane on storage in a frozen state. For example, early-harvest Sendai virus triggers hemolysis only when the virus has undergone a freeze–thaw cycle before incubation with the red blood cells.[10] This observation also implies that the absence of hemolysis does not necessarily reflect a lack of virus fusion activity.

A typical protocol for virus-induced hemolysis is as follows. Approximately $10-30$ μg of virus (protein) is suspended with human erythrocytes (10^8 cells) in a final volume of 0.6 ml buffer, consisting of either 120 mM KCl/30 mM NaCl/10 mM Na_2HPO_4, pH 7.4, or 150 mM NaCl/5 mM Hepes, pH 7.4. Eppendorf tubes are convenient incubation vials. The incubation temperature is 37°. After appropriate time intervals the reaction is stopped by adding 0.4 ml cold buffer, followed by centrifugation (Eppendorf centrifuge, 3 min, maximal speed) of the samples at 2°. Hemoglobin in the supernatant is determined by measuring the absorbance at 540 nm. Total release is accomplished by suspending the same amount of erythrocytes in 1% (v/v) Triton X-100.

Both virus-induced cell–cell fusion and hemolysis may give some rough indication as to the pH and temperature dependence of virus fusion activity. No insight is obtained into relevant questions such as number of virus particles that bind, relative to the number that fuse, with a cellular membrane, nor is it possible to obtain an accurate estimation of the kinetics of these processes, relative to the kinetics of endocytosis, the latter process mediating the entry of many families of enveloped viruses. As exemplified by the strategies developed in studies of the fusion of artificial membranes,[11] answers to such questions are pertinent in order to obtain insight into the molecular details of the mechanism of virus fusion. As a consequence, efforts were undertaken to apply a technology similar to that used in artificial membrane fusion in studies dealing with virus fusion.

[10] M. Homma, Y. K. Shimizu, and N. Ishida, *Virology* **71**, 41 (1976).
[11] D. Hoekstra and N. Düzgüneş, this volume [2].

Both electron spin resonance probes and fluorescent probes have been used to study the kinetics of early interactions between viruses and target membranes.

Electron Spin Probes as Markers for Virus Fusion Activity

Using assays based on electron spin resonance (ESR), fusion of viruses with cellular membranes can be detected by registering the deposition of the nucleocapsid into the cytoplasm of the cell or the mixing of viral and cellular membrane (i.e., plasma or endosomal membrane). The former approach relies on using TEMPO choline[12] as a fusion reporting molecule, the latter on applying spin-labeled (phospho-)lipids. Viruses can be loaded with TEMPO choline (Molecular Probes, Inc.) by incubating the virus with the probe for several hours. The probe seems to attach preferentially to the viral nucleocapsid. At a sufficiently high binding density, the ESR spectrum displays an exchange-broadened signal, which is converted to a sharp, increased signal when the nucleocapsid gains access into the cytoplasm, causing release and dilution of the probe. A similar principle holds for viruses labeled with spin-labeled lipids at a relatively high concentration (10–20 mol% with respect to total lipid). When such labeled viruses fuse with a nonlabeled target membrane, lipid dilution will result in elimination of the exchange-broadened signal, as obtained for the labeled virus particles per se.[13,14]

An alternative approach for monitoring lipid mixing as a measure of virus fusion, based on lipid spin probes, is to monitor physical changes in the environment of the probe, as reflected by changes in spectral properties. With suitable target membranes such as erythrocytes, a clear distinction may exist in bilayer fluidity relative to that of the virus membrane, the latter being more rigid. Spectral changes can be detected when the probe, initially incorporated into the viral membrane (at much lower concentrations than in the case of monitoring changes in spin broadening), randomizes into a target membrane of different fluidity. These changes can be quantitatively related to the fractions of spin-labeled virions that bind to and fuse with target membranes.[15]

A major drawback of these procedures is that they are quite laborious, requiring sampling at different time intervals after initiation of the fusion reaction, followed by recording and analysis of the spectra. This precludes continuous monitoring of the fusion process with these procedures. These disadvantages can generally be overcome by using assays based on fluorescence.

[12] T. Maeda, K. Kuroda, S. Toyama, and S.-I. Ohnishi, *Biochemistry* **20**, 5340 (1981).
[13] M. Umeda, S. Nojima, and K. Inoue, *J. Biochem. (Tokyo)* **97**, 1301 (1985).
[14] K. Kuroda, K. Kawasaki, and S.-I. Ohnishi, *Biochemistry* **24**, 4624 (1985).
[15] D. S. Lyles and F. R. Landsberger, *Biochemistry* **18**, 5088 (1979).

Fluorescence Assays to Monitor Virus Fusion

As noted elsewhere,[16] the application of fluorescence assays to monitor fusion offers a number of advantages over other assays. These advantages include convenient accessibility of the techniques, high sensitivity, relative ease in obtaining quantitative data, and possibilities to monitor or detect fusion by both fluorimetry and fluorescence microscopy.

A number of assays based on fluorescence have been employed to measure the fusion between viral envelopes and target membranes. Some of these methods allow continuous monitoring of the fusion process, whereas others require analysis of samples taken from the incubation mixture after certain time intervals. For measurement of rapid, initial kinetics, the latter procedures are therefore less attractive, but in spite of this drawback they are superior over other, nonfluorescence procedures as far as quantitation and molecular characterization of virus fusion activity are concerned.

One of these procedures involves the application of fluorescence photobleaching recovery (FPR), which measures the lateral mobility of fluorescently labeled membrane components. The labeled membrane components can be either proteins or lipids. One approach has been to label the viral proteins with fluorescein isothiocyanate (FITC)[17] by incubating intact virus with the fluorophore (1.5 mg virus/mg FITC) for several hours. Little, if any, labeling of the lipids is obtained, whereas virus-induced agglutination and hemolysis are not affected by the probe. The occurrence of fusion of the labeled virus can then be detected by measuring the fraction of fluorophore that becomes laterally mobile in the plane of the cell membrane after merging and random mixing of viral and cell membrane constituents.[17a] Such measurements are done by focusing a laser beam on a relatively small spot on the cell surface. After high-energy bleaching, the recovery of fluorescence in the bleached area is monitored at a lower intensity, and the fraction of fluorescence recovered, relative to that prior to bleaching, gives a measure of virus particles that are bound (nonmobile, i.e., the fluorescence of this fraction is not recovered) and those actually fused (laterally mobile fraction).

Instead of monitoring lateral mobility of viral proteins, exogenously inserted fluorescent lipids can be employed for this purpose. A nonexchangeable fluorescent lipid analog, N-(7-nitrobenz-2-oxa-1,3-diazol-4-yl)phosphatidylethanolamine (N-NBD-PE), can be used, and the probe has

---

[16] D. Hoekstra, in "Membrane Fusion" (J. Wilschut and D. Hoekstra, eds.), p. 289. Dekker, New York, 1991.

[17] Y. I. Henis and T. M. Jenkins, *FEBS Lett.* **151**, 134 (1983).

[17a] Y. I. Henis, this volume [26].

been incorporated into viral envelopes by a reconstitution procedure.[18] The FPR techniques have provided valuable insight as to the role of viral protein mobility in virus-induced and reconstituted envelope-induced fusion of cells, as well as whether interactions between viral proteins are relevant for triggering virus fusion activity.[17-19]

In combination with another fluorescent lipid analog, N-(lissamine rhodamine B sulfonyl)phosphatidylethanolamine (N-Rh-PE), N-NBD-PE has also been used to monitor the fusion of viruses or their reconstituted envelopes in a continuous fashion, by means of resonance energy transfer. This procedure is described elsewhere in this volume.[11] Because the probes cannot be readily incorporated into intact viral membranes, that is, by exogenous addition, liposomes containing 0.8–1.0 mol% of each of the fluorescent analogs (with respect to total lipid) are commonly used as target membranes.[20-22] When using reconstituted viral envelopes, the probes can be incorporated into the envelope by solubilizing the lipid analogs in the detergent used for viral envelope reconstitution.[23] The probes are subsequently mixed with detergent–solubilized envelope constituents at a concentration of 0.8–1.0 mol% each with respect to total viral lipid. On reconstitution of the viral envelope during detergent dialysis, the probes become an integral part of the viral membrane at concentrations that allow efficient energy transfer. In this case, fusion studies can be carried out with both artificial and biological target membranes. Fusion is revealed as a relief of energy transfer efficiency, monitored as an increase of NBD fluorescence. Finally, it should be noted that between pH 5.0 and 9.0, application of this assay reliably reflects the efficiency and occurrence of virus fusion. It has also been reported,[22] however, that at very low pH values (pH 4.0) artifacts may arise, owing to the decreased absorption properties of N-Rh-PE under these conditions, thus artificially enhancing NBD fluorescence and therefore requiring appropriate corrections.

Pyrene-labeled lipids can also be profitably used to monitor fusion, albeit not in a continuous fashion, of intact viruses with liposomes and biological membranes. As described for fusion of liposomal systems,[11] this approach relies on changes in the excimer/monomer fluorescence intensity ratio, which occur when the probe dilutes on fusion between labeled and

[18] B. Aroeti and I. Henis, *Biochemistry* **25,** 4588 (1986).
[19] Z. Katzir, O. Gutman, and Y. I. Henis, *Biochemistry* **28,** 6400 (1989).
[20] O. Eidelman, R. Schlegel, T. S. Tralka, and R. Blumenthal, *J. Biol. Chem.* **259,** 4622 (1984).
[21] T. Stegmann, D. Hoekstra, G. Scherphof, and J. Wilschut, *Biochemistry* **24,** 3107 (1985).
[22] R. I. MacDonald, *J. Biol. Chem.* **262,** 10392 (1987).
[23] M. C. Harmsen, J. Wilschut, G. Scherphof, C. Hulstaert, and D. Hoekstra, *Eur. J. Biochem.* **149,** 591 (1985).

unlabeled membranes. With liposomes as target membranes, the probe is usually incorporated at a relatively high concentration (~ 10 mol%) in the liposomal bilayer. After initiation of fusion, samples are taken at various time intervals, and emission spectra are recorded between approximately 360 and 520 nm, at an excitation wavelength of 330–350 nm. The fluorescence intensities of excimer and monomer emission are then determined at 470 and 385 nm, respectively.

An attractive and more versatile approach is to label the viral lipids with pyrene, enabling one to measure fusion of the intact virus with biological target membranes as well (see Fig. 6). Such a derivatization of the viral lipids can be accomplished[24] by growing the virus on cells that have been metabolically labeled with pyrene fatty acids. Because the virus acquires its lipids from the host cell, the pyrene-labeled phospholipids thus become an integral part of the viral membrane.

An assay that has found wide application[25-30] in studies of virus fusion is that based on the relief of fluorescence self-quenching (Fig. 2). This assay[25] involves the exogenous insertion of a fluorescent probe, octadecylrhodamine B chloride[31] (R_{18}), in the viral bilayer by briefly incubating a virus suspension with an ethanolic solution ($\leq 1\%$, v/v) of the probe. The concentration of the probe is taken such that it will cause efficient quenching of fluorescence, once inserted in the viral bilayer. When the labeled viruses fuse with nonlabeled target membranes, the lipidlike probe becomes diluted and hence its surface density decreases. Concomitantly an increase in fluorescence is observed which increases proportionally when fusion proceeds, thus allowing kinetic and quantitative measurements of fusion of intact viruses with both artificial and biological membranes (see e.g., Fig. 3).

A typical protocol for labeling the virus and measuring its fusion activity is as follows. A stock solution of R_{18} (obtained from Molecular Probes, Inc., Eugene, OR) is prepared (1–2 mM) in chloroform/methanol (1:1, v/v). An aliquot of this solution, stored at −20°, is taken and dried under a stream of argon gas. Subsequently, the probe is solubilized in ethanol, and this solution is then rapidly injected with a Hamilton syringe into the

[24] R. Pal, Y. Barenholz, and R. R. Wagner, *Biochemistry* **27**, 30 (1988).

[25] D. Hoekstra, T. de Boer, K. Klappe, and J. Wilschut, *Biochemistry* **23**, 5675 (1984).

[26] D. Hoekstra and K. Klappe, *J. Virol.* **58**, 87 (1986).

[27] A. Puri, J. Winick, R. J. Lowy, D. Covell, O. Eidelman, A. Walter, and R. Blumenthal, *J. Biol. Chem.* **263**, 4749 (1988).

[28] R. K. Scheule, *Biochim. Biophys. Acta* **899**, 185 (1987).

[29] T. Stegmann, D. Hoekstra, G. Scherphof, and J. Wilschut, *J. Biol. Chem.* **261**, 10966 (1986).

[30] F. Sinangil, A. Loyter, and D. Volsky, *FEBS Lett.* **239**, 88 (1988).

[31] P. M. Keller, S. Person, and W. Snipes, *J. Cell Sci.* **28**, 167 (1977).

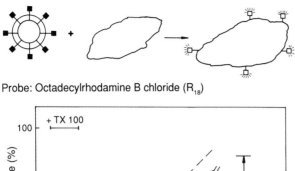

Probe: Octadecylrhodamine B chloride (R_{18})

FIG. 2. Schematic representation of the principle of the R_{18} assay. The virus is labeled with a self-quenching concentration of R_{18} and suspended in the fusion medium (time a). When appropriate target membranes are added (time b), fusion will lead to a decrease in surface density of the probe, causing a concomitant increase of fluorescence. The relief of self-quenching is monitored continuously as a function of time. Parameters such as the lag time (c, see also Fig. 3), initial rate, and extent of fluorescence increase after a specified time are commonly determined as a measure of fusion. For further details, see Refs. 16 and 25.

virus-containing buffer solution, under vigorous vortexing. For labeling of Sendai virus, 1 to 2 mg of viral protein is labeled with $10-20$ nmol R_{18} (in 10 μl ethanol solution). The final incubation volume is 1 to 1.5 ml, that is, the final ethanol concentration is 1% (v/v) or less. The mixture is incubated at room temperature for 30 min in the dark. The efficiency of probe incorporation in the viral bilayer is of the order of 70%. Therefore, noninserted probe has to be removed, which is done by chromatography on Sephadex G-75 (1 × 15 cm). Nonincorporated R_{18} strongly adsorbs to the top of the column, whereas R_{18}-labeled virus is recovered in the void volume fraction. In principle, free probe can also be removed by centrifugation. However, column chromatography is preferred because this procedure is rapid (10 min) and removes the free probe very efficiently from the preparation, whereas centrifugation is more laborious (at least 1 hr) and may lead to cosedimentation of the free probe. An alternative procedure for removal of free probe is sucrose density-gradient centrifugation of the labeled virus.[32]

[32] N. Düzgüneş, M. C. Pedroso de Lima, L. Stamatatos, D. Flasher, D. Alford, D. S. Friend, and S. Nir, *J. Gen. Virol.* **73**, 27 (1992).

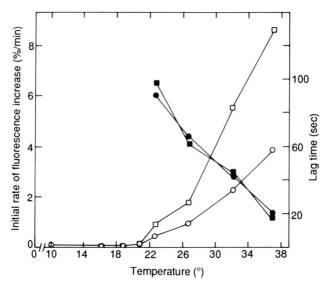

Fig. 3. Fusion of Sendai virus with erythrocyte membranes and effect of polyethylene glycol. R_{18}-labeled Sendai virus was incubated with erythrocyte membranes ("ghosts"), and fusion was monitored in the presence or absence of 4% polyethylene glycol as a function of temperature. Lag times (filled symbols) and initial rates (open symbols) were determined as described in Fig. 2. Note that significant virus fusion is observed only at temperatures above approximately 21°, that the initiation of fusion occurs faster (lag time decreases) when the temperature increases, and that the initial rate of fusion increases considerably in the presence (□) of the dehydrating agent polyethylene glycol. [For further details, see D. Hoekstra, K. Klappe, H. Hoff, and S. Nir, J. Biol. Chem. **264**, 6786 (1989), from which this figure was reproduced with permission.]

The amount of R_{18} that becomes incorporated into the membrane can be determined by extracting the virus with chloroform/methanol/0.1 N HCl and measuring the fluorescence of an aliquot of the extract (in chloroform). A calibration curve is constructed using known concentrations of R_{18}. Up to 1000 pmol/ml, the fluorescence intensity increases linearly with R_{18} concentration.

Because the virus contains the fusion-reporting molecule, its merging with both artificial and biological membranes can now be monitored. Before starting the experiment, the fluorescence scale is calibrated, using Triton X-100 (1%, v/v). The detergent does not interfere with rhodamine fluorescence, and the only correction factor required is that due to dilution. The R_{18}-labeled virus is suspended into the fusion medium, and the residual fluorescence, measured at excitation and emission wavelengths of 560 and 590 nm, respectively, is taken as the zero level. Subsequently, the fluorescence of a Triton X-100-treated sample is read, and its fluorescence, corrected for sample dilution, is then set at 100% (infinite dilution). After calibration, a fresh virus sample is mixed in the fusion medium (the fluorometer sample chamber is equipped with a magnetic stirring device to ensure optimal mixing during the experiment), the chart recorder is activated, and after the zero level has been marked the target membrane preparation is added by injecting the preparation into the cuvette with a Hamilton syringe through a small hole in the cover of the sample chamber.

The occurrence of fusion is then monitored continuously by following the increase in fluorescence. Parameters usually used to characterize the fusion process are the initial rate of fluorescence increase, determined from the slopes of the fluorescence readings at time zero, and the final extent of fluorescence increase (see Fig. 2). For Sendai virus, we routinely incubate the virus with target membranes overnight to determine this level. For viruses that fuse more efficiently than Sendai virus, this level can be reached much more rapidly, as reflected by leveling off of the fluorescence tracings. The extent of fusion reached will depend on the surface area available for dilution. With mammalian cells as target membranes, the conditions can be taken such that if all viruses would fuse, infinite dilution is obtained (i.e., a density of the probe in the fusion product of less than 0.1–0.2 mol%, representing conditions at which self-quenching of the probe is essentially negligible). Under such conditions the increase in fluorescence is directly proportional to the extent of fusion.[25] With liposomes such a condition can also be accomplished, provided that a large excess of liposomes is added. At lower concentrations the extent of fusion can be derived from a calibration curve, constructed by incorporating known amounts of R_{18} in the liposomal bilayers and measuring its fluorescence as a function of the mole percent R_{18}. Up to 9 mol%, the fluorescence increases linearly.[33,34] It is worthwhile to note that the R_{18} assay provides the possibility of obtaining simultaneously the fraction of viruses that actually fuses and the fraction that, under certain conditions, remains bound to the target membrane without engagement in fusion.

When virus fusion experiments are carried out at 37° (i.e., the virus is added to the target membrane at this temperature without a low-temperature preincubation), using assays as described above, the fluorescence increase essentially reflects the overall fusion reaction. This overall process involves an initial binding step of adjacent membranes and the actual fusion reaction per se. A theoretical model, based on a simple mass action kinetic model, has been developed, which, in conjunction with such

[33] K. Klappe, J. Wilschut, S. Nir, and D. Hoekstra, *Biochemistry* **25**, 8252 (1986).
[34] S. Nir, K. Klappe, and D. Hoekstra, *Biochemistry* **25**, 8261 (1986).

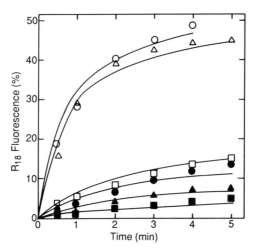

FIG. 4. Kinetic analysis of Sendai virus–liposome fusion. R_{18}-labeled Sendai virus was suspended in a medium of pH 7.4 (filled symbols) or 5.0 (open symbols), and fusion was initiated on injection of liposomes of various compositions. The curves were obtained by continuous monitoring of the R_{18} fluorescence increase. The data points were obtained by simulation of the fusion reaction, applying the mass action kinetic model (see text). The aggregation and fusion rate constants are given in Table I. The liposomes consisted of cardiolipin (CL, 10 μM lipid, O, ●), CL/dioleoylphosphatidylcholine (DOPC) (molar ratio 1:1, 25 μM lipid, △, ▲), and phosphatidylserine (PS, 50 μM lipid, □, ■). Note that the fusion event as a function of pH differs between these liposomes and erythrocyte membranes (see Fig. 5). (Reproduced from Ref. 34, with permission.)

fluorescence assays, provides a means to distinguish between these sequential steps and to determine the separate rate constants.[7,34] Thus, parameters that affect viral attachment can be determined by estimating the aggregation rate constant, whereas those that affect the molecular events involved in the fusogenic destabilization of virus and target membrane can be appreciated by determining the fusion rate constant (Fig. 4, Table I).

Lipid Mixing Assays: Some Critical Notes

When membrane fusion is measured using exogenous probes, it is essential to bear in mind that the probe itself may interfere with a particular step in the overall fusion process. With lipid mixing assays, an additional hazard may arise from the transfer or exchange of probe between membranes, in which case the probe may not report a merging step. For each system under study, it is therefore imperative to include appropriate control experiments. For fusion of pure liposomal systems,[11] control experiments should include, whenever possible, the application of contents

TABLE I
INTERACTION OF SENDAI VIRUS WITH LIPOSOMES[a]

Liposome composition	pH	$f(\text{sec}^{-1})$	$C\ (M^{-1}\ \text{sec}^{-1})$
CL	7.4	0.018	9×10^7
CL/DOPC	7.4	0.009	8×10^7
PS	7.4	0.005	7×10^7
CL	5.0	1	1.8×10^8
CL/DOPC	5.0	0.2	1.2×10^8
PS	5.0	0.03	5×10^7

[a] Rate constants of fusion (f) and aggregation (C) were determined as a function of liposomal membrane composition and pH. Experimental conditions were as described in the legend to Fig. 4. Data were obtained by kinetic simulation. Note that in all cases the liposome composition and pH markedly affect the fusion reaction itself, rather than the aggregation step. For details, see Ref. 34.

mixing assays. For biological membrane systems, contents mixing is commonly impossible. Other criteria have to be applied therefore, usually relying on a comparison with other lipid mixing assays, in conjunction with specific, known fusion properties of the membrane preparation as

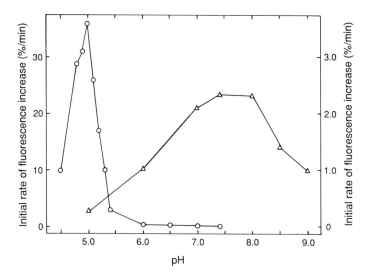

FIG. 5. Fusion of viruses with erythrocyte membranes as a function of pH. With certain mammalian cell membranes, Sendai virus (△) displays optimal fusion activity in the neutral pH region. Influenza virus (○) is known to enter cells by receptor-mediated endocytosis and to fuse with the endosomal membrane from within at mildly acidic pH. As shown, these events can be simulated using erythrocyte membranes and monitoring the fusion of R_{18}-labeled viruses as a function of pH. The initial rates are plotted.

derived from other experiments. For viruses, this can be done quite conveniently.

The amount of probe required for profitable use of the R_{18} assay as discussed above is fairly high. This is dictated by the need to obtain a reasonable degree of self-quenching (see also discussion in Ref. 16). However, at this concentration, hemolysis induced by labeled or unlabeled virus is virtually indistinguishable.[35] Also, the fusion of R_{18}-labeled virus with liposomes and the fusion of unlabeled virus with liposomes labeled with a much lower concentration of lipid derivatives capable of undergoing resonance energy transfer[11] display very similar fusion kinetics.[25] Furthermore, viruses can be bound to target membranes under conditions (i.e., low temperatures) where fusion does not occur. Under those conditions there should be no relief of fluorescence self-quenching (cf. Fig. 3), thus excluding transfer of single probe molecules. Neither should such a transfer occur under conditions where fusion is inhibited, for example, on specific proteolytic treatment of virus that causes the removal of the protein that mediates viral fusion.[1,3,32] Incubation of the labeled virus and target cells with an excess of unlabeled virus should inhibit fluorescence dequenching and cause no probe transfer to the unlabeled virus.[26,36] Many viruses display a mildly acidic pH-dependent fusion activity (Fig. 5). This criterion can also be used to verify the reliability of a lipid mixing assay registering virus fusion activity. Apart from providing a wealth of detailed insight into the overall fusion process itself, kinetic analysis of the fusion data should indicate that the attachment of virus to a target membrane follows second-order kinetics while the fusion reaction itself is first-order.[7,34,37]

Finally, it is also possible to compare different fusion assays in the same system. For Sendai virus such a comparison has been made using the R_{18} assay, a fluorescence photobleaching recovery technique, monitoring of the lateral mobility of N-NBD-PE, and a chemical assay that distinguishes bound from fused virus particles.[18] With erythrocyte ghosts as target membranes, very similar results in terms of fused and bound virus fractions were obtained for all three assays. A close similarity in the kinetics of the fusion of vesicular stomatitis virus with phospholipid vesicles has been reported in a comparative study with exogenously labeled virus, using R_{18} and pyrene-labeled fatty acid, respectively[24] (Fig. 6). More recently, another important means of monitoring viral protein-induced fusion has been described, based on aqueous contents mixing.[38] The approach

[35] D. Hoekstra, K. Klappe, T. de Boer, and J. Wilschut, *Biochemistry* **24**, 4739 (1985).
[36] M. C. Pedroso de Lima, S. Nir, D. Flasher, K. Klappe, D. Hoekstra, and N. Düzgüneş, *Biochim. Biophys. Acta* **1070**, 446 (1991).
[37] S. Nir, K. Klappe, and D. Hoekstra, *Biochemistry* **25**, 2155 (1986).
[38] D. P. Sarkar, S. J. Morris, O. Eidelman, J. Zimmerberg, and R. Blumenthal, *J. Cell Biol.* **109**, 113 (1989).

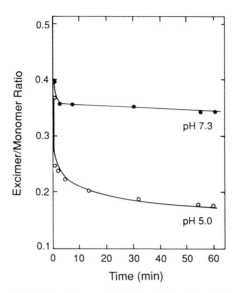

FIG. 6. pH-Dependent fusion of pyrene-labeled vesicular stomatitis virus with PS vesicles. The virus was metabolically labeled with pyrene fatty acid, yielding a virus containing pyrene-labeled phospholipids in its envelope. The labeled viruses were incubated (at 37°) with PS vesicles in a medium of pH 7.3 (●) or pH 5.0 (○). At various time intervals an aliquot of the incubation mixture was taken, and the excimer/monomer ratio was measured [λ_{ex} 330 nm; λ_{em} 385 nm (monomer) and 470 nm (excimer)]. The sharp decline in the ratio at pH 5.0, but not at pH 7.3, is consistent with penetration of the virus into cells by endocytosis, and fusion, at mildly acidic pH, in the endosomal compartment. (Data were redrawn from Ref. 24.)

relies on fusing red blood cells, loaded with a water-soluble fluorescent dye, N-(7-nitrobenzofurazan-4-yl)taurine (NBD-taurine), with influenza hemagglutinin-expressing fibroblasts. The fluorescent dye is partially quenched by hemoglobin. On fusion with the fibroblasts, the dye is diluted, causing a concomitant increase of fluorescence. The kinetics of the cell fusion process thus monitored was closely matched by the kinetics of fusion followed by relief of R_{18} fluorescence self-quenching that occurred when R_{18}-labeled red blood cells fused with the unlabeled hemagglutinin-expressing fibroblasts (Fig. 7).

In summary, a number of fusion assays, based on fluorescence, are available that allow the continuous monitoring of fusion of viruses with a variety of target membranes. Provided that careful control experiments are carried out that exclude potential hazards in the system under study (i.e., control experiments should be done for each virus and target membrane used), these assays can be valuable tools to elucidate and understand the mechanism of virus fusion.

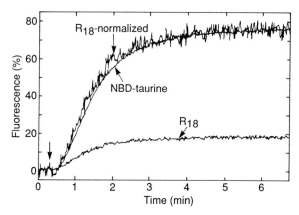

FIG. 7. Fusion of influenza hemagglutinin-expressing fibroblasts with erythrocytes, monitored by contents and lipid mixing. The erythrocytes were loaded with the water-soluble fluorescent dye NBD-taurine, and R_{18} was incorporated into their membranes. Labeled cells were incubated with hemagglutinin-expressing fibroblasts, and fusion was triggered by lowering the pH to 5.0 (arrow). The relief of NBD-taurine fluorescence quenching (as a measure of fusion by contents mixing, see text) and the relief of self-quenching of the membrane marker R_{18} (as a measure of fusion by lipid mixing) was continuously monitored as a function of time. The contents mixing curve was corrected for leakage. Note that, after normalizing the R_{18} curve to the maximal extent of NBD-taurine dequenching, the kinetics of fusion obtained with both assays are very similar. [Data were redrawn from Ref. 38 (reproduced with permission).]

Acknowledgments

Parts of the work cited in this chapter were supported by a grant from the National Institutes of Health (AI 25534) and Grant 86-00010 from the United States–Israel Binational Science Foundation. The secretarial assistance of Mrs. Rinske Kuperus is gratefully acknowledged.

[21] Kinetics of Fusion of Enveloped Viruses with Cells

By Anu Puri, Michael J. Clague,
Christian Schoch, and Robert Blumenthal

Introduction

Fusion of viral envelopes with cell membranes has been assessed by means of biological assays[1] (infectivity, syncytium formation, hemolysis), by electron microscopic techniques,[2] and by determination of the transfer of biochemical markers from virus to cells.[3] These methods are often limited by inaccurate quantitation of the extent of fusion, and they do not allow continuous monitoring of the fusion reaction. Therefore, attempts have been made to generate direct fusion assays which allow continuous, sensitive, and quantitative monitoring of the fusion process. Fluorescence[4] and electron spin resonance techniques[5] have been developed to continuously monitor the fusion of intact virions with their targets. These assays can be performed without removal of unfused viruses and allow analysis of initial steps in the fusion process.

In this chapter we describe an assay for fusion of intact viruses with biological targets that is based on the principle of fluorescence dequenching. According to the method, which was originally used in studies of liposome fusion,[6] the fluorophore is incorporated into one of the fusing partners in such a way that its fluorescence is self-quenched. Once fusion has occurred, the fluorophore diffuses into the larger target, resulting in relief of self-quenching. Consequently an increase in the fluorescence signal is observed. Fluorescence dequenching assays have been developed for monitoring the mixing of lipids as well as aqueous spaces.[7] Core-mixing assays for viral fusion require reconstitution of viral envelopes,[8] whereas membrane mixing assays can be performed with intact virions.

[1] Y. Okada, Curr. Top. Membr. Transp. 32, 297 (1988).
[2] K. N. J. Burger, L. J. Caller, P. M. Frederik, and A. J. Verkleij, this volume [27].
[3] M. Marsh, this volume [19].
[4] D. Hoekstra and K. Klappe, this volume [20].
[5] S. Ohnishi and K. Kuroda, this volume [24].
[6] J. N. Weinstein, S. Yoshikami, P. Henkart, R. Blumenthal, and W. A. Hagins, Science 195, 489 (1977).
[7] S. J. Morris, D. Bradley, G. C. Gibson, P. D. Smith, and R. Blumenthal, in "Spectroscopic Membrane Probes" (L. Loew, ed.), p. 161. CRC Press, Boca Raton, Florida, 1988.
[8] A Loyter, V. Citovsky, and R. Blumenthal, Methods Biochem. Anal. 33, 128 (1988).

This chapter deals with a lipid mixing assay which utilizes the lipophilic fluorescent dye octadecylrhodamine B (R18).[4,9-11] R18 was originally developed by Keller et al.[9] to monitor cell fusion. Subsequently it was found that R18 could be incorporated into intact virions under self-quenching conditions without redistribution of the probe into the target membrane under conditions where no fusion occurs.[10] Fusion of viral envelopes with cell membranes has been studied using the R18 dequenching assay with a variety of different virus strains, including those that infect cells by acid-activated fusion following endocytosis [e.g., influenza virus[12,13] and vesicular stomatitis virus (VSV)[14,15]] and those that fuse directly with the plasma membrane at neutral pH (e.g., Sendai virus,[4,11] human immunodeficiency virus,[16] mumps virus,[17] vaccinia virus,[18] respiratory syncytial virus,[19] and Epstein-Barr virus[20]). According to the R18 dequenching assay, fluorescence changes associated with membrane fusion in a virus–cell suspension are measured directly using a spectrofluorometer. This method requires relatively simple instrumentation, and experimental conditions (e.g., temperature, incubation time, pH) can be easily manipulated.

Labeling of Intact Virions with Octadecylrhodamine B

Biophysical measurements of viral fusion require pure virus preparations.[21] Contamination of the virus stocks by cellular proteins either leads to underestimation of the fusion yield or complicates the analysis by nonspecific dequenching of the probe (see below). The procedures described below are for labeling of VSV or influenza virus, but they can be applied for labeling of other viruses. Stocks of R18 (Molecular Probes, Inc., Eugene, OR) are made to 1 mg/ml (1.28 mM) in absolute ethanol and

[9] P. M. Keller, S. Person, and W. Snipes, J. Cell Sci. 28, 167 (1977).

[10] D. Hoekstra, T. de Boer, K. Klappe, and J. Wilschut, Biochemistry 23, 5675 (1984).

[11] Y. I. Henis, this volume [26].

[12] T. Stegmann, H. W. Morselt, J. Scholma, and J. Wilschut, Biochim. Biophys. Acta 904, 165 (1987).

[13] O. Nussbaum and A. Loyter, FEBS Lett. 221, 61 (1987).

[14] R. Blumenthal, A. Bali-Puri, A. Walter, D. Covell, and O. Eidelman, J. Biol. Chem. 262, 13614 (1987).

[15] A. Puri, J. Winnick, R. J. Lowy, D. Covell, O. Eidelman, A. Walter, and R. Blumenthal, J. Biol. Chem. 263, 4749 (1988).

[16] F. Sinangil, A. Loyter, and D. J. Volsky, FEBS Lett. 239, 88 (1988).

[17] C. Di Simone and J. D. Baldeschwieler, Biophys. J. 57, 490a (1990) [Abstr.].

[18] R. W. Doms, R. Blumenthal, and B. Moss, J. Virol. 64, 4884 (1990).

[19] N. Srinivasakumar, P. L. Ogra, and T. D. Flanagan, J. Virol. 65, 4063 (1991).

[20] N. Miller and L. M. Hutt-Fletcher, J. Virol. 62, 2366 (1988).

[21] B. W. J. Mahey, "Virology: A Practical Approach." IRL Press, Oxford, 1985.

stored at $-70°$ in 0.5-ml aliquots in amber-colored vials with Teflon–rubber septums. The R18 stocks can be used up to 15 days when stored in the dark at 4°. The fluorescent properties of R18 in different environments have been described previously.[4,8,11]

An aliquot of 5.0–7.5 μl R18 solution is added with vortex mixing to 1 ml of a virus suspension (0.5 mg viral protein) in isotonic buffer at pH 7.4. The sample is allowed to stand at room temperature for 10–20 min in the dark. The ratio of R18 to virus may need to be varied to incorporate the probe at optimal quenching. Addition of an insufficient quantity of R18 may result in insufficient quenching. Addition of too much R18, on the other hand, may result in "supersaturation" of the viral membrane with the dye and consequently nonspecific transfer of R18 (in the absence of fusion). The percent quenching as a fusion of membrane concentration of R18 has been determined in liposomes[10] and intact virus.[22]

To remove unincorporated R18, the incubation mixture is loaded on a Sephadex PD-10 column (Pharmacia Fine Chemicals, Piscataway, NJ). The R18-labeled virus is eluted in the void volume, while the unincorporated R18 is retained in the column. Virus recovery from the column is 70–75%, as determined by viral protein assays. Unincorporated R18 can also be removed by centrifugation (60,000 g, 30 min, 4°) of the virus preparation onto a 25% sucrose cushion.[23] Although the latter method requires more time, it offers the advantages of concentrating R18-labeled virus from a dilute incubation mixture and of removing any enzymes which may be used to modify the virus in fusion studies. Sucrose and any R18 remaining unincorporated is removed by gel chromatography. Recovery of R18-labeled virus is 65–75%, as determined by viral protein assays. The extent of quenching in the samples is assessed by measuring the fluorescence of the R18-labeled virus before and after addition of Triton X-100 (0.05%, v/v, final concentration).

R18-labeled virus loses fusion activity after storage at 4° presumably owing to aggregation of virus particles. It is possible to observe single virions as well as large aggregates (4–8 particles) and "superclusters" (greater than 9 particles) by intensified quantitative fluorescent video optical microscopy.[15,24] Viral aggregates can be broken up by brief sonication or by repeated passage through a 26-gauge needle. Large viral aggregates, which interface with the fusion assay, may be removed by centrifugation (600 g for 3 min),[15] or by filtration through a 0.22-μm Millipore (Bedford,

[22] B. Aroeti and Y. I. Henis, *Exp. Cell Res.* **170**, 322 (1987).
[23] A. Puri, F. Booy, R. W. Doms, J. M. White, and R. Blumenthal, *J. Virol.* **64**, 3824 (1990).
[24] R. J. Lowy, D. P. Sarkar, Y. Chen, and R. Blumenthal, *Proc. Natl. Acad. Sci. U.S.A.* **87**, 1850 (1990).

TABLE I
BIOLOGICAL AND MODEL MEMBRANES AS TARGETS FOR VIRAL FUSION

Target	Advantage	Disadvantage
Liposomes	Well-defined composition Easy handling Availability Modification of membrane components and size Incorporation of receptors	Low fusion efficiency Fluorescence dequenching artifacts
Erythrocytes	Well-defined composition and structure Manipulation of cytoskeleton and lipid distribution Availability Handling	Limited virus range (receptors for only para- and orthomyxoviruses)
Cultured cells	Appropriate receptors Metabolic processes like endocytosis can be studied Results have direct implication for *in vivo* system	Structure and composition of the membrane not well-defined Not readily adapted to stopped-flow mixing chambers

MA) filter (Millex GV).[25] R18-labeled influenza virus appears to be less prone to loss of fusogenicity with time at 4° than R18-labeled VSV. In the case of Epstein-Barr virus, it has been shown that incorporation of R18 into the viral membrane does not affect either binding of virus to its cell surface receptor or viral infectivity.[20] For VSV, whose cell surface receptor is not well-defined, labeling with R18 did appear to enhance binding to cells.[14]

Target Membrane

The target membrane plays an important role in viral fusion. We briefly discuss the advantages and drawbacks of various targets commonly used to study viral fusion (see Table I). Liposomes are the simplest targets in which to study viral fusion. Their composition can be easily manipulated, and appropriate receptors can be incorporated in the liposomal surface. However, their use is restricted since, in some cases, the observed fluorescence dequenching is unrelated to biological fusion,[8] or their fusion efficiency is very low as compared to biological membranes.[26] A great

[25] M.-T. Paternostre, R. J. Lowy, and R. Blumenthal, *FEBS Lett.* **243**, 251 (1989).
[26] D. P. Sarkar and R. Blumenthal, *Membr. Biochem.* **7**, 231 (1988).

number of viral fusion studies have also been performed with erythrocytes and erythrocyte ghosts. Sialoglycolipids on erythrocyte membranes serve as receptors for Sendai[27] and influenza virus.[28] Methods are available to manipulate the distribution of phospholipids between the outer and inner monolayers of the erythrocyte membrane,[29-31] as well as procedures to control the interaction of cytoskeletal elements with the membrane.[32] These advantages make erythrocyte membranes attractive targets for viral fusion, but use of such membranes is limited because they lack receptors for most of the viruses. On the other hand, a variety of virus receptors have been identified on the surface of cultured cells.[33] Moreover, cultured cells are used to study entry of viruses via the endocytic pathway[12-14] (see Fig. 1). Another advantage of using the cultured cells is the possibility to study mutants or to modify components of the target membrane under different culture conditions.

Virus – Cell Association

In our studies the measured fusion is dominated by virus bound to the membrane at the time of mixing.[14,15,31] Thus, we can neglect consideration of diffusion limited virus – cell association in our analysis.[34] Figure 2 illustrates the fusion of VSV with erythrocyte ghosts: Fusion triggered when virus is prebound to erythrocyte ghosts is compared with fusion when virus and ghosts were added together without prebinding.[31] In the latter case virus and ghosts have to associate by random collisions before undergoing the process of fusion itself. If virus is prebound to ghosts, 30 – 40% fluorescence dequenching is observed after 400 sec (Fig. 2). If virus and ghosts are added separately, the final level of fluorescence dequenching is less than 5% of the prebound value, which is probably due to virus bound in the time before pH reduction (2 – 5 min).

[27] M. A. Markwell, L. Svennerholm, and J. C. Paulson, *Proc. Natl. Acad. Sci. U.S.A.* **78**, 5406 (1981).

[28] J. C. Paulson, J. E. Sadler, and R. L. Hill, *J. Biol. Chem.* **254**, 2120 (1979).

[29] P. Williamson, L. Algarin, J. Bateman, H. R. Choe, and R. A. Schlegel, *J. Cell. Physiol.* **123**, 209 (1985).

[30] S. Grimaldi, R. Verna, A. Puri, S. J. Morris, and R. Blumenthal, *in* "Advances in Biotechnology of Membrane Ion Transport" (P. L. Jorgensen and R. Verna, eds.), Serono Symposium Publications, Vol. 51, p. 197. Raven, New York, 1988.

[31] M. J. Clague, C. Schoch, L. Zech, and R. Blumenthal, *Biochemistry* **29**, 1303 (1990).

[32] P. S. Low, *Biochim. Biophys. Acta* **864**, 145 (1986).

[33] R. W. Compans, A. Helenius, and M. Oldstone (eds.) "Cell Biology of Virus Entry, Replication and Pathogenesis." Alan R. Liss, New York, 1989.

[34] S. Nir, this volume [28].

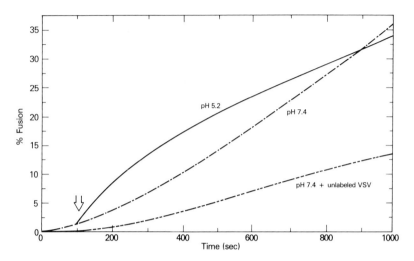

FIG. 1. Kinetics of fusion of R18-labeled VSV with Vero cells. R18-Labeled VSV was preincubated with Vero cells at 4°, pH 7.4, for 60 min. The cells were washed by centrifugation. Fifty microliters of the R18-labeled VSV–Vero complex was injected into a prewarmed cuvette at 37°, pH 7.4. The increase in fluorescence at pH 7.4 is due to endocytosis of the virus by the nucleated cells and subsequent fusion of the membranes of virus and endosomes as the pH is lowered in the endocytic compartment. It is blocked in the presence of a 10-fold excess of unlabeled VSV at pH 7.4. Fusion at the plasma membrane is triggered by lowering the pH in the medium (arrow) by addition of a small volume of a low pH buffer. Percent fusion is calculated from fluorescence dequenching values according to Eq. (1) (From Blumenthal *et al.*[14])

The R18 dequenching assay depends on dilution of the dye as a result of fusion of viral and cell membranes. To obtain maximum dequenching, relatively low amounts of virus need to be bound to the cell surface. However, a sufficient amount needs to be bound to obtain the fluorescence intensity necessary for spectrofluorometric measurement. In experiments of VSV with erythrocyte ghosts a virus to cell ratio of 100:1, which corresponds to a surface area ratio of about 1:37, allows full R18 dequenching while maximizing the signal observed.[31] The number of virus particles bound per cell should be optimized for each system being studied.

To estimate the amount of virus bound to the target membrane, the R18-labeled virus is incubated with a known concentration of cells at 4°. Free virus is separated from virus–cell complexes by centrifugation at low speed (300 *g*, 5 min, 4°), and the pellets are washed with cold buffer. To estimate the amount of cell-associated virus, the fluorescence in the pellet and supernatant is measured after addition of Triton X-100 (0.05%, v/v final concentration) and compared with a known amount of the same R18-labeled virus after solubilization with Triton X-100.

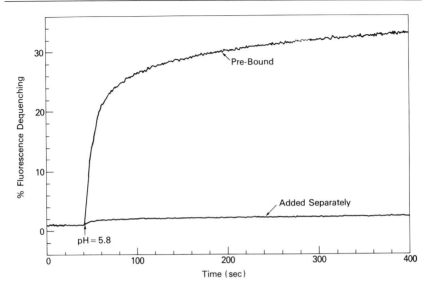

Fig. 2. Effect of association on VSV–erythrocyte ghost fusion kinetics. R18-Labeled VSV prebound to ghosts, or R18-labeled VSV and ghosts added separately, were incubated in 2 ml phosphate-buffered saline, pH 7.4, at 37°. At about 150 sec (only 40 sec shown), the pH in the medium was changed (arrow) to 5.83 by adding a small volume of low pH buffer. (From Clague et al.[31])

Kinetics of Viral Fusion

Fluorescence increases that occur as a result of fusion of R18-labeled virus with cells are monitored spectrofluorometrically after manual injection of the virus–cell complexes into a thermostatted cuvette containing a stirred buffer solution. For viruses which fuse at neutral pH (e.g., Sendai virus,[4,11] human immunodeficiency virus,[16] mumps virus,[17] vaccinia virus,[18] respiratory syncytial virus,[19] and Epstein-Barr virus[20]) the buffer solution is at neutral pH, whereas for viruses which require acid activation for fusion (e.g., influenza virus[12,13] and VSV[14,15]) the buffer solution is set at the desired pH. The mixing time according to this protocol is about 5 sec, and fluorescence changes may occur before the spectrofluorometric measurement. Alternatively, for acid-activated viruses, fusion may be triggered by injection of a small volume of low pH buffer into the cuvette containing the stirred suspension of virus–cell complexes at neutral pH. This method allows observation of the changes in the fluorescence signal with respect to the fluorescence at neutral pH. However, addition of a low pH solution may create overshoots in the local H^+ concentration, and thus trigger fusion in a subpopulation of virus–cell complexes. This effect may be significant when fusion is studied at pH values close to the threshold pH for fusion. The extent of fusion may be affected by possible dissociation of virus from the target membrane during preincubation or during the fusion process. Therefore, it is important to maintain the same preincubation conditions prior to triggering fusion.

Figures 1 and 2 show fluorescence dequenching on fusion of VSV with cells triggered by the low pH buffer injection methods. A 2 ml suspension of R18-labeled VSV–cell complexes at pH 7.4 and 37° is placed in a disposable plastic cuvette and stirred with a 2×8 mm Teflon-coated magnetic stir bar for about 20 sec. The pH of the incubation mixture is lowered by addition of 20 μl of a low pH buffer. Fluorescence is followed for 400 to 1000 sec. At the end of the run, 10 μl of 10% (v/v) Triton X-100 is added to obtain R18 fluorescence at infinite dilution. The pH of the solution in the cuvette is always measured at the end of the experiment.

To relate the fluorescence dequenching to the extent of fusion[8] we assume that the R18 is diluted into an infinite reservoir when a single virus fuses with the cell membrane. The total fluorescence change of a population of fused viruses then equals the fluorescence change associated with fusion of a single virus times the number of viruses that have fused. It is assumed that the fluorescence change associated with the fusion of a single virus does not depend on how many viruses have fused with a target membrane. The relationship between percent fusion and the fluorescence dequenching[8] is then given by Eq. (1):

$$\% \text{ Fusion} - \frac{(F - F_0)}{(F_t - F_0)} \tag{1}$$

where F, F_0, and F_t are fluorescence values at a given time, at zero time, and after disruption of virus–cell complexes with Triton X-100, respectively. The total fluorescence of the dye in the detergent micelle may differ from that in the cell membrane under conditions of "infinite dilution." For R18 in Vero cells, a correction factor of 1.56 has been used to account for this difference.[14]

Initial events of viral fusion may be submersed in the mixing time in experiments where small volumes of virus–cell suspensions or low pH buffer solutions are manually injected into the cuvette. A high time resolution of those initial events can be obtained by means of stopped-flow mixing.[31] This technique allows analysis of any changes in the fluorescence signal within a time scale of 50 msec. Figure 3 shows data on rapid kinetics of fusion of VSV with erythrocyte ghosts using stopped-flow mixing. The measurement is performed with an SFA-11 rapid kinetics accessory (Hi-tech Scientific Ltd., Salisbury, England). One syringe contains the R18-labeled VSV and ghost suspension in a pH 7.4 buffer, whereas the other is filled with a low pH buffer. Fusion is triggered by mixing the contents of

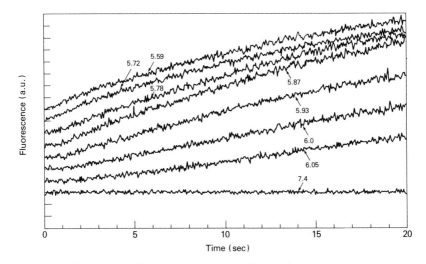

FIG. 3. Rapid kinetics of fluorescence changes on fusion of R18-labeled VSV with erythrocyte ghosts. The reaction was triggered by rapid mixing of equal volumes of an R18-labeled ghost suspension and a low pH buffer solution at 37°. Ten data sets were averaged for each pH indicated. The fluorescence values at zero time, when the suspension was mixed, were the same for all pH values but have been offset for clarity. (From Clague et al.[31])

the two syringes 1:1 into the observation cuvette within 50 msec. The stopped-flow mixing technique provides excellent resolution of the initial stages of viral fusion. It is particularly applicable to pH-triggered viral fusion. Those viruses that fuse at neutral pH need to be triggered in alternate ways (e.g., a temperature jump). Stopped–flow measurements with cultured cells may be limited by technical problems such as sedimentation of cells in the chamber tubing and shear in the mixing chamber.

Specificity of Octadecylrhodamine B Fluorescence Dequenching

Various control experiments have been designed to ascertain that changes in R18 fluorescence are associated with fusion. Although fluorescence of R18 is by itself not pH-dependent, changes in fluorescence of R18-labeled virus have been observed in the absence of target membranes. These effects probably reflect changes in the quantum yield owing to rearrangement of R18 within the plane of the viral membrane. Such effects are variable and are dependent on the efficiency of labeling. The fluorescence of the R18-labeled virus as a function of pH should be checked prior to binding to the cells. If the fluorescence changes are significant, the virus preparation should be discarded. Otherwise, a correction factor can be applied by subtraction of fluorescence changes observed prior to binding to the target.

Experimental protocols which involve inhibition of fusion without interference with binding are the most appropriate. A good example of such a control involves experiments with the uncleaved precursor of influenza hemagglutinin, which binds to sialoglycoproteins/lipids on target membranes with the same affinity as the cleaved form but is not fusogenic.[35] Antibodies against viral envelope proteins which inhibit fusion without affecting binding (e.g., VSV,[14] Epstein-Barr virus,[20] and vaccinia virus[18]) have been used in control experiments. Alternatively, virions have been pretreated at temperatures above 56° or at a pH below 5 to inactivate fusion activity.[8] However, some virus strains (e.g., VSV[15] or the Japan strain of influenza virus[23]) appear to be resistant to inactivation by low pH pretreatment. Viral fusion activity may also be inhibited by treatment of virus with chemical reagents[8] [e.g., dithiothreitol (DTT), hydroxylamine] and enzymes (e.g., trypsin). It is then necessary to determine the effects of these treatments on binding versus fusion activity.

Analysis of Fusion Kinetics

The parameters to be considered for the analysis of fusion kinetics are lag times, the rise time of the fluorescence change, and the extent of the fusion reaction. The underlying assumption in the interpretation of the fusion kinetics is that the fluorescence changes reflect the kinetics of activation of the viral fusion process, and not the redistribution kinetics of the probe itself. One can show by a simple calculation based on particle size and lipid diffusion that a lipid probe would diffuse from an intact virus within 100 msec once membrane continuity has been established.[36] Video microscopic observations indicate a slower redistribution time of R18.[24] However, a theoretical calculation indicates that, even if the redistribution rates are of the order of magnitude of the activation reactions, the kinetics (specifically the delay times) are not significantly affected.[36]

At the optimum temperature of fusion, lag times for intact VSV range between 0.5 and 2 sec,[31] and they are only resolved by stopped-flow kinetics. At lower temperatures it is possible to observe the lag times by manual injection of virus–cell complexes. A decrease in lag time appears to be well correlated with an increase in the slope of the curves[31] (see Fig. 3). This fact is consistent with the notion that the same activation processes determine lag times and slopes. Longer lag times and smaller slopes are

[35] H.-D. Klenk, R. Rott, M. Ohrlich, and J. Blodorn, *Virology* **68**, 426 (1975).
[36] Y. Chen and R. Blumenthal, *Biophys. Chem.* **34**, 283 (1989).

indicative of a slow activation reaction. Lag times in fusion with the plasma membrane should not be confused with the slow rise in fluorescence seen on incubation of VSV[14] or influenza virus[12,13] with living cells at neutral pH (see Fig. 1). The latter effect is due to endocytosis of the virus by the nucleated cells and subsequent fusion of the membranes of virus and endosomes as the pH is lowered in the endocytic compartment. Because there is no endocytosis of intact virus by erythrocytes, we observed no fluorescence changes at pH 7.4 (see Fig. 3).

Figure 3 also shows variation in the extent of fusion at suboptimal pH values. Similar observations have been made with other acid-activated viruses.[37] Our interpretation of this pH dependence is that the viral proteins undergo pH-dependent conformational transitions which are not associated with fusion while the virus is bound to the target membrane. This will then give rise to "desensitization." If the rates of activation reactions are rapid relative to those of the desensitization pathway, optimal extents will be reached. This interpretation has become apparent after application of detailed curve-fitting techniques [e.g., the simulation analysis and modeling (SAAM) package used to analyze kinetic data by compartmental analysis[38]] to the body of kinetic data acquired by rapid-mixing assays.[31]

Conclusions

Using a fluorescence assay for continuous monitoring of fusion, the kinetics and extent of fusion of intact virions with a variety of targets can be studied. Control experiments show that R18 dequenching is specific for the fusion reaction. Because the assay does not require cells to be intact,[14] experimental conditions (e.g., pH, temperature, osmotic pressure) can be manipulated easily to shed light on various aspects of the fusion mechanism. The pathway of viral entry (e.g., fusion with the plasma membrane versus entry via the endocytic pathway) can be readily assessed based on kinetics and the effects of inhibitors. Analysis of the fusion kinetics using rapid mixing techniques (50 msec time resolution) reveal that the final viral fusion event is preceded by a complex multistep process.

Acknowledgments

We thank Drs. D. Dimitrov and R. J. Lowy for many helpful suggestions.

[37] S. Ohnishi, *Curr. Top. Membr. Transp.* **32,** 257 (1988).
[38] M. Berman, W. F. Beltz, P. C. Greif, R. Chabay, and R. C. Boston, "CONSAM User's Guide," DHHS, NIH, Bethesda, Maryland. (The package can be obtained from Dr. L. Zech, Bldg. 10, Rm 6B13, National Institutes of Health, Bethesda, MD 20892.)

[22] Metabolic Labeling of Viral Membrane Lipids by Fluorescent Fatty Acids: Studying Virus Fusion with Target Membranes

By Yechezkel Barenholz, Ranajit Pal, and Robert R. Wagner

Introduction

Enveloped virions introduce their nucleocapsid into eukaryotic host cells by two different routes, both of which are fusion-dependent. The common denominator is that the fusion is mediated by envelope viral proteins, although the two routes differ in their target membrane within the host cell.[1] In most systems studied so far (such as togá-, rhabdo-, and myxoviruses), the virions are introduced into cellular endosomes by receptor-mediated endocytosis, which is followed by low pH-dependent fusion triggered by acidification of the endosome lumen.[1,2] Fusion between the endosomal and viral membranes causes release of the transcribable viral genome and all the viral accessory proteins into the host cell cytosol.[1,3]

The second fusion pathway, for which paramyxoviruses are the main representatives, is characterized by receptor-mediated binding of the virions to the host cell plasma membrane and direct fusion between the virions and the host cell plasma membrane, which results in the introduction of the nucleocapsid into the cytosol. This fusion is pH-independent. The entry mechanism of the human immunodeficiency virus (HIV, a retrovirus) genome into its host cells involves the CD4 receptor of the host cells, although the exact mechanism is still a controversial issue.[4–7a]

Knowledge of the mechanism of fusion of virions with their target membranes is obtained from two main sources: virus–cell systems[1] and

---

[1] T. Stegmann, R. W. Doms, and A. Helenius, *Annu. Rev. Biophys. Chem.* **18**, 187 (1989), and references therein.

[2] J. White, M. Kielian, and A. Helenius, *Q. Rev. Biophys.* **16**, 151 (1983).

[3] S. Ohki, D. Doyle, T. D. Flanagan, S. W. Hui, and E. Mayhew (eds.), "Molecular Mechanisms of Membrane Fusion." Plenum, New York, 1988.

[4] P. J. Maddon, A. G. Dalgeish, J. S. McDougal, P. R. Clapham, R. A. Weiss, and R. Axel, *Cell (Cambridge, Mass.)* **47**, 333 (1986).

[5] J. M. McCune, L. B. Rabin, M. B. Feinberg, M. Lieberman, J. C. Kosek, G. R. Reyes, and I. L. Weissman, *Cell (Cambridge, Mass.)* **53**, 55 (1988).

[6] C. D. Pauza and T. M. Price, *J. Cell Biol.* **107**, 959 (1988).

[7] B. S. Stein, S. D. Gowda, J. D. Lifson, R. C. Penhallow, K. G. Bensh, and E. G. Engleman, *Cell (Cambridge, Mass.)* **49**, (1987).

[7a] N. Düzgüneş (ed.), "Mechanisms and Specificity of HIV Entry into Host Cells." Plenum, New York, 1991.

virus–liposome systems.[1-3, 8-10] Like most other biological fusion processes, the fusion of virions with the biological target membrane is characterized as a process which neither affects host cell membrane asymmetry nor involves leakage of macromolecules into the external medium. It is, however, characterized by the intermixing of components of the viral envelope and the host cell target plasma or endosomal membrane.[1-3,8,9] The intermixing of contents and of membrane components serves as the basis of most fusion assays, particularly quantitative ones. These fusion assays are based on a variety of biochemical and physical methods.[1-11] Among the assays used to determine the mixing of viral membrane and target membrane components, the fluorescent methods are the most popular. The extensive use of various fluorescent assays is justified by the following: (1) high sensitivity; (2) the ability to perform continuous on-line measurements; and (3) the ability to use kinetic models to obtain direct information on the fusion mechanism.[11,11a]

Two general approaches are used to measure mixing of virion membrane components with those of its target membrane. In the first type of assays, both membranes are labeled with different probes, so that the mixing of the two probes can be followed. These assays are referred to as "mixing of probes" assays. Alternatively, only one of the membranes is labeled, and the dilution of this probe by the unlabeled membrane is then used to follow the fusion. These assays are classified as "dilution of probes" assays.

Both types of assays can also be categorized according to the method of measuring fluorescence: (1) resonance energy transfer (RET) between two fluorescent lipid analogs to measure the change in the average distance between the two different fluorophores (quenching and dequenching for "mixing of probes" and "dilution of probes" assays, respectively);[1-11a] (2) Change in excimer to monomer ratio (E/M ratio) owing to dilution of concentration-dependent excimer-forming fluorophores, such as 2-pyrenyldodecanoylphosphatidylcholine (applied to "dilution of probes" assays only)[12,13]; or (3) dequenching by fusion-dependent dilution of con-

[8] N. Düzgüneş, Subcell. Biochem. 11, 195 (1985).

[9] J. Wilschut and D. Hoekstra, Chem. Phys. Lipids 40, 145 (1986).

[10] D. Hoekstra, K. Klappe, T. Stegmann, and S. Nir, in "Molecular Mechanisms of Membrane Fusion" (S. Ohki, D. Doyle, T. D. Flanagan, S. W. Hui, and E. Mayhew, eds.), 339. Plenum, New York, 1988.

[11] A. Loyter, V. Citovsky, and R. Blumenthal, Methods Biochem. Anal. 33, 120 (1988).

[11a] N. Düzgüneş and J. Bentz, in "Spectroscopic Membrane Probes," (L. M. Loew, ed.) Vol. 1, p. 117. CRC Press, Boca Raton, Florida, 1988.

[12] S. Amselem, Y. Barenholz, A. Loyter, S. Nir, and D. Lichtenberg, Biochim. Biophys. Acta 860, 301 (1986).

[13] R. Pal, Y. Barenholz, and R. R. Wagner, Biochemistry 27, 30 (1988).

centration-dependent self-quenched fluorophores, such as the fluorescent fatty acid octadecylrhodamine (ODR) (applied to "dilution of probes" assays only).[14] For all assays, the simplest approach is to use fluorescent labeled liposomes as target membranes and then to follow induced changes in RET, E/M ratio, or dequenching, using the above methods.

The use of these methods is prone to various artifacts such as lipid exchange, which is especially problematic for probes such as ODR. As ODR is a fatty acid, it may undergo relatively fast, spontaneous transmembrane exchange. This possibility can be ruled out, however, by means of several control experiments.[14-14c] Other artifacts for assays based on RET are referred to as "rapid artifacts" and "slow artifacts." The rapid artifacts result from the change in the lipid environment or pH in the vicinity of the fluorophores, and they differ from the dilution factor itself. The effect of change in environment near the fluorophore can be corrected by establishing a calibration curve or, preferably, by introducing fluorophores whose effect on the environment is minimal, and which are not going to interfere with the effect of lipid dilution. These probes are characterized by polar fluorophores located at carbon atom 12 of the long-chain acyl chain at position 2 of phosphatidylcholine.[15]

Overcoming the rapid artifacts reveals the problems of the slow artifacts. The latter are related to interactions between liposomes, which do not involve fusion with the virions. Examples include liposome aggregation, liposome collapse and "hemifusion," and redistribution of fluorescent probes in the liposome membrane induced by liposome aggregation and/or liposome collapse.[15] The correction for slow artifacts is not simple, especially for viral-induced RET; however, this is a lesser problem for the change in excimer monomer ratio.[12] In most of the studies on viral fusion, liposomes were used as target membranes, and, in most cases, the liposomes, and not the virions, were fluorescently labeled.[1,3,8-10,14-16]

A complementary and more biologically relevant approach is to use fluorescently labeled virions. The method used thus far has been to label

[14] D. Hoekstra, T. de Boer, K. Klappe, and J. Wilschut, *Biochemistry* **23**, 5675 (1984).

[14a] N. Düzgüneş, M. C. Pedroso de Lima, L. Stamatatos, D. Flasher, D. Alford, D. S. Friend, and S. Nir, *J. Gen. Virol.* **73**, 27 (1992).

[14b] M. C. Pedroso de Lima, S. Nir, D. Flasher, K. Klappe, D. Hoekstra, and N. Düzgüneş, *Biochim. Biophys. Acta* **1070**, 446 (1991).

[14c] D. Hoekstra and K. Klappe, this volume [20].

[15] J. R. Silvius, R. Leventis, and P. M. Brown, in "Molecular Mechanisms of Membrane Fusion" (S. Ohki, D. Doyle, T. D. Flanagan, S. W. Hui, and E. Mayhew, eds.), p. 531. Plenum, New York, 1988.

[16] N. Düzgüneş, T. M. Allen, J. Fedor, and D. Papahadjopoulos, in "Molecular Mechanisms of Membrane Fusion" (S. Ohki, D. Doyle, T. D. Flanagan, S. W. Hui, and E. Mayhew, eds.), p. 543. Plenum, New York, 1988.

purified virions with ODR, a fluorescent fatty acid which spontaneously inserts into lipid bilayers and biological membranes.[10,11,14-14c] Although this approach is very attractive because of its simplicity, some important questions regarding this assay remain unanswered. The most important questions are related to ODR organization in the viral envelope and how well the dequenching of ODR represents the mixing of viral membrane with the target membrane. An alternative approach to the dequenching of ODR is based on the use of metabolically fluorescent-labeled virions. The main advantage of this approach is that the metabolic labeling reflects the actual transmembrane and lateral organization of membrane components.

This approach requires the labeling of the virion host cells by growing them in the presence of fluorescent precursors (such as fluorescent fatty acids). During cell division and growth, the fluorescent precursor will be incorporated into the cells and will become part of the host cell membrane lipids through biosynthetic routes and intracellular traffic. After viral infection of the prelabeled cells and viral replication, the virions that bud from these cells will acquire part of the cell plasma membrane and thereby will include fluorescent lipids in their membrane. The use of this approach is demonstrated here for fusion of vesicular stomatitis virions (VSV) with liposomes. VSV is used because its protein and lipid composition, distribution of lipids between the two faces of the membrane, lipid–protein interactions, and lipid dynamics are well characterized.[17,18] The metabolic prelabeling of the baby hamster kidney (BHK-21) host cells is accomplished by their growth in the presence of the ω-pyrenyl fatty acids. The isolated and purified virions contain pyrenyl phospholipids in their envelope.

The pyrenyl moiety is the label of choice for three main reasons. (1) It is possible to follow the fusion and to quantify the mixing of labeled virion membranes with target membranes through the dilution-dependent change in excimer to monomer ratio.[12] (2) The combination of using the hydrophobic pyrenyl group as the fluorophore together with the location of this moiety at position ω of the phospholipid acyl chain minimizes local perturbation to the lipid bilayers (Y. Barenholz, unpublished data, 1985). (3) The versatility in the physical properties of the pyrenyl moiety, as fluorophore and chromophore, make it a very attractive candidate not only for fusion or transmembrane transfer studies, but also for obtaining basic information on physical changes occurring in the membrane during and after fusion.[19,20]

[17] E. J. Patzer, N. F. Moore, Y. Barenholz, J. M. Shaw, and R. R. Wagner, *J. Biol. Chem.* **253**, 4544 (1978).

[18] R. Pal, Y. Barenholz, and R. R. Wagner, *Biochim. Biophys. Acta* **906**, 175 (1987).

[19] H. J. Pownall and L. C. Smith, *Chem. Phys. Lipids* **50**, 191 (1989).

[20] Y. Barenholz, T. Cohen, R. Korenstein, and M. Ottolenghi, *Biophys. J.* **60**, 110 (1991).

Materials and Methods

Chemicals

16-(9-Anthroyloxy)palmitate (C16AP), 2-(9-anthroyloxy)palmitate (C2AP), 9-(1-pyrenyl)nonanoic acid (PyC9), 10-(1-pyrenyl)decanoic acid (PyC10), 12-(1-pyrenyl)dodecanoic acid (PyC12), and 16-(1-pyrenyl)hexadecanoic acid (PyC16) are obtained from Molecular Probes (Junction City, OR); 10-(1-pyrenyl)-10-oxodecanoic acid (PyC10-keto) was obtained from Sigma Chemical Co. (St. Louis, MO). For best results the pyrene fatty acids should be purified by HPLC.[21] The structures of the fluorescent fatty acids are compared to that of palmitic acid in Scheme 1. 1-Palmitoyl-2-oleoyl-phosphatidylserine (POPS) is obtained from Avanti Polar Lipids (Birmingham, AL).

Cells and Virus

BHK-21 cells are trypsinized gently and passed directly into Dulbecco's modified Eagle's medium (GIBCO, Grand Island, NY) containing 5% fetal calf serum, 10% tryptose phosphate broth, 1% penicillin, and 1% streptomycin. Cells are grown as monolayers at 37° in 20 ml medium. When confluent cells are passaged 1 to 10 in 75-cm²-surface Falcon plastic flasks; passaged cells reach confluency within 3 days (about 4×10^7 cells/flask). The BHK-21 cells used for the preparation of fluorescent virions are grown in the presence of the desired fluorescent fatty acid. The desired concentration (in the range of 5–40 μg/ml) of fluorescent fatty acid in dimethyl sulfoxide (DMSO) is added to the growth medium; the final concentration of DMSO is kept below 1%. When cell growth is complete, generally after 3 days, the medium is removed, and the cell layer is washed once with phosphate-buffered saline (PBS) (pH 7.4) and infected with VSV (Indiana serotype, San Juan strain) at a multiplicity of 0.1 plaque-forming units (pfu)/cell. No fluorescent fatty acid is added to the medium after infection. The supernatant fluids are collected 16–18 hr postinfection, pooled, and centrifuged at 900 g for 20 min at 4°. All subsequent steps are performed at 4° or on ice.

The harvested supernatant is then centrifuged at 80,000 g for 90 min through a 2-ml pad of 50% glycerol in Earle's balanced salt solution (EBSS) in an SW27 rotor. The resulting pellet is gently resuspended into EBSS, layered over a preformed 0–40% linear sucrose gradient containing 50 mM tris(hydroxymethyl)aminomethane (Tris), 0.25 M NaCl, and

[21] R. C. Heresko, T. C. Markello, Y. Barenholz, and T. E. Thompson, *Chem. Phys. Lipids* **38**, 263 (1985).

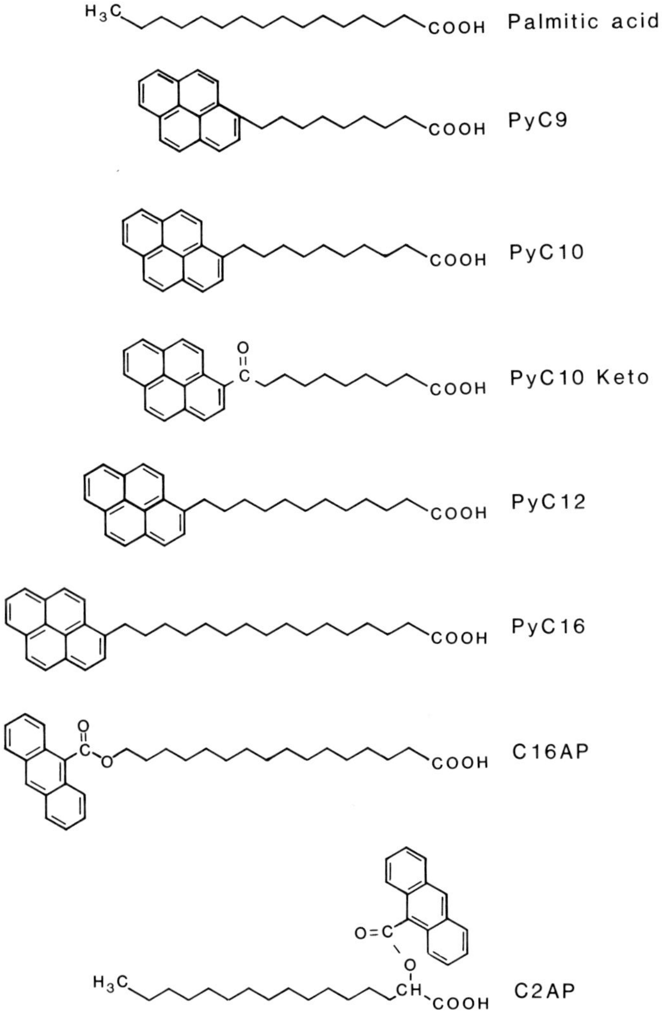

SCHEME 1. Structures of fluorescent fatty acids added to the BHK cell growth medium: 9-(1-pyrene)nonanoic acid (PyC9), 10-(1-pyrene)decanoic acid (PyC10), 10-(1-pyrene)-10-ketodecanoic acid (PyC10-keto), 12-(1-pyrene)dodecanoic acid (PyC12), 16-(1-pyrene)hexadecanoic acid (PyC16), 16-(9-anthroyloxy)palmitic acid (C16AP), and 2-(9-anthroyloxy)palmitic acid (C2AP).

0.5 mM ethylenediaminetetraacetic acid (EDTA) (pH 7.6), and centrifuged for 90 min at 35,000 g in an SW25.1 rotor. The clearly visible band of B virions is harvested, diluted approximately 5-fold with PBS, and pelleted through a 2-ml pad of 50% glycerol in PBS at 60,000 g for 90 min in an SW25.1 rotor. The virus pellets are drained and gently resuspended in PBS, and the suspension is overlaid on a gradient of 0–40% potassium tartrate and 0–20% glycerol containing 20 mM Tris (pH 7.6). Centrifugation to equilibrium is for 18 hr at 35,000 g in an SW25.1 rotor. The visible band is harvested, diluted, and pelleted as before. The virus pellets are drained and carefully washed with PBS before resuspension in PBS at a concentration of 2–10 mg/ml (viral protein). Virus is stored at 4° for use within 2 days or at −80° for longer storage periods.[20,22]

Viral Infectivity

The infectivity of the virus is measured by assaying for plaques formed on monolayers of L-929 cells.[23] For this L-929 cells are grown as monolayers at 37° in 20 ml of Eagle's basal medium (BME) containing Hanks' salts supplemented with 10% fetal calf serum and 1% each of penicillin and streptomycin. Cells are passaged at confluency 1 to 3 in 75-cm² Falcon plastic flasks and reach confluency within 3 days (about 2×10^7 cells/flask).

Removal of Free Fluorescent Fatty Acids from Virions

Most (> 90%) of the fluorescent lipids obtained during metabolic labeling of the virions are phospholipids (see Table I). As demonstrated below, a small fraction of the viral fluorescence is due to unreacted free fluorescent fatty acid. Fluorescent fatty acids undergo a fast intermembrane transfer ($t_{1/2}$ 7 min) (Fig. 3 below) that complicates data interpretation. This problem is overcome by taking advantage of the difference in $t_{1/2}$ values of spontaneous transfer of fluorescent fatty acids and of fluorescent phospholipids ($t_{1/2}$ 12 hr) (see Fig. 3). Routine removal of most free fluorescent fatty acids is done by incubating VSV (5 mg protein) in 5 ml 5% heat-inactivated fetal calf serum in PBS for 30 min (4 times $t_{1/2}$) at 37°. This procedure removes more than 90% of the viral fluorescent fatty acid with less than 5% reduction in viral fluorescent phospholipids. The virions are then purified by 0–40% sucrose density-gradient centrifugation as described above. These virions, referred to as "free pyrenyl fatty acid-depleted virions," are the only virions used for fusion studies.[13]

Lipid Analysis

The total lipids of the virions are extracted by the method of Folch et al.[24] The virions are extracted by 19 volumes of chloroform–methanol

---

[22] Y. Barenholz, N. F. Moore, and R. R. Wagner, *Biochemistry* **15**, 3563 (1976).

[23] J. J. McSharry and R. R. Wagner, *J. Virol.* **7**, 59 (1971).

[24] J. Folch, M. Lees, and G. H. Sloan-Stanley *J. Biol. Chem.* **266**, 497 (1957).

solution (2:1, v/v), then water is added to give a final 8:4:3 volume ratio of chloroform–methanol–water. Two phases are formed after mixing and centrifugation. The lower (chloroform) phase contains more than 99% of the total viral lipids. Total phospholipids are quantified using the procedure of Bartlett.[25] The cholesterol level is determined with the aid of cholesterol oxidase as described elsewhere[26] using sodium taurocholate as detergent to make cholesterol available for the enzyme.

Phospholipid Composition of Vesicular Stomatitis Virus

The phospholipids and neutral lipids are separated by thin-layer chromatography (TLC) on 250 μm silica gel G plates (Analytical Techniques, Newark, DE) using a solvent system of chloroform–methanol–NH_4OH (65:25:5, v/v). The fluorescent lipids are detected under UV light. Then, for detection of all spots the TLC plates are sprayed with primuline solution (1 mg per 100 ml acetone–water, 80:20, v/v). After solvent evaporation the lipid spots are detected using UV light. The advantages of the primuline spray include the high sensitivity of the assay and the lack of interference with other assays such as ninhydrin spray and phosphorus determination. The lipid spots are scraped from the TLC plates and analyzed for phosphorus content using the Bartlett procedure as described by Barenholz et al.[27] It is worth noting that the above specified TLC silica plates have a very low phosphorus content, and, therefore, they are suitable for quantitative determination of phospholipid by TLC.

Protein Determination

Protein is determined by the method of Lowry et al.[28] using crystalline bovine plasma albumin as the standard.

Phospholipase C Treatment

Phospholipase C (Clostridium perfringens welchii type I EC 3.1.4.3) from Sigma (St. Louis, MO) is dissolved in 20 mM Tris and 100 M NaCl (pH 7.5) buffer. Virions are suspended at a concentration of 0.5–1.0 mg/ml in the same buffer containing 1.3 mM $CaCl_2$ and are treated with 0.075 unit/ml phospholipase C at 37° for 3 hr until all the phospholipids accessible to phospholipase C are hydrolyzed.[19] The reaction is stopped by

[25] G. R. Bartlett, J. Biol. Chem. 234, 466 (1959).
[26] Y. Barenholz, E. Patzer, N. F. Moore, and R. R. Wagner, in "Enzymes in Lipid Metabolism" (S. Gatt, L. Freysz, and P. Mandel, eds.), p. 45. Plenum, New York, 1978.
[27] Y. Barenholz, E. Yechiel, R. Cohen, and R. Deckelbaum, Cell Biophys. 3, 115 (1981).
[28] O. H. Lowry, N. F. Rosebrough, A. L. Farr, and R. J. Randall, J. Biol. Chem. 193, 265 (1951).

extraction of the lipids. The lipid composition before and after phospholipase C hydrolysis is analyzed and quantified by TLC as described above.

Quantitation of Extent of Viral Labeling

Total viral lipids are extracted by the Folch procedure.[24] Individual phospholipids are analyzed by extraction of the spots from the TLC plate by chloroform–methanol (2:1, v/v) and chloroform–methanol–water (1:2:1, v/v) solutions, and the extracts are processed by the Folch procedure.[24] The lower phase is protected from light and dried, either under a stream of nitrogen or by using flash evaporation, under reduced pressure. The dried lipids are then dissolved in ethanol. The total, or individual, pyrenyl lipids in the virus is estimated by reading the optical density of the lipid extract in ethanol at 345 nm and comparing this with standard curves obtained with known amounts of fatty acids in ethanol.

Liposome Preparation

Small unilamellar vesicles (SUV) composed of 1-palmitoyl-2-oleylphosphatidylserine (POPS) are used as the target membrane for fusion studies. It is important to use Ca^{2+}-free POPS.[29] POPS is dissolved in chloroform–methanol (2:1, v/v). Then 0.1 M sodium EDTA (pH 7.2) in double-distilled water is added to a final volume ratio of 8:4:3 chloroform–methanol–0.1 M EDTA.[24,29] After vortexing and centrifugation, the aqueous methanolic upper phase is removed and the lower chloroform phase washed with synthetic Folch upper phase (chloroform–methanol–water 6:94:98, v/v). More than 98% of the Ca^{2+}-free POPS is recovered in the lower, chloroform phase. The SUV are prepared as described by Barenholz et al.[29,30] with minor modifications. In short, Ca^{2+}-free POPS in chloroform is dried using a combination of flash evaporation under reduced pressure followed by vacuum pump evaporation. Hydration of the thin lipid film is carried out in unbuffered 0.9% NaCl solution, to give multilamellar vesicles (MLV). The SUV are produced from MLV by ultrasonic irradiation.[26] SUV are fractionated by differential centrifugation.[30] The SUV are diluted to give a final concentration of 5–10 mM POPS.

Fluorescence Measurements

Fluorescence emission intensity is measured in quartz cells with a fluorescence spectrophotometer (e.g., Perkin-Elmer MPF3, MPF-44, or

[29] D. Lichtenberg and Y. Barenholz, Methods Biochem. Anal. 33, 337 (1988).
[30] Y. Barenholz, D. Gibbes, B. J. Litman, T. E. Thompson, and F. D. Carlsson, Biochemistry 16, 2806 (1977).

LS-5). The pyrene-labeled lipids in virions are excited at 330 nm, and the uncorrected spectra are recorded at wavelengths from 360 to 530 nm. The excimer to monomer fluorescence intensity ratio (E/M) is measured by exciting the pyrene-labeled virions at 330 nm; the fluorescence intensities of excimer and monomer emissions are determined at 470 and 385 nm, respectively.[19,20]

Characterization of Viral Labeling

Effect of Chain Length of ω-(1-Pyrenyl) Fatty Acids and of Pyrene Fluorophore Modification

The degree of incorporation of pyrene fatty acids into VSV membranes is shown in Table I. The structures of the pyrenyl fatty acids are compared to palmitic acid in Scheme 1. Pyrenyl fatty acids with chain lengths of 9, 10, 12, and 16 carbons are all significantly incorporated in the VSV membrane, whereas the incorporation is markedly less for the PyC10-keto fatty acid. This may be due to greater polarity in the ω-pyrene moiety (see Scheme 1). Total lipids extracted from virions are analyzed by TLC, and the amount of fluorophore present in phospholipids and neutral lipids is measured. As shown in Table I, most of the pyrenyl fluorescence (>90%) is found in phospholipids, whereas only 8% of the pyrene fluorescence is detected in neutral lipid fractions, primarily as free fatty acids. It is worth

TABLE I
INCORPORATION OF PYRENE LIPIDS IN VESICULAR STOMATITIS VIRUS[a]

Type of fatty acid in growth medium	Total pyrene lipids[b]	Pyrene phospholipids[b]	Pyrene neutral lipids[b]
PyC9	25.0	21.3	1.8
PyC10	27.5	21.6	1.6
PyC12	22.8	21.2	1.7
PyC16	26.1	21.2	1.8
PyC10-keto	6.8	5.9	1.0

[a] Virions were released from BHK-21 cells grown in the presence of ω-pyrenyl fatty acids of different chain lengths. BHK-21 cell monolayers grown to confluency in the presence of various pyrene fatty acids (30 μg/ml of growth medium) were infected with VSV at a multiplicity of 0.1 pfu/cell. The virions released in the medium were harvested 21 hr after infection and purified. Lipids were extracted and separated into phosphlipids and neutral lipids by thin-layer chromatography (TLC). The amount of pyrene incorporated in each fraction was measured by optical density at 345 nm with suitable calibration curves for each fatty acid in which the molar extinction was determined (see text).

[b] Total pyrene lipids, phospholipids, and neutral lipids were calculated as nanomoles fatty acid equivalent per milligram of viral protein.

noting that when cells are labeled in the presence of 30 μg fatty acid/ml medium, and VSV is produced under conditions described in Table I, about 8.0% of viral phospholipid acyl chains contain a pyrenyl moiety.

Effect of Concentration of ω-Pyrenyl Fatty Acids During Cell Growth

The amount of pyrenyl lipids incorporated into the VSV membrane is also measured as a function of the amount of pyrenyl-labeled fatty acid added to the cell growth medium. In this experiment, BHK cells are grown to confluency in monolayer cultures in the presence of different amounts of PyC16 fatty acid. The cells are then infected with VSV in a medium devoid of pyrenyl fatty acids, and virions released after incubation for 20 hr at 37° are purified and assayed for their content of pyrenyl lipids and total protein. As shown in Fig. 1, the pyrenyl content in the VSV envelope

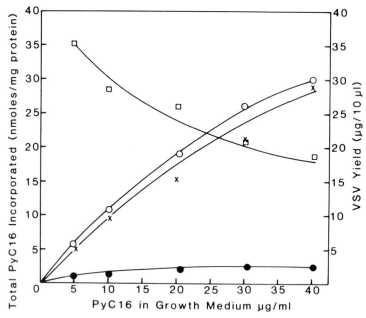

FIG. 1. Uptake of fluorescent lipids by VSV as a function of the amount of pyrenylhexadecanoic acid (PyC16) added to the cell growth medium, and the effect on virus yield. BHK cell monolayers were grown to confluency in the presence of various amounts of PyC16 and then infected with VSV (multiplicity of infection ∼0.1 pfu/ml). The virions were harvested at 20 hr postinfection, and the lipids were extracted and separated by TLC into neutral lipids and phospholipids as described in the text. Total pyrene incorporation in each fraction was measured by optical density at 345 nm. Symbols indicate total pyrene lipid uptake (○), neutral pyrene lipids (●), pyrene phospholipid (×), and total VSV protein content harvested from BHK cells grown in different amounts of PyC16 (□). (From Ref. 13 with permission.)

increases progressively when greater amounts of pyrenyl fatty acid are present in the preinfection BHK cell growth medium. The pyrenyl moiety incorporated in the viral membrane is associated primarily with the phospholipids.[13]

Figure 1 also shows that increasing the concentration of pyrenyl fatty acids in the BHK growth medium, from 5 to 40 μg/ml medium, reduces viral production by 50%. However, the metabolically labeled virions have identical infectivity with unlabeled virions. The infectivity of the PyC16-containing virions released from BHK cells grown in a medium containing 40 μg/ml of PyC16 is measured by plaque assays on L cell monolayers (see above) and compared with the infectivity of an identical amount of VSV released from BHK cells grown in the absence of pyrene fatty acids. Both control and pyrene lipid-labeled virions have an identical infectivity titer of 3×10^{10} pfu/ml, which suggests that the presence of pyrene lipids in the virion membrane has no effect per se on the infectivity of the virus.[13]

Distribution of Pyrene Lipids among Vesicular Stomatitis Virus Membrane Lipids and between Two Faces of Viral Membrane

The major phospholipids present in the membrane of VSV grown in BHK cells[17] are phosphatidylcholine (PC), sphingomyelin (SPM), phosphatidlyethanolamine (PE), and phosphatidylserine (PS) in approximate ratios of 24:24:31:18.[17] Table II shows the overall composition and bilayer distribution of pyrene fluorescence in the various phospholipids in the membrane of VSV harvested from BHK cells grown in the presence of PyC16 fatty acid. PyC16 constitutes 8% of total VSV acyl chains. PC and PE are the major lipids labeled with pyrene under such conditions, representing more than 85% of the total pyrene content in VSV. Only 5.3% of the total pyrene fluorophore is found in SPM and PS combined, although SPM and PS constitute 24% and 18%, respectively, of the total phospholipids in the VSV membrane.[17] As noted previously, neutral lipids (mainly free fatty acid) account for only a small amount (8.7%) of the total fluorescence. In some experiments, a degraded product of pyrene lipid represents less than 5% of the total VSV pyrene fluorescence.

The phospholipids in the envelope of intact VSV are present in two pools, based on availability to phospholipase C.[18] The accessible pool is considered to reside in the outer monolayer of the viral membrane and the inaccessible pool is presumed to be in the inner monolayer.[17] This analysis reveals that the choline-containing phospholipids are usually located in the outer layer, whereas the inner layer of the VSV membrane is mainly composed of aminophospholipids.[17] The bilayer distribution of VSV phospholipids labeled with PyC16 acyl chains is examined by comparing, by

TABLE II
COMPOSITION AND BILAYER DISTRIBUTION OF VESICULAR STOMATITIS
VIRUS MEMBRANE LIPIDS[a]

Lipid class	Lipid composition (%)	Total PyC16 lipids of intact VSV (%)	Lipids hydrolyzed by phospholipase C (%)	
			Total lipids	PyC16 lipids
SPM	24	} 5.3	0	0
PS	18			
PC	24	42.2	95	95
PE	31	43.6	47	35
NL	—	8.7	—	—

[a] Lipids were labeled with 16-(1-pyrenyl)hexadecanoic acid (PyC16). Virions (1 mg/ml) grown in BHK-21 cells prelabeled with PyC16 fatty acid and in unlabeled BHK-21 cells were treated with phosphlipase C as described in the text. The lipids were extracted from untreated and phospholipase C-treated virions and separated into various phospholipid species and neutral lipids (NL) by TLC. The amount of pyrene incorporated in each species was measured by optical density at 345 nm. The amount of each phospholipid hydrolyzed was calculated by expressing the phospholipid content of the enzyme-treated virus as a percentage of an untreated control run in parallel. The percent hydrolyzed was then obtained by subtracting this value from 100%. Percent composition was calculated from the amount of each pyrene phospholipid in the untreated control. Eight mole percent of the acyl chains of the virions grown on fluorescently labeled cells as fluorescent.

TLC, the fluorescent lipids extracted from virions exposed and unexposed to hydrolysis by phospholipase C (see above). Although no hydrolysis of SPM or PS is observed following exposure of virions to phospholipase C, 95% of pyrenyl-PC and 35% of pyrenyl-PE are hydrolyzed by phospholipase C (Table II). A similar observation can be made with VSV grown in BHK-21 cells in the absence of pyrenyl fatty acids, where 95% of unlabeled PC and 47% of PE are found to be accessible to phospholipase C hydrolysis.[17] It is evident from the data shown in Table II that pyrenyl-PC and pyrenyl-PE have an asymmetrical distribution, similar to that of unlabeled PC and PE. It is not possible to assess the distribution of pyrenyl-SPM and pyrenyl-PS, but, collectively, they represent only 5–8% of the total viral pyrene–lipid content. Moreover, the PS head group resists hydrolysis by *Clostridium welchii* phospholipase C.[17]

Fluorospectral Characterization of Metabolically Pyrenyl-Labeled Intact Vesicular Stomatitis Virus

Pyrenyl-containing molecules have a typical fluorescence excitation and emission spectrum, as demonstrated in Fig. 2 for VSV metabolically

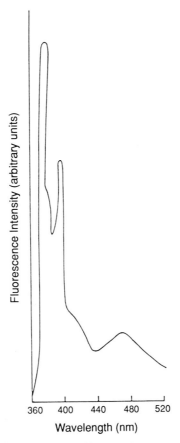

FIG. 2. Fluorescence emission spectrum of intact vesicular stomatitis virions. Virions were produced on BHK-21 cells after their metabolic labeling by PyC16 (see text and Table I). The intensity of fluorescence emission was recorded between 360 and 520 nm. For more details, see text.

labeled by PyC16 fatty acid. This spectrum is slightly different from that of free pyrene in membranes. It is characterized by only two sharp monomer peaks at fluorescence emissions of 385 and 410 nm, compared with three sharp peaks for free pyrene.[20] Similar to that of free pyrene, the spectrum is also characterized by a concentration-dependent broad excimer emission centered at 470 nm. The latter, being an excited state reaction, has no counterpart in the excitation spectrum.[19,20] The excimer is a result of collision in the lipid bilayer between one excited and one nonexcited monomer to give a dimer (excimer) with a typical broad emission fluorescence.[19,20] This is an excited state diffusion-controlled reaction, and, therefore, the excimer cannot be accounted for in the excitation spectrum. The fluorescence spectrum of PyC16-containing lipids in the VSV membrane is identical to free PyC16 or phospholipids containing ω-pyrenyl acyl chains in liposomes.[20] No monomer three-peak pattern representing the presence of free pyrene is observed in the VSV spectrum. The E/M ratio increases concurrently with increasing temperature. The E/M ratio in the viral envelope increases from 0.13 to 0.32 by raising the temperature from 4° to 37°, owing to the increase in diffusion rate with temperature.

Potential Artifacts of the Fusion Assay

The use of fluorescence techniques for fusion quantification has inherent pitfalls that may lead to misinterpretation. For VSV metabolically labeled with pyrenyl phospholipids, the extent of viral lipid dilution into the target membrane during fusion is determined from the reduction in the excimer to monomer ratio.[12] In order for the assay to be a quantitative fusion monitor it should meet the following criteria.

Stability of Labeling

The labeled phospholipid has to be a real and stable component of the VSV membrane. This criterion is met, as more than 90% of the labeling is in viral envelope phospholipids (mainly PC and PE), and their distribution between the two faces of the membrane resembles that of the normal phospholipids (Table II).

Intermembrane Transfer of Lipids

The relative contribution of spontaneous intermembrane transfer through the aqueous phase has to be known.[30] Phospholipids are transferred spontaneously between interacting vesicles, with a half-time ranging from 2 to 72 hr, depending on the physical state of the lipids in the bilayer and the lateral organization of the transferable species in the bilayer plane.[19] Similar spontaneous transfer of phospholipids is also noted among lipoproteins.[19] Spontaneous transfer may lead to a net one-way transfer of lipids from one membrane to another, thereby causing a first-order fusion-independent dilution of the pyrene fluorophore in both the donor VSV membrane and the acceptor target membrane which will result in the reduction of E/M independent of fusion.

It is important to determine whether pyrene lipids present in the VSV envelope can also undergo spontaneous transfer to the target membrane, and if the kinetics of this transfer are relevant to the time scale of the fusion. Virions containing pyrenyl lipids in the membrane are incubated

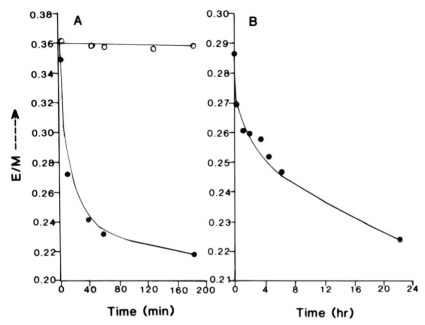

FIG. 3. Spontaneous rates of transfer of pyrene lipids from virion membranes to acceptor serum lipoprotein molecules before (A) and after (B) removal of free fatty acids from VSV. (A) VSV (200 μg) labeled with 16-(1-pyrenyl)hexadecanoic acid (PyC16) was added to 2 ml of PBS containing either no serum or 5% fetal calf serum at 37°, and the E/M ratio was measured at various times thereafter. Symbols denote virions treated with PBS alone (O) and virions treated with 5% serum in PBS (●). (B) VSV (500 μg) labeled with PyC16 fatty acid was incubated with 1 ml of 5% serum in PBS at 37°C for 10 min. The virions were then purified by sucrose density-gradient centrifugation as described in the text. The fatty acid-depleted virions (200 μg) were then added to 2 ml of PBS containing 5% serum at 37° and the E/M ratio measured at various times. (From Ref. 13 with permission.)

with PBS alone or with PBS containing 5% heat-inactivated fetal calf serum. The E/M ratio is measured at various time intervals after incubation at 37° (Fig. 3A). The E/M ratio of VSV incubated with PBS alone shows no change with time, whereas the virions incubated with PBS containing 5% serum show a marked initial drop in the E/M ratio, which slowly decreases with time. The initial rapid drop in E/M ratio ($t_{1/2}$ ~7 min) is due to the transfer of free fatty acids present in the viral membrane (see Table I and methods section), which are known to be transferred very rapidly.[19] The kinetics of spontaneous transfer are then reexamined after the free fatty acids are removed from the viral envelope (see methods section). To this end, virions are first incubated at 37° with

5% serum in PBS for 10 min, then purified by centrifugation through a 0–40% sucrose density gradient. Such preincubation removes the free fatty acids from the viral membrane, as determined by TLC of extracted viral lipids. The pyrene fatty acid-depleted virions are then incubated with 5% serum in PBS and the E/M ratio measured at different time intervals. As shown in Fig. 3B, the E/M ratio decreases with time at a much slower rate ($t_{1/2} \geq 12$ hr).

These experiments show that the interaction of virions with lipid acceptor leads to a spontaneous transfer of free pyrene fatty acids at a rapid rate and to transfer of pyrene phospholipids at a much slower rate. The rate of the latter is slow enough that it does not interfere with the fast fusion-related E/M reduction with a $t_{1/2}$ under 5 min (compare Figs. 3B and 4). Therefore, when pyrenyl free fatty acid-depleted virions are used, no correction for spontaneous transfer is needed.

Effect of Medium Composition

Medium composition variables such as pH and the presence of calcium ions are known to affect fluorescence properties of certain fluorophores. These artifacts are referred to as fast artifacts.[15,16] In our system the E/M ratio is pH-independent over the pH range of 4.0–7.4. At pH values lower than 3.0, the E/M ratio is reduced. The presence of Ca^{2+} in 0.9% NaCl also does not affect the E/M ratio. The absence of medium-dependent changes is explained by the fact that the pyrenyl moiety is located in the bilayer center,[19,20] which is characterized by its very low dielectric constant and low degree of order.[31,32] This also explains why the fluorophore, which is not present on the membrane surface, serves only as a fusion monitor, without affecting the fusion process itself.

Slow Artifacts

Slow artifacts do not exist in this study, because the assay is based on a bimolecular collision reaction in which excited pyrenyl monomers collide with nonexcited pyrenyl monomers. Therefore, results are not affected by aggregation or collapse of the membrane.

Other artifacts

Certain artifacts, such as the contribution of hemifusion,[11a] in which membranes intermix lipids only between their outer monolayers, have not

---

[31] E. Sackman, in "Biophysics" (W. Hoppe, et al., eds.), p. 425. Springer-Verlag, Berlin, (1983).

[32] M. K. Jain, "Introduction to Biological Membranes," 2nd Ed., Wiley (Interscience), New York, 1988.

been studied for the pyrenyl excimer dilution assay.[15,16] In general, it should be stressed that, this being a probe dilution assay, it is less prone to artifacts than the probe mixing assays.[16]

Fusion Studies

Virion–Liposome Fusion

An application of metabolically pyrenyl-labeled virions to study virus fusion with target membranes is given here for the case of VSV fusion with POPS vesicles. Only labeled VSV depleted of free pyrenyl fatty acid is used in these studies. POPS SUV are chosen as the target membranes because it has been reported that PS, but not PC, is the putative membrane component for VSV adsorption.[33] This has been confirmed by the finding that the fusion of G protein vesicles (virosomes) and liposomes is dependent on the amount of PS in the target vesicles, and on acidic pH.[34] The relevance of this pH dependency to the biological intracellular fusion of VSV with the endosomal membrane is still a controversial issue, which is, however, beyond the scope of this chapter.

During fusion of liposomes with VSV at pH 5.0, the excimer fluorescence decreases markedly with time, whereas the monomer emission intensity increases, which results in time-dependent reduction in the E/M ratio. At neutral pH, however, the spectra are unaltered with the time of incubation. Figure 4 shows that pyrenyl phospholipid in the VSV membrane undergoes a very sharp decline in E/M ratio as a result of mixing with POPS SUV at pH 5.0 and 37°, indicating that membrane fusion is essentially completed in 10 min. By comparison, the pyrene E/M ratio shows very little change in interaction of VSV with POPS vesicles at pH 7.3 for a period as long as 60 min. The E/M ratio of the VSV pyrenyl phospholipid is not significantly affected at pH 5.0 in the absence of POPS receptor vesicles.

The importance of the G protein in the fusion process is suggested by the finding that pyrenyl phospholipid, when present in liposomes formed by lipids extracted from VSV lipids in the absence of G protein, exhibits almost no alteration in the E/M ratio (Fig. 4). Moreover, the fusion reaction at pH 5.0 with POPS vesicles is markedly reduced when virions were rendered free of glycoprotein spikes by exposure ot the proteolytic

---

[33] R. Schlegel, R. B. Dickson, M. C. Willingham, and I. Pastan, *Cell (Cambridge, Mass.)* **32,** 639 (1983).

[34] O. Eidelman, R. Schlegel, T. S. Tralka, and R. Blumenthal, *J. Biol. Chem.* **259,** 4622 (1984).

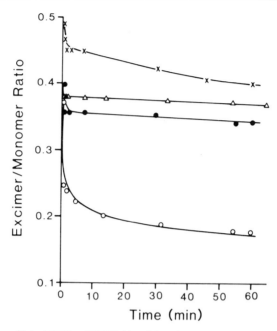

FIG. 4. Fusion of intact VSV or VSV lipid vesicles with 1-palmitoyl-2-oleoylphosphatidyl-serine (POPS) small unilamellar vesicles (SUV) at pH 5.0 or 7.3 as measured by E/M ratios. VSV (5 mg) labeled metabolically with PyC16 fatty acid was incubated with 5 ml of 5% fetal calf serum (heat inactivated) in PBS at 37° for 15 min to remove free pyrene fatty acids. The virions were then purified by 0–40% sucrose gradient centrifugation. Fatty acid-depleted virus (100 μg) was then added to 3 ml of 10 mM citrate buffer, pH 5.0 or 7.3 at 37°, and virus–vesicle fusion was initiated by adding 150 nmol of POPS vesicles. The E/M ratio was determined over a period of 1 hr by exciting the sample at 330 nm; emission was measured at 385 and 470 nm for monomer and excimer, respectively. VSV lipid vesicles were prepared by extracting the lipids from pyrene-labeled virions (200 μg) by the method of Folch et al.[24] and the protein-free vesicles were prepared by sonicating the VSV lipid suspension in 10 mM N-[tris(hydroxymethyl)methyl]glycine (Tricine), pH 7.5 Fusion of viral lipid vesicles with POPS vesicles was performed as described above. Symbols denote VSV alone at pH 5.0 (△), VSV with POPS vesicles at pH 5.0 (○), VSV with POPS vesicles at pH 7.3 (●), and VSV lipid vesicles with POPS vesicles at pH 5.0 (✕). (From Ref. 13 with permission.)

enzyme thermolysin.[35] These studies suggest strongly that the membrane of intact virions undergoes rapid fusion with the target membrane of POPS vesicles at pH 5.0, but not at neutral pH, and this reaction is modulated by the G protein, which may have to be in a specific low pH-dependent conformation. It is also clear that excimerization of an endoge-

[35] N. F. Moore, Y. Barenholz, and R. R. Wagner, J. Virol. 19, 126 (1976).

nous pyrenyl phospholipid probe provides a useful and biologically relevant method for monitoring fusion of a viral membrane with a model target membrane.

The effect on the assay of medium temperature has been studied by comparing the reduction in E/M ratio for the fusion of metabolically pyrenyl-labeled VSV with unlabeled POPS vesicles at 4° and 37°. Figure 5 reveals that the E/M ratio undergoes no significant change when virions are incubated with POPS vesicles at pH 5.0 at 4° for 30 min. When VSV labeled with somewhat greater amounts of pyrenyl phospholipid was preincubated at pH 5.0 with POPS vesicles at 4°, and then warmed to 37°, a rapid decline in the E/M ratio was observed, reaching a plateau at approximately 5 min. It seems likely, therefore, that the pyrenyl phospholipid-labeled virions adsorb to POPS vesicles at 4°, but the fusion event is triggered by the increase in the reaction temperature. This is in agreement with what has been observed for other viral systems, such as Sendai virus.[12] The fact that the fusion process has a higher energy of

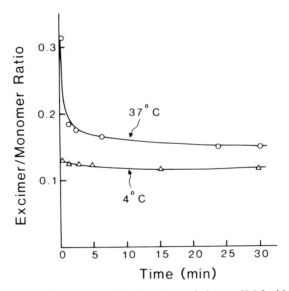

FIG. 5. Temperature dependence of VSV membrane fusion at pH 5.0 with POPS vesicles. Vesicular stomatitis virions biologically labeled with PyC16 fatty acid were depleted of free fatty acid as described for Fig. 3. Fatty acid-depleted virions (100 μg) were incubated with 150 nmol of POPS vesicles at 4° for 30 min. The suspension of virions was then added to 3 ml of 10 mM citrate buffer, pH 5.0 at 37° (O), and the E/M ratio was measured at intervals for 30 min as described for Fig. 3. Fusion of VSV with POPS vesicles at 4° (△) was monitored by adding fatty acid-depleted VSV (100 μg) to 3 ml of 10 mM citrate, pH 5.0 at 4° and fusion was initiated by adding 150 nmol of POPS vesicles. (From Ref. 13 with permission.)

activation than the viral–target membrane attachment may be related to the effect of temperature on the fraction of active virions.[12]

Effect of Antibodies to G Protein on Viral Membrane Fusion

The evidence is quite conclusive that the G protein is responsible for host–membrane recognition, as well as being the fusion factor.[1,34–36] It is necessary, therefore, to determine to what extent G protein-specific monoclonal antibodies (MAb) affect the reduction of E/M of the pyrenyl-labeled virions on fusion of VSV and POPS vesicles. Monoclonal antibodies to four epitopes were found to neutralize the infectivity of VSV, whereas MAbs to seven other epitopes did not.[37]

Two of the monoclonal antibodies, MAb6, which neutralizes VSV infectivity, and MAb17, which does not, have been tested to determine their capacity to inhibit the fusion activity of VSV. VSV is incubated with MAb6 or MAb17 [150 μg of immunoglobulin G (IgG) of each] for 60 min at room temperature, and these virions, as well as control virions, each containing pyrenyl phospholipids, are incubated with POPS vesicles at pH 5.0 and 37°. As shown in Fig. 6, both MAb6 and, to a somewhat lesser degree, MAb17 inhibit the reduction in E/M ratio of PyC16 phospholipid exhibited by control VSV on interaction at pH 5.0 with POPS vesicles. Because these two monoclonal antibodies to two separate G protein antigenic determinants have a dramatic effect on the VSV fusion reaction, the inhibition of fusion does not account for the markedly different capacities of the two monoclonal antibodies to neutralize viral infectivity. It is possible that binding of antibodies to VSV spikes might also induce aggregation of the virions, and thus prevent virus–vesicle association, a prerequisite step for virus–vesicle fusion.

Comparative Fusion Studies with an Exogenous Fluorophore

It is important to compare the fusion kinetics of the metabolically pyrenyl-labeled virions with the popular fusion assay of virions labeled exogenously with ODR (see Introduction). Octadecylrhodamine readily partitions into biological membranes at self-quenched concentrations. The fusion-dependent dequenching of the exogenous probe can be monitored continuously to measure the rate of membrane fusion.[11,14,14c]

VSV is labeled with a quenching concentration of ODR by incubating purified virions with the fluorophore for 60 min at 37°. Sonicated POPS vesicles are added to labeled virions in citrate buffer at pH 5.0 or pH 7.3,

[36] D. H. L. Bishop, P. Repik, J. J. Obijeski, N. F. Moore, and R. R. Wagner, *J. Virol.* **16**, 75 (1975).

[37] W. A. Volk, R. M. Snyder, D. C. Benjamin, and R. R. Wagner, *J. Virol.* **42**, 220 (1982).

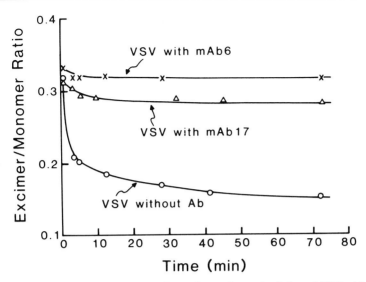

FIG. 6. Effect of monoclonal antibodies to G protein on the fusion of VSV with POPS vesicles. Virions labeled with PyC16 were depleted of free fatty acids as described for Fig. 3. Fatty acid-depleted virions (100 μg) were then incubated with purified IgG (150 μg) of neutralizing (MAb6) or nonneutralizing (MAb17) monoclonal antibody to G protein at room temperature for 60 min. The virion–antibody mixture was then added to 3 ml of 10 mM citrate buffer (pH 5.0), and the fusion reaction was performed at 37° by adding 150 nmol of POPS vesicles. Symbols denote control VSV without antibody (O), VSV pretreated with MAb6 (X), and VSV pretreated with MAb17 (Δ). (From Ref. 13 with permission.)

and the fluorescence intensity at 590 nm is monitored continuously. Figure 7 reveals very rapid dequenching of the rhodamine probe in virions incubated at 37° with POPS vesicles at pH 5.0. In contrast, only a minimal amount of the ODR fluorescence is dequenched on incubation with POPS vesicles at pH 7.3. In both cases, plateaus of unquenched fluorophores are reached within 3 min. The data obtained with the exogenous ODR agree with results obtained with the metabolically labeled virions (Fig. 4). In both systems, fusion of VSV with target POPS vesicles is pH-dependent. At pH 5.0, fusion occurs very rapidly with a short half-time ($t_{1/2} \sim 2$ min) and is virtually completed within 10 min. Therefore, the comparison between the two assays further validates the ODR assay.

Potential Applications of Metabolically Fluorescently Labeled Vesicular Stomatitis Virions

Pyrenyl lipid-labeled VSV is a useful tool to investigate the kinetics and factors influencing the intermixing of the viral membrane with the target membrane during membrane fusion. The biological labeling does not

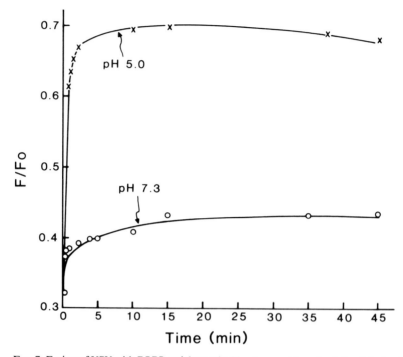

FIG. 7. Fusion of VSV with POPS vesicles probed by dequenching of octadecylrhodamine B chloride (R_{18}) as measured by the ratio of quenched fluorescence intensity (F) to completely unquenched fluorescence intensity (F_0). VSV (1 mg) was incubated with R_{18} (1:100 probe to lipid molar ratio) in PBS at 37° for 45 min. (The R_{18} probe was added to the viral suspension in ethanol so that the final ethanol concentration in solution was less than 0.5%.) The free probe molecules were removed by pelleting the virions in an SW50.1 rotor at 25,000 rpm for 60 min through a 5% glycerol pad. The R_{18}-labeled virions (100 μg) were then added to 3 ml of 10 mM citrate buffer, pH 5.0 or pH 7.3 at 37°, and fusion was initiated by adding 150 nmol of POPS vesicles. The emission intensity (F) was measured at 590 nm, with excitation at 560 nm. The total unquenched fluorescence (F_0) was obtained by complete solubilization of R_{18}-labeled virions with 1% Triton X-100. (From Ref. 13 with permission.)

inhibit viral infectivity. The pyrenyl acyl chain constitutes approximately 8% of the viral envelope acyl chains. PC and PE contain more than 85% of the label; the distribution of the labeled lipids between the two faces of the viral membrane reflects the distribution of total PE and PC.

The fact that the half-time of spontaneous transfer of phospholipids from virions is very slow ($t_{1/2} > 12$ hr) and that for fusion it is very fast ($t_{1/2} \sim 2$ min) makes it possible to use the reduction in E/M ratio to follow the kinetics of fusion of enveloped virions with various membrane targets. Using this assay, it has been confirmed that the VSV G protein plays an important role in mediating fusion of VSV with model membranes. The fusogenic activity of the G protein is implicated by the finding that viral lipid vesicles, free of viral proteins, do not undergo fusion with PS vesicles at pH 5.0. Furthermore, monoclonal antibodies to G protein, of both neutralizing and nonneutralizing types, inhibit virus–liposome fusion markedly. The metabolic labeling by pyrenyl acyl chains has many other applications.

Quantification of Fusion Process

The mass action kinetic model and procedure described by Amselem et al.[12] provide analytical solutions for the final levels of probe dilution and numerical solutions for the kinetics of the overall fusion process, in terms of rate constants for the virus–target membrane adhesion, dissociation, and fusion. According to this model, fusion is a two-step process: adherence of the virions to the target membrane, followed by the merging of the viral and target membrane. The use of this model requires the establishment of a calibration curve describing the E/M ratio as a function of dilution.[12] The extent of viral lipid dilution in each experiment is then determined from this calibration curve. The basic equation for the probe dilution (D_p) is

$$D_p = (V_F + T_L)/T_L = V_F/T_L + 1$$

where T_L is the total concentration of target membrane lipid and V_F is the lipid concentration of the *fused* virions. When all virus particles fuse, then all viral lipids $(V = V_F)$ are involved in the fusion, and the maximal probe dilution (D_L) is given by $D_L = (V + T_L)/T_L = V/T_L + 1$. The fraction of virions which fuse with the target membrane depends on the molar concentration ratio of virions and membrane particles. This ratio is given by $(A_V/A_T) \times (T_L/V)$; where A_V and A_T are the surface area of virions and target membrane, respectively, which can be determined from particle size using one of the methods described elsewhere.[29] For further details of this analysis, see Amselem et al.[12] and references therein.

Involvement of Viral G Proteins in Fusion Process

There is extensive circumstantial evidence that the VSV G protein is not only the viral fusion protein, but also the ligand to which cell surface receptors bind during receptor-mediated endocytosis.[1] The G protein is acylated, via a thioester bond, to the cysteine residue at position 14 from the G protein carboxy terminus, with preference for palmitic acid.[19,36,38] Although the role of the acyl chain is not fully understood,[18] one can take advantage of the metabolic covalent attachment of acyl chains to the VSV G protein and use it to study the role of this protein in the various stages of

---

[38] J. K. Rose, G. A. Adams, and C. J. Gallione, *Proc. Natl. Acad. Sci. U.S.A.* **81,** 2050 (1984).

the fusion process, especially the aggregation state and dynamics of the protein. The feasibility of labeling this protein with fluorescent fatty acids has already been demonstrated.[39] The labeling of the VSV G protein with fluorescent fatty acids occurs during the infection and replication of the VSV in BHK cells prelabeled with fluorescent fatty acid, as described above. As most of the label is present in viral envelope phospholipids, the viral proteins have to be purified and undergo complete removal of the phospholipids and free fatty acids.

Studies similar to those described in Table I have been carried out using palmitic acid containing the anthroyloxy fluorophore attached, via an ester bond, to either C-2 or C-16 of the palmitic acid moiety (C2AP and C16AP, respectively). The C16AP incorporates well into phospholipids, with only low levels remaining as free fatty acid (13.3 and 2.15 nmol/mg protein, respectively). The C2AP shows very low incorporation into viral phospholipids, probably owing to steric inhibition of the anthroyloxy moiety at position 2 to the activation of this fatty acid (see Scheme 1). Also, viral labeled C16AP and C2AP contain relatively large amounts of biodegradation compounds, yet unidentified, which result from the hydrolysis of the ester bond through which the anthroyloxy moiety is attached to the acyl chain. Using metabolic labeling by fluorescent 16-(9-anthroyloxy)palmitate, it is found that among the VSV proteins only the G protein is fluorescently labeled by a covalent thioester bond. The labeled G protein is then reconstituted in dipalmitoylphosphatidylcholine (DPPC) vesicles to form G protein proteoliposomes. Steady-state fluorescence anisotropy of the fluorescent labeled VSV G protein in the proteoliposomes (G virosomes) indicates that the rotational mobility of G protein 16-(9-anthroyloxy)palmitoyl moiety is restricted, although it does not sense the DPPC gel-to-liquid crystalline phase transition.

The G protein is also acylated by ω-(1-pyrenyl) fatty acids. The use of the acyl pyrenyl protein may allow following of protein aggregation in the G virosomes. G virosomes and intact virions undergo similar fusion processes with target membranes. The use of G virosomes in which the G protein is labeled with a pyrenyl acyl chain may enable one to get better insight into the changes occurring in the lateral organization of the G protein in the G virosome membrane, and to evaluate various models which have been proposed to explain the exact mechanism by which viral fusion proteins induce virus–target membrane fusion.[40]

[39] A. W. Petri, R. Pal, Y. Barenholz, and R. R. Wagner, *J. Biol. Chem.* **256,** (1980).

[40] R. Blumenthal, A. Puri, A. Walter, and O. Eidelman, *in* "Molecular Mechanisms of Membrane Fusion" (S. Ohki, D. Doyle, T. D. Flanagan, S. W. Hui, and E. Mayhew, eds.), p. 367. Plenum, New York, 1988.

[23] Preparation, Properties, and Applications of Reconstituted Influenza Virus Envelopes (Virosomes)

By ROMKE BRON, ANTONIO ORTIZ, JAN DIJKSTRA,
TOON STEGMANN, and JAN WILSCHUT

Introduction

Enveloped viruses utilize a membrane fusion strategy to introduce their genome into the cytoplasm of the host cell.[1,2] This infectious entry, in principle, can occur through direct fusion of the viral envelope with the cell plasma membrane, or it may occur from within endosomes after cellular uptake of intact virions through receptor-mediated endocytosis. In the latter route of entry, the target membrane for fusion of the viral envelope is the limiting membrane of the endosomal cell compartment.

It is well established that influenza virus utilizes the endocytic route of infection.[1-3] Expression of the fusion activity of the virus is strictly dependent on a mildly acidic pH, precluding direct fusion of the viral envelope with the plasma membrane. After endocytic uptake, the virus encounters an acidic environment in the lumen of the endosomes, activating its fusion capacity. Accordingly, inhibitors of vacuolar acidification, such as chloroquine or NH_4Cl, block cellular infection by influenza virus. In model systems, influenza virus can be induced to fuse with various artificial or biological target membranes, including the plasma membrane of cultured cells, by lowering the pH of the medium.

The membrane fusion activity of influenza virus is mediated solely by the major envelope glycoprotein, hemagglutinin (HA). HA represents the best-characterized membrane fusion protein so far.[1-5] The HA spike, protruding some 13.5 nm from the viral surface, is a homotrimeric molecule. Each monomer consists of two disulfide-linked subunits, HA1 (47 kDa) and HA2 (28 kDa). The N terminus of HA2 plays a crucial role in the fusion activity of HA. This so-called fusion peptide is a conserved stretch of some 20 amino acid residues that are mostly hydrophobic in nature. At

[1] T. Stegmann, R. W. Doms, and A. Helenius, *Annu. Rev. Biophys. Biophys. Chem.* **18,** 187 (1989).

[2] J. M. White, *Annu. Rev. Physiol.* **52,** 675 (1990).

[3] R. W. Doms, J. White, F. Boulay, and A. Helenius, *in* "Membrane Fusion" (J. Wilschut and D. Hoekstra, eds.), p. 313. Dekker, New York, 1990.

[4] D. C. Wiley and J. J. Skehel, *Annu. Rev. Biochem.* **56,** 365 (1987).

[5] J. Wilschut and R. Bron, *in* "Viral Fusion Mechanisms" (J. Bentz, ed.), p. 133. CRC Press, Boca Raton, Florida, 1993.

neutral pH the fusion peptides are buried within the stem of the HA trimer at about 3.5 nm from the viral surface. However, at low pH a conformational change in the HA results in their exposure,[6] activating the fusion capacity of the molecule. There is good evidence to indicate that the HA fusion peptides mediate the fusion reaction via penetration into the target membrane.[3,7,8]

Ultimate elucidation of the molecular mechanisms involved in viral membrane fusion processes requires the isolation, purification, and functional reconstitution of the viral spike glycoproteins. Thus, the isolated fusion proteins can be studied in a well-defined lipid environment at controlled surface densities. In addition, in studies on viral fusion and cell entry, the use of reconstituted viral envelopes (virosomes) offers distinct advantages over the use of native virus in that specific reporter molecules can be introduced in the membrane of the virosomes or in their aqueous lumen. Finally, virosomes provide a promising carrier system for the introduction of foreign substances, either water-soluble or membrane-associated, into living cells.

Reconstitution of Influenza Virus Envelopes (Virosomes)

Background

Many studies have been aimed at the functional reconstitution of viral envelopes. It is beyond the scope of this chapter to discuss these in detail (for a review, see Ref. 9). Unfortunately, from the available data, it is impossible to extrapolate a consensus procedure that would yield predictable results in a variety of systems. This may in part be due to the fact that different parameters have been used to evaluate the structural and functional characteristics of the reconstitution products. Obviously, the criteria for evaluation of the properties of viral reconstitution products will depend on the specific aims of the reconstitution experiment. For example, if reconstituted viral envelopes are being produced for immunization purposes,[10] membrane fusion activity of the products is usually not essential.

---

[6] J. M. White and I. A. Wilson, *J. Cell Biol.* **105**, 2887 (1987).

[7] C. Harter, P. James, T. Bächi, G. Semenza, and J. Brunner, *J. Biol. Chem.* **264**, 6459 (1989).

[8] T. Stegmann, J. M. Delfino, F. M. Richards, and A. Helenius, *J. Biol. Chem.* **266**, 18404 (1991).

[9] A. Walter, O. Eidelman, M. Ollivon, and R. Blumenthal, *in* "Membrane Fusion" (J. Wilschut and D. Hoekstra, eds.), p. 395. Dekker, New York, 1990.

[10] N. El Guink, R. M. Kris, G. Goodman-Snitkoff, P. A. Small, Jr., and R. J. Mannino, *Vaccine* **7**, 147 (1989).

In other cases, where the reconstitution was carried out with the purpose of generating a model system for studying the membrane fusion activity of the particular virus involved, the functional characteristics of the reconstitution products have not always been rigorously evaluated.[11-17] For example, reconstitution of hemagglutination or hemolytic activity does not necessarily imply concomitant reconstitution of membrane fusion activity. In our view, direct demonstration of biologically relevant membrane fusion activity represents the only rigorous criterion for the functional reconstitution of viral envelopes.

Use of $C_{12}E_8$ or Triton X-100. Several examples of functional reconstitution of viral envelopes that meet the above rigorous criterion have been described recently. These are all based on the use of a nonionic detergent with a relatively low critical micellar concentration (CMC). Both octaethyleneglycol mono(n-dodecyl)ether ($C_{12}E_8$) and Triton X-100 have been applied successfully in the reconstitution of envelopes derived from vesicular stomatitis virus (VSV),[18] influenza virus,[19-21] Sendai virus,[22] and Semliki Forest virus (SFV).[22a] In all cases the virosomes exhibited biologically relevant fusion activity. The reconstitution procedure involved solubilization of the viral envelope in an excess of the detergent, removal of the viral nucleocapsid by ultracentrifugation, and subsequent controlled removal of the detergent from the supernatant by a two-step treatment with BioBeads SM2, a hydrophobic resin, consisting of a styrene–divinylbenzene copolymer.[23] We note that the solubilization of phospholipid bilayers by $C_{12}E_8$ and the subsequent reconstitution of vesicular systems through removal of the detergent with BioBeads SM2 has recently been studied in detail.[24]

[11] W. H. Petri and R. R. Wagner, *J. Biol. Chem.* **254**, 4313 (1979).

[12] D. K. Miller, B. I. Feuer, R. Vanderoef, and J. Lenard, *J. Cell Biol.* **84**, 421 (1980).

[13] R. T. C. Huang, K. Wahn, H. D. Klenk, and R. Rott, *Virology* **104**, 294 (1980).

[14] K. Kawasaki, S. Sato, and S.-I. Ohnishi, *Biochim. Biophys. Acta* **733**, 286 (1983).

[15] O. Eidelman, R. Schlegel, T. S. Tralka, and R. Blumenthal, *J. Biol. Chem.* **259**, 4622 (1984).

[16] Y. Hosaka, Y. Yasuda, and K. Fukai, *J. Virol.* **46**, 1014 (1983).

[17] P. Sizer, A. Miller, and A. Watt, *Biochemistry* **26**, 5106 (1987).

[18] K. Metsikkö, G. van Meer, and K. Simons, *EMBO J.* **5**, 3429 (1986).

[19] T. Stegmann, H. W. M. Morselt, F. P. Booy, J. F. L. van Breeman, G. Scherphof, and J. Wilschut, *EMBO J.* **6**, 2651 (1987).

[20] O. Nussbaum, M. Lapidot, and A. Loyter, *J. Virol.* **61**, 2245 (1987).

[21] M. Lapidot, O. Nussbaum, and A. Loyter, *J. Biol. Chem.* **262**, 13736 (1987).

[22] A. Vainstein, M. Hershkovitz, S. Israel, S. Rabin, and A. Loyter, *Biochim. Biophys. Acta* **773**, 181 (1984).

[22a] R. Bron, C. N. van der Veere, W. Eisenga, M. Sulter, J. Scolma, and J. Wilschut, submitted for publication.

[23] P. W. Holloway, *Anal. Biochem.* **55**, 304 (1973).

[24] D. Levy, A. Gulik, M. Seigneuret, and J. L. Rigaud, *Biochemistry* **29**, 9480 (1990).

Use of Octyl Glucoside. The use of low-CMC detergents has a disadvantage in that they cannot be readily removed from the system by dialysis. For this reason, many reconstitution procedures, not only those involving viral fusion proteins, have relied on the use of detergents with a relatively high CMC that can be completely removed by dialysis. A widely used detergent in this category is the nonionic detergent n-octyl-β-D-glucopyranoside (octyl glucoside, OG), which has a CMC of about 20 mM.

With respect to the reconstitution of viral envelopes based on the use of OG, results have been variable. First, in an OG dialysis method, involving the formation of protein–lipid cochleates after or during the removal of the detergent, large proteoliposomes have been produced that were capable of fusion-mediated introduction of encapsulated macromolecules into cells.[25,26] The formation of the cochleate structures in this procedure required the presence of an excess of exogenous phospholipid (phosphatidylserine) and cholesterol. The method has been applied for generating proteoliposomes carrying the fusion proteins of influenza virus or Sendai virus. As for Sendai virus, reconstitution of fusogenic virosomes based on slow OG dialysis in the absence of exogenous lipids has also been described,[27] but in the case of VSV the virosomes did not display biologically relevant fusion activity.[11,12,15,18] As for influenza, reconstituted viral envelopes, produced via OG dialysis in the absence of exogenous lipids, were either not tested for fusion activity,[10] or reportedly exhibited fusion activity at neutral pH.[13] In our hands, attempts to reconstitute influenza virus envelopes from OG-solubilized virus were frustrated by the formation of nonuniform virosome preparations consisting of lipid-rich and protein-rich fractions.[19]

Conclusion. We conclude that, at this point, the use of a low-CMC detergent, preferably the well-defined detergent $C_{12}E_8$, to solubilize the viral membrane, followed by controlled detergent removal using BioBeads SM2,[18,19,24] represents the method of choice in attempts aimed at the functional reconstitution of viral envelopes from the initial envelope proteins and lipids.

Procedure for Functional Reconstitution of Influenza Virus Envelopes

Virus and Reagents. Influenza virus can be grown to high titers in the allantoic cavity of 10-day-old embryonated chicken eggs, and it is readily

[25] S. Gould-Fogerite and R. J. Mannino, *Anal. Biochem.* **148**, 15 (1985).
[26] S. Gould-Fogerite, J. E. Mazurkewicz, K. Raska, K. Voelkerding, J. M. Lehman, and R. J. Mannino, *Gene* **84**, 429 (1989).
[27] M. C. Harmsen, J. Wilschut, G. Scherphof, C. Hulstaert, and D. Hoekstra, *Eur. J. Biochem.* **149**, 591 (1985).

purified from the allantoic fluid.[28] Briefly, harvested allantoic fluid is centrifuged at 1000 g for 15 min in the cold to remove debris, after which the virus is sedimented from the supernatant at 75,000 g for 90 min at 4°. The virus pellet is resuspended in HNE buffer (150 mM NaCl, 0.1 mM EDTA, and 5 mM HEPES, adjusted to pH 7.4) and subjected to sucrose gradient centrifugation (10–60%, w/v, linear sucrose gradient in HNE) at 100,000 g for 16 hr at 4°. The virus equilibrates as a single band at approximately 45% (w/v) sucrose. The band is collected, then frozen in small aliquots at −80°. The protein content of the virus preparation is determined according to Peterson.[29] and the phospholipid content, after quantitative extraction of the lipids from a known amount of virus,[30] according to Böttcher et al.[31]

The detergent $C_{12}E_8$ (Nikko Chemicals, Tokyo, Japan; $C_{12}E_8$ can also be purchased from Fluka, Buchs, Switzerland, or Calbiochem, San Diego, CA) is dissolved in HNE at a concentration of 100 mM. BioBeads SM2 (from Bio-Rad, Richmond, CA) are washed with methanol and subsequently with water, according to Holloway,[23] and stored under water. Just before use the beads are drained on filter paper and weighed. Sucrose solutions for gradient centrifugation are made in HNE on a weight per volume basis.

Reconstitution Procedure. Influenza virus (the equivalent of 1.5 μmol membrane phospholipid) is diluted in HNE and sedimented for 30 min at 50,000 g (e.g., in a Beckman Ti50 rotor) at 4°. To the pellet, 0.7 ml of 100 mM $C_{12}E_8$ in HNE is added. The pellet is gently resuspended by repeatedly moving the suspension through a 1-ml syringe with a 25-gauge needle, avoiding the formation of air bubbles. When the pellet is completely resuspended, solubilization is allowed for another 15 min on ice. Based on the assumption that the phospholipid-to-cholesterol molar ratio in the viral membrane is 2:1, the detergent-to-lipid ratio will be approximately 30:1. Subsequently, the viral nucleocapsid is removed by centrifugation for 30 min at 85,000 g at 4°. This step is conveniently carried out in a Beckman TL100 tabletop ultracentrifuge, using 1.3-ml vials plus adapters in the TL100-3 rotor at 50,000 rpm. A small sample of the supernatant can be taken at this stage for protein and phospholipid analysis. Of the initial viral protein and phospholipid, 35% (representing almost all of the membrane protein) and over 90%, respectively, is recovered in the supernatant.

[28] Y. Okada, *Exp. Cell Res.* **26,** 98 (1962).
[29] G. L. Peterson, *Anal. Biochem.* **83,** 346 (1977).
[30] E. G. Bligh and W. J. Dyer, *Can. J. Biochem. Biophysiol.* **37,** 911 (1959).
[31] C. J. F. Böttcher, C. M. van Gent, and C. Fiers, *Anal. Chim. Acta* **24,** 203 (1961).

The supernatant (0.63 ml) is transferred to a 1.5-ml Eppendorf vial containing 180 mg (wet weight) pre-washed BioBeads SM2. The supernatant is shaken in a Vibrax-VXR shaker (IKA Labortechnik, Staufen, Germany) at 1400 rpm for 60 min at room temperature. At this stage the suspension is still clear. Subsequently, an additional amount of 90 mg wet BioBeads is added, and shaking is continued for 10 min at 1800 rpm. The suspension becomes turbid at this point, indicating the formation of vesicular structures. Subsequently, the virosome suspension is centrifuged on a 10–40% (w/v) discontinuous sucrose gradient for 90 min at 130,000 g at 4° (e.g., in Beckman SW50.1 tubes, containing 1.0 ml of 40% and 3.0 ml of 10% sucrose). The virosomes appear as a thin opalescent band and are collected from the interface in 0.5 ml.

Evaluation of Structural and Functional Characteristics of Virosomes

As a model system for mimicking the native viral envelope, virosomes should resemble the viral envelope as much as possible. This implies that the virosomes should have a structure and composition similar to that of the viral envelope. The preparation should consist of a relatively uniform population of vesicles in terms of size and protein-to-lipid ratio. Residual detergent should be minimal and not interfere with virosome function. With respect to functional characteristics, the virosomes should mimic the biological activity of the native viral envelope. In the case of influenza virosomes, this implies that they should exhibit pH-dependent membrane fusion activity in a biologically relevant model system as well as in cells. The characteristics of this activity should resemble those of the native virus.

Equilibrium Density-Gradient Analysis

Spike protein incorporation in reconstituted vesicles is readily assessed by equilibrium density-gradient analysis. The virosome preparation, collected from the discontinuous sucrose gradient (see above), is diluted 3-fold with HNE, and 0.6 ml is applied to a 10–60% (w/v) linear sucrose gradient in HNE (e.g., in Beckman SW50.1 tubes). The gradient is centrifuged at 170,000 g for 30 hr at 4°, after which fractions of 0.5 ml are collected and analyzed for protein[29] and phospholipid[31] content. Figure 1 shows the results of such an analysis.[19] The virosomes appear as a single band, containing both protein and phospholipids. The density of the virosomes is approximately 1.12 mg/ml.

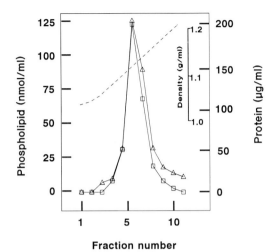

FIG. 1. Equilibrium density-gradient analysis of reconstituted influenza virosomes. The analysis was carried out as described in the text. Squares, phospholipid phosphate; triangles, protein; dashed line, buoyant density (determined from the refractive index of the fractions).

Virosome Composition

Analysis of influenza virosomes by sodium dodecyl sulfate–polyacrylamide gel electrophoresis (SDS-PAGE) (not shown) has demonstrated that the virosomes contain only the spike proteins. The viral nucleoprotein NP and the matrix protein M1 are not detectable. Western blot analysis has revealed that the minor integral membrane protein of influenza virus, M2, is not reconstituted in the virosomes either.[31a] The virosomes have a protein-to-phospholipid ratio that is similar to the ratio in the solubilization mixture after sedimentation of the nucleocapsid (1.42 mg/μmol as determined for the X-47 strain, Ref. 19).

Recoveries of viral membrane protein and phospholipid in the virosome preparation range from 30 to 50% relative to the initially solubilized material. Residual detergent in influenza virosomes has been determined to be 7.5 mol % relative to the total virosomal lipid, using ^{14}C-labeled $C_{12}E_8$ (CEA, Saclay, France) as a marker.[19] This level does not appear to affect the fusogenic activity of the virosomes to any significant extent.

[31a] R. Bron, A. P. Kendal, H. D. Klenk, and J. Wilschut, submitted for publication.

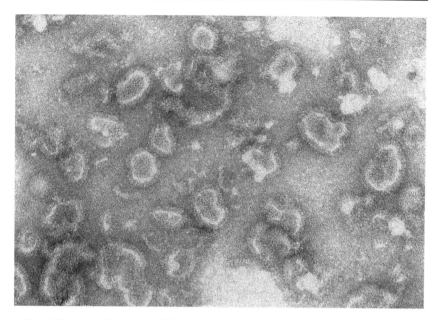

FIG. 2. Electron micrograph of influenza virus (X-47) strain) virosomes, negatively stained with 2% phosphotungstate, pH 7.4. Magnification: ×72,210. For further details, see text.

Virosome Morphology

Negative-stain electron microscopy (EM) is the most widely applied and accessible technique for obtaining information with respect to the structure and size of artificial and biological membrane vesicles in the submicron size range. The staining solution should be chosen with care and, in the case of influenza virosomes, be preferably of neutral pH, so as to avoid acid-induced conformational changes of the spike proteins.

Briefly, a droplet of the virosome suspension, after dialysis against isotonic ammonium acetate buffered to neutral pH with 5 mM HEPES, is applied to a grid with a carbon-coated Formvar film, after glow-discharge of the grid. The specimen is placed upside down for 1 min on a droplet of 2% phosphotungstic acid (PTA) at neutral pH (1% sodium silicotungstate of neutral pH provides a good alternative[32]), drained and dried in air.

Figure 2 presents a negative-stain electron micrograph of influenza virosomes, prepared according to the procedure described above.[19] In this particular case, the specimen was viewed in a Philips EM 400 electron microscope at 80 kV.

[32] R. W. H. Ruigrok, N. G. Wrigley, L. J. Calder, S. Cusack, S. A. Wharton, E. B. Brown, and J. J. Skehel, *EMBO J.* **5**, 41 (1986).

The most reliable way to visualize the structure of viral reconstitution products is by means of cryo-EM.[18,19,33] However, this technique requires a rapid-freezing device and an electron microscope equipped with a cryo transfer stage, which are not usually found among the standard equipment of an EM facility. We have applied cryo-EM techniques in the characterization of influenza virosomes[19] and observed that they are unilamellar vesicles of approximately 100 nm in diameter, which is similar to the size of the native virus. The virosomes are densely covered with spikes, which appear to be present on both sides of the membrane.

Fusion Activity of the Virosomes

Fluorescence Lipid Mixing Assays. Fusion of either intact viruses or reconstituted viral envelopes with biological or artificial target membranes can be followed with the R_{18} assay,[34] which relies on the fluorescence self-quenching properties of the fatty acid probe octadecylrhodamine B (R_{18}). The R_{18} assay has been proved to reveal reliably the general fusion characteristics of several enveloped viruses in a number of model systems. However, we have observed recently[34a] that the probe is not homogeneously distributed in the labeled viral or virosomal membrane. Possibly as a result of this, R_{18} may, after fusion, dilute incompletely and/or in a retarded fashion into the target membrane. In addition, even though in several cases control experiments have been carried out to exclude the possibility of transfer of individual R_{18} molecules between virus and target membrane,[34,34b-d] under specific conditions such exchange of the probe may well occur and has, in fact, been observed.[34e]

Reconstitution of viral envelopes offers the option to coreconstitute fluorescent reporter molecules in the virosomal membrane. However, labeling of virosomes with R_{18} via coreconstitution, rather than postlabeling, does not solve the problems of inhomogeneous probe distribution and molecular exchange eluded to above. Therefore, it is preferable to coreconstitute fluorescently labeled phospholipid analogs in the viral envelopes. In this respect, there are essentially two possibilities: first, the application of a

[33] J. Lepault, F. P. Booy, and J. Dubochet, *J. Microsc.* **129,** 89 (1983).

[34] D. Hoekstra, T. De Boer, K. Klappe, and J. Wilschut, *Biochemistry* **23,** 5675 (1984).

[34a] T. Stegmann, P. Schoen, R. Bron, J. Wey, I. Bartoldus, A. Ortiz, J. L. Nieva and J. Wilschut, submitted for publication.

[34b] T. Stegmann, D. Hoekstra, G. Scherphof, and J. Wilschut, *J. Biol. Chem.* **261,** 10966 (1986).

[34c] V. Citovsky, R. Blumenthal, and A. Loyter, *FEBS Lett.* **193,** 135 (1985).

[34d] N. Düzgüneş, M. C. Pedroso de Lima, L. Stamatatos, D. Flasher, D. Alford, D. S. Friend, and S. Nir, *J. Gen. Virol.* **73,** 27 (1992).

[34e] H. Wunderli-Allenspach, M. Günthert, and S. Ott, *Biochemistry* **32,** 900 (1993).

fluorescence resonance energy transfer (RET) pair, and, second, the use of fluorescent excimer-forming lipids.

The most widely used assay in the RET category is that involving the donor probe N-(7-nitrobenz-2-oxa-1,3-diazol-4-yl)phosphatidylethanolamine (N-NBD-PE) and the acceptor N-(lissamine rhodamine B sulfonyl)PE (N-Rh-PE).[35,35a] A variant of this assay, utilizing the same donor N-NBD-PE but a different acceptor, cholesterol-anthracene-9-carboxylate (CAC), has also been used.[36] These probes are incorporated in the virosomal membrane during reconstitution. After fusion, the two fluorophores dilute into the target membrane, resulting in a decrease of their overall surface density and a concomitant decrease of the RET efficiency. This decrease can be followed as an increase of the donor (N-NBD-PE) fluorescence. On the basis of this assay, pH-dependent fusion of influenza virosomes with erythrocyte ghosts and BHK cells has been demonstrated.[19]

The excimer assay, involving pyrene-labeled lipids, relies on the capacity of the pyrene fluorophore to form excited dimers (excimers) between a probe molecular in the excited state and a probe molecule in the ground state.[37] The fluorescence emission of the excimer is shifted to higher wavelengths by about 100 nm relative to the emission of the monomer. Excimer formation is dependent on the distance between the probe molecules. Thus, coupled to one of the acyl chains of a phospholipid molecule, such as phosphatidylcholine (PC), the pyrene probe provides a sensitive measure of the surface density of the labeled molecules in a lipid bilayer membrane. On fusion of a pyrene-labeled membrane with an unlabeled membrane, the pyrene surface density decrease can be monitored as a reduction of the excimer fluorescence.

We prefer the pyrene assay over the RET assay, for the following reasons. First, the pyrene assay is more sensitive. Particularly in measurements of intracellular fusion, a higher signal-to-noise ratio is achieved. Second, the low-pH-induced conformational change of influenza HA has an appreciable effect on the fluorescence properties of N-Rh-PE.[36] Third, the use of pyrene-labeled lipids offers the possibility to compare directly the fusogenic activity of whole virus and corresponding reconstituted virosomes on the basis of the same assay, as pyrene-labeled intact virus can be produced from cultured cells whose phospholipids are labeled metabolically by growth in the presence of pyrene fatty acids.[38]

[35] D. K. Struck, D. Hoekstra, and R. E. Pagano, *Biochemistry* **20**, 4093 (1981).
[35a] D. Hoekstra and N. Düzgüneş, this volume [2].
[36] S. A. Wharton, J. J. Skehel, and D. C. Wiley, *Virology* **149**, 27 (1986).
[37] H.-J. Galla and W. Hartmann, *Chem. Phys. Lipids* **27**, 199 (1980).
[38] R. Pal, Y. Barenholz, and R. R. Wagner, *Biochemistry* **27**, 30 (1988).

In the following, we briefly describe the procedures involved in the R_{18} and pyrene assays. Fusion of the labeled virosomes can be conveniently measured using resealed human erythrocyte ghosts[39] as a model biological target membrane system. Alternatively, fusion activity toward liposomes can be assessed, in which case it is important to avoid liposomes consisting primarily of negatively charged phospholipids, such as cardiolipin, as these appear to support a fusion reaction with influenza virus[40] or virosomes,[19] whose characteristics deviate from those of fusion with biological membranes. In our hands, fusion with liposomes consisting of a 2:1 mixture of PC and PE (Avanti Polar Lipids, Alabaster, AL), and containing 5 mol % of the ganglioside G_{D1a} or total bovine brain gangliosides (Sigma Chemical Co., St. Louis, MO) serving as sialic acid-containing receptors for the virus/virosomes, gives satisfactory results. It is to be noted that the ganglioside receptors are not essential for fusion of the virus/virosomes with the liposomes, but merely enhance the binding of the virus/virosomes to the target liposomes.[40] Finally, fusion can be monitored in an on-line fashion using cultured cells as targets. Here, one has the option of following the endocytic uptake of the virosomes at neutral pH and their fusion from within endosomes. Alternatively, the virosomes may be induced to fuse with the cell plasma membrane by lowering the extracellular pH.

Octadecylrhodamine B Assay. Intact influenza virus or reconstituted virosomes (the equivalent of 500 nmol viral phospolipid, in 1.0 ml HNE) are labeled by addition of 30 nmol R_{18} (Molecular Probes, Eugene, OR) in 10 μl ethanol and subsequent incubation at room temperature in the dark for 1 hr. The virions/virosomes are separated from unincorporated probe by gel filtration on a 15-cm Sephadex G-50 column in HNE, during which the free probe remains at the top of the gel.

Fusion is assessed, with either erythrocyte ghosts or liposomes as target membranes, in the cuvette of a fluorometer under continuous stirring at an excitation wavelength of 560 nm and emission wavelength of 590 nm. The zero fusion level is set to the R_{18} fluorescence of the unfused virus/virosomes, while maximal fusion is determined after the addition of Triton X-100 to 0.5% (v/v).

Pyrene Assay. C_{10}-Pyrene-PC (Molecular Probes) is incorporated in the virosomal membranes by coreconstitution, as follows. The supernatant obtained after solubilization of the viral membrane and sedimentation of the nucleocapsid (see above) is added to a dry film of the probe (10 mol %

[39] T. L. Steck and J. A. Kant, this series, Vol. 31, p. 172.
[40] T. Stegmann, S. Nir, and J. Wilschut, *Biochemistry* **28**, 1698 (1989).

with respect to the viral lipid). The mixture is lightly shaken (not stirred in order to avoid foam formation) to allow mixing of the probe with the viral lipids, and detergent is removed as described above.

Fusion of pyrene-labeled virosomal membranes with unlabeled target membranes (erythrocyte ghosts or liposomes, see above) is monitored as a decrease of the excimer fluorescence at 480 nm (excitation wavelength, 343 nm). If fusion is monitored continuously, the excitation slit width should be set as narrow as possible, to limit photobleaching of the probe. The zero fusion level is set to the initial excimer fluorescence and the 100% level to the background intensity of the target membranes at 480 nm.

Examples. Figure 3A gives an example of fusion of pyrene-PC-labeled influenza virosomes with human erythrocyte ghosts. For comparison, fusion of R_{18}-labeled intact virus is also shown. Fusion was measured after prebinding of the virus/virosomes to the ghosts in the cold at neutral pH, as outlined in the legend to Fig. 3. (It is also possible to inject virus into a cuvette containing the ghosts at the desired pH, although this will produce slower fusion kinetics and smaller extents of fusion.) Figure 3A shows that the fusion of both virus and virosomes is strictly dependent on low pH. Clearly, the fusion of the pyrene-PC-labeled virosomes is faster and more extensive than that of the R_{18}-labeled intact virus. However, in control studies (not shown), R_{18}-labeled virosomes (labeled either by coreconstitution of the probe or by postlabeling) was found to also exhibit a slower and less extensive fusion reaction with ghosts than pyrene-PC-labeled virosomes, indicating that the differences in fusion between virus and virosomes shown in Fig. 3A are due to the use of the R_{18} assay in the case of the virus, rather than a reflection of a real difference in fusion potential between virus and virosomes.

Figure 4 presents on-line intracellular fusion recordings of influenza virosomes in BHK-21 cells. Pyrene-PC-labeled virosomes were allowed to bind to the plasma membrane of trypsinized cells at 4°. The virosome–cell mixture was subsequently injected in the fluorescence cuvette containing prewarmed (37°) Hanks'/HEPES buffer. After a lag phase of about 3–4 min, explained by the time required for the virosomes to reach the endosomes, the excimer fluorescence decreases in 1 hr to less than 60% of the initial value (curve C), indicating that over 40% of the cell-associated virosomes have fused. Under the conditions of the experiments, this corresponds to approximately 2000–2500 virosomes fused per cell. If the experiment is performed in the presence of 20 mM NH$_4$Cl, an inhibitor of endosomal acidification, the fusion is blocked (Fig. 4, curve A), or interrupted in case the NH$_4$Cl is added during the course of virosome internalization (Fig. 4, curve B).

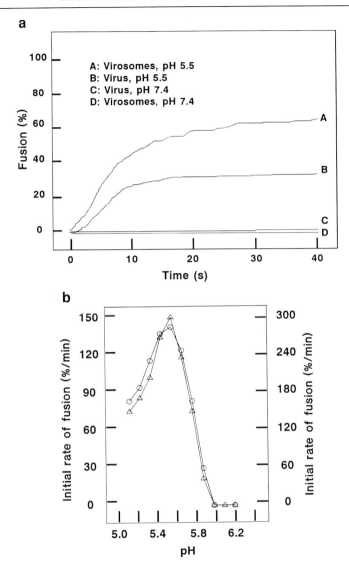

FIG. 3. Fusion of R_{18}-labeled influenza virus (X-99 strain) and pyrene-PC-labeled viro-somes with human erythrocyte ghosts. Virions/virosomes were allowed to bind to the ghosts for 15 min at 4° and pH 7.4; after washing, an aliquot of a concentrated suspension of the complex was injected into fusion medium (135 mM NaCl, 15 mM sodium citrate, 10 mM MES, 5 mM HEPES), adjusted to the desired pH, at 37° in the cuvette of the fluorimeter. Virus fusion was monitored as an increase of the R_{18} fluorescence, virosome fusion as a decrease of the pyrene-PC excimer fluorescence. (A) Real-time fusion recordings. (B) Initial rates of fusion as a function of pH (triangles, virus fusion, left-hand scale; circles, virosome fusion, right-hand scale). Note that the X-99 virus has a relatively high pH threshold (pH 6.0) and optimum (pH 5.6) for fusion.

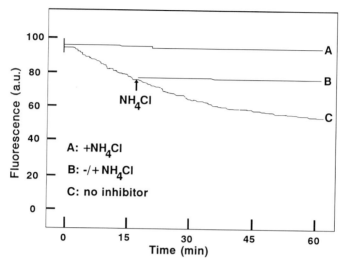

FIG. 4. Intracellular fusion of pyrene-PC-labeled influenza virosomes (X-99 strain) in BHK-21 cells in suspension. The virosomes were allowed to bind to the cells for 1 hr at 4° in Hanks'/HEPES buffer (137 mM NaCl, 5.4 mM KCl, 0.44 mM KH$_2$PO$_4$, 0.41 mM MgSO$_4$, 0.40 mM MgCl$_2$, 1.3 mM CaCl$_2$, 5.6 mM glucose, and 10 mM HEPES, pH 7.4). After washing, an aliquot of the cell suspension was injected into Hanks'/HEPES buffer at 37° in the cuvette of the fluorimeter. Fusion was monitored as a decrease of the pyrene excimer fluorescence. The NH$_4$Cl concentration (curves A and B) was 20 mM (pH 7.4).

Hemolytic Activity

A readily accessible alternative for the direct assessment of the fusion activity of influenza virosomes is given by the determination of their hemolytic activity. It is our experience that the fusion activity of influenza virosomes, produced according to the procedure described above, is invariably paralleled by hemolytic activity, exhibiting a pH dependence identical to that of fusion. One should, however, bear in mind that reconstitution of hemolytic activity does not necessarily reflect reconstitution of fusion activity.

Hemolytic activity of influenza virosomes is determined by adding the virosomes (the equivalent of 1 nmol of phospholipid, in a volume of 25 μl) to 4 × 10^7 washed human erythrocytes in 975 μl fusion buffer (135 mM NaCl, 15 mM sodium citrate, 10 mM MES, 5 mM HEPES), set to various pH values. After incubation at 37° for 30 min, the mixture is centrifuged for 3 min at 1350 g. Lysis of erythrocytes is quantified by the measurement of absorbance of the hemoglobin in the supernatant at 541 nm. Maximal hemolysis is determined after lysis of the erythrocytes in distilled water.

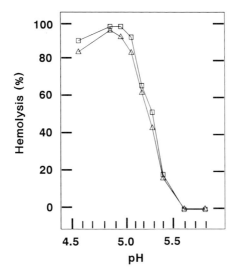

FIG. 5. Lysis of human erythrocytes after pH-dependent fusion with X-31 influenza virions (triangles) or virosomes (squares). For details, see text.

Figure 5 presents the results of a typical determination of the hemolytic activity of influenza virosomes as a function of pH, in comparison with that of the native virus. The virus strain used in this example (X-31) has a lower pH threshold for fusion and hemolysis than the X-99 strain (see Fig. 3B). Invariably, the pH dependence of fusion and that of hemolysis of virus or virosomes of the same strain are found to be the same.

Virosome-Mediated Introduction of Foreign Molecules into Cells

Not only do virosomes provide a useful model system for studying the membrane fusion activity of viral spike glycoproteins or the cellular entry mechanisms of enveloped viruses, they also represent a powerful carrier system for the introduction of foreign molecules into living cells. In these applications of virosomes, as opposed to their use as a model for the native virus, the most prominent criterion for functionality is the capacity of the vesicles to retain encapsulated or membrane-associated substances and to deliver them efficiently to cells, whereas optimal correspondence of the virosomes to the native viral envelope in terms of composition and size and/or the presence of residual detergent are less important.

Both water-soluble and membrane-associated substances can be incorporated in virosomes, the former by encapsulation in the aqueous interior of the virosomes, the latter by coreconstitution in the virosomal membrane. Examples of virosome-mediated delivery of water-soluble macromolecules to cells include the transfer of genes.[26,41] With respect to membrane-associated molecules, efficient insertion of specific receptors into the plasma membrane of receptor-deficient cells has been achieved.[42] Below, we describe briefly two examples of virosome-mediated delivery of foreign molecules to cells.

Delivery of Water-Soluble Molecules

The A chain of diphtheria toxin (DTA) has been used as an aqueous-phase marker of the virosomes in order to explore the capacity of influenza virosomes to deliver water-soluble molecules to the cytoplasm of cultured cells. DTA is an enzyme that efficiently inhibits cellular protein synthesis by ADP-ribosylation of elongation factor 2 (Ref. 43). However, when added to cells, DTA is not toxic, as it lacks the capacity to bind to cellular receptors. In the intact toxin, this binding capacity is located on the B subunit. When isolated DTA is encapsulated in virosomes during their preparation, it is delivered efficiently to the cytoplasm of cells via endocytic uptake of the virosomes and fusion at the level of the endosomes.[43a]

Briefly, 250 μg of lyophilized DTA is added to the solubilized viral envelope, after the sedimentation of the nucleocapsid, and reconstitution is carried out as described above. To ensure complete separation from non-encapsulated DTA, the virosomes are subjected to flotation on a discontinuous metrizamide gradient in HNE after the discontinuous sucrose gradient step. The virosome fraction from the sucrose gradient is diluted 10-fold with HNE, and the virosomes are sedimented for 1 hr at 170,000 g in a SW 50.1 rotor. Subsequently, the virosome pellet is resuspended in 0.5 ml HNE and mixed with 1.0 ml of 52.5% (w/v) metrizamide. On top of this solution (in a Beckman SW 50.1 tube), 3.0 ml of 30% (w/v) metrizamide in HNE and 0.5 ml HNE are layered, and the gradient is centrifuged for 2 hr at 170,000 g. The virosomes are collected from the top of the gradient in 0.5 ml and appropriately diluted in Hanks'/HEPES. The virosomes are allowed to bind for 1 hr at 4° to BHK-21 cells grown in 96-well titer plates. The nonbound virosomes are washed away. Subsequently, endocytosis is induced by the addition of Hanks'/HEPES buffer, pH 7.4 (see legend to Fig. 4) at 37°, followed by incubation at 37° for 1 hr. After a 16-hr incubation in culture medium, ^{3}H-labeled leucine is added to the wells, and 6 hr later proteins are precipitated with ice-cold 5% trichloroa-

[41] D. J. Volsky, T. Gross, F. Sinangil, A. Kuszynsky, R. Bartzatt, T. Dambaugh, and E. Kieff, *Proc. Natl. Acad. Sci. U.S.A.* **81**, 5926 (1984).

[42] D. J. Volsky, I. M. Shapiro, and G. Klein, *Proc. Natl. Acad. Sci. U.S.A.* **77**, 5453 (1980).

[43] A. M. Pappenheimer, Jr., *Annu. Rev. Biochem.* **46**, 69 (1977).

[43a] R. Bron, A. Ortiz, and J. Wilschut, submitted for publication.

cetic acid (TCA). The protein precipitates are dissolved in 0.1 N NaOH. Incorporation of [³H] leucine is assessed by liquid scintillation counting.

The results, presented in Table I, indicate that DTA encapsulated in fusogenic virosomes induces a complete inhibition of cellular protein synthesis. On the other hand, free DTA or empty virosomes have no effect, while the effect of virosome-encapsulated DTA can be blocked completely with NH₄Cl or by a pretreatment of the virosomes alone at low pH, causing an irreversible inactivation of their fusion capacity. Delivery of DTA to the cellular cytoplasm can also be accomplished through fusion of the virosomes at the cellular plasma membrane, by a brief incubation (1–2 min) of the cell-bound virosomes in a low pH buffer.

Insertion of Membrane-Associated Molecules

As an example of a biologically active membrane-associated compound, we have studied bacterial lipopolysaccharide (LPS). LPS is a potent

TABLE I

DELIVERY OF DIPHTHERIA TOXIN A CHAIN BY INFLUENZA
VIROSOMES TO BHK-21 CELLS[a]

Condition	Protein synthesis (% of control)
Free DTA (0.1 mg/ml)	99
Empty virosomes	97
DTA in virosomes, 60 min, pH 7.4	3
DTA in virosomes, 2 min, pH 5.4	5
DTA in virosomes, low-pH pretreated, 60 min, pH 7.4	97
DTA in virosomes, + 20 mM NH₄Cl, 60 min, pH 7.4	98

[a] DTA-containing influenza virosomes (derived from the X-97 strain of the virus) were prepared as described in the text. Virosomes (1 nmol of phospholipid) were allowed to bind to 5×10^4 BHK-21 cells in 96-well microtiter plates for 1 hr at 4°. The unbound virosomes were washed away. Fusion of the virosomes with the plasma membrane was induced by addition of fusion buffer (135 mM NaCl, 15 mM sodium citrate, 10 mM MES, 5 mM HEPES), pH 5.4, and incubation for 2 min at 37°. Fusion from within endosomes was induced by addition of Hanks'/HEPES buffer and incubation for 1 hr at 37°. In either case, after washing, the cells were incubated for 16 hr in culture medium containing 20 mM NH₄Cl, after which protein synthetic capacity was assessed on the basis of [³H]leucine incorporation in TCA-insoluble material.

activator of, among other cells, B lymphocytes.[44] The mechanism of action of LPS is not known, but it is believed to act at the level of the plasma membrane. When free LPS is added to cells, it, in part, partitions into the plasma membrane and/or interacts with a putative receptor, eventually resulting in cell activation. LPS, incorporated in the membrane of influenza virosomes that are subsequently fused to the plasma membrane of murine splenic B lymphocytes, is much more potent in activating the cells than free LPS.[45] Because fusion-inactivated LPS virosomes induce little cell activation, the potentiation of the action of LPS is due to its fusion-mediated insertion into the cell plasma membrane.

Concluding Remarks

The establishment of a procedure for the functional reconstitution of viral envelopes has set the stage for further delineation of the membrane fusion characteristics of viral envelope glycoproteins. Now that the conditions for generating virosomes, which closely mimic the corresponding native viral envelope in terms of composition, structure, and biological activity, have been defined, purified viral fusion proteins can be reconstituted in a controlled lipid environment at a controlled surface density on the basis of an otherwise similar reconstitution procedure. We believe that ultimate elucidation of viral fusion mechanisms will require the investigation of isolated fusion proteins under such well-defined conditions and the combined application of advanced biophysical, biochemical, and morphological techniques.

In addition, virosomes represent an efficient carrier system for the introduction of foreign molecules into cells, with numerous potential applications in areas ranging from cell biological research and membrane biotechnology to drug delivery and vaccine development. Further optimization of the virosomes as delivery vehicles should concentrate primarily on improvement of their encapsulation efficiency. For example, the entrapment of DTA in the procedure described above is relatively low, since the enclosed aqueous volume of the virosomes, at the relatively low concentration at which they are produced, represents a small fraction of the total volume. The production of fusogenic proteoliposomes, involving the intermediate formation of protein-lipid cochleates, yields vesicles with a much higher relative encapsulation efficiency.[25,26] Alternatively, one may consider the possibility of first generating liposomes, applying procedures that result in very high relative encapsulation efficiencies. Viral fusion

[44] D. C. Morrison and J. L. Ryan, *Adv. Immunol.* **28**, 293 (1979).
[45] J. Dijkstra, A. de Haan, R. Bron, J. Wilschut, and J. L. Ryan, submitted for publication.

proteins might then be postinserted into the liposomal membrane at sub-CMC detergent concentrations. Successful postinsertion of a viral fusion protein in preformed lipid vesicles has been reported.[46] However, whether encapsulated macromolecules are retained during such a procedure remains to be established.

Acknowledgments

We acknowledge the support of The Netherlands Organization for Scientific Research (fellowship to R. Bron), the European Molecular Biology Organization (long-term fellowship to A. Ortiz), the U.S. National Institutes of Health (Research Grant AI-25534 to N. Düzgüneş and J. Wilschut), the Du Pont–Merck Pharmaceutical Company (grant to the University of Groningen), and ImClone Systems, Inc., New York (grant to the University of Groningen). We thank Solvay-Duphar B. V., Weesp, The Netherlands, for ample supplies of virus, and Jan van Breemen and Frank Booy for their contribution to the EM work shown in Fig. 2.

[46] R. K. Scheule, *Biochemistry* **25**, 4223 (1986).

[24] Electron Spin Resonance Methods for Studying Virus–Cell Membrane Fusion

By Shun-ichi Ohnishi and Kazumichi Kuroda

Disappearance of Exchange Broadening of Spin Labels by Fusion

When two membranes fuse, the lipid molecules should mix with each other by lateral diffusion. Therefore, when one membrane containing spin-labeled phospholipids at a high concentration is fused with other, nonlabeled membranes, dilution should dramatically change the electron spin resonance (ESP) spectrum. The ESR spectrum broadens in membranes containing more spin labels, and the three-line ESR peak heights decrease, owing to the spin–spin interactions. When the concentration of

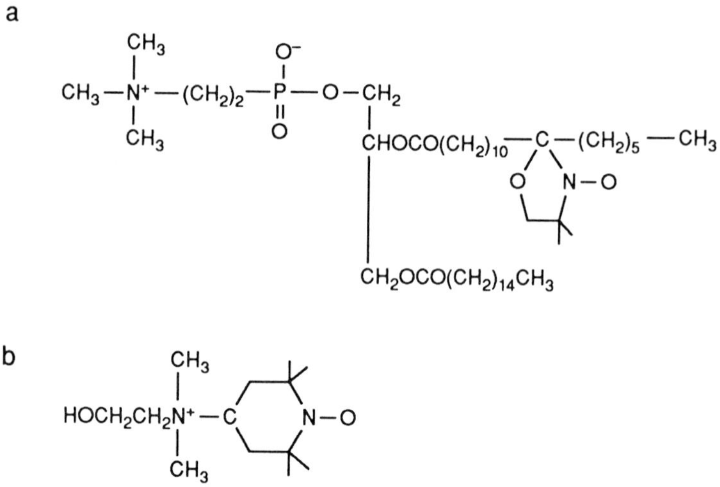

a

b

FIG. 1. Structure of spin-labeled phosphatidylcholine (PC*) and tempocholine.

labeled phospholipids is higher than 50 mol %, the three-line ESR spectrum becomes a single line, its line width decreasing with the concentration.[1]

To study fusion of viruses with cells, first the envelope of the virus is tagged with spin-labeled phospholipids. Hemagglutinating virus of Japan (HVJ; Sendai virus) is labeled by incubation with vesicles of spin-labeled phosphatidylcholine (PC*) (Fig. 1), at a high concentration, namely, 10 mol % of the envelope phospholipids. Then spin-labeled HVJ* is added to unlabeled red blood cells (RBC), and the ESR spectrum of the virus–cell aggregates is measured. The spectrum greatly changes from the exchange-broadened spectrum (solid line, Fig. 2a) to the sharp three-line spectrum (dotted line, Fig. 2a).[2] The peak height increases approximately three times, which clearly indicates the dilution of PC* by mixing with RBC lipids.

Model spectra of PC* are measured at various concentrations in erythrocyte total lipid liposomes (see Fig. 3A). The peak height decreases with the concentration of PC*. Figure 3B shows the concentration dependence of the normalized central peak height, that is, the central peak height divided by the double integrated are of the spectrum. The double integrated area is parallel to the number of spins in the sample. The peak height for HVJ* at 10% spin labels is 0.125, while that for a dilute erythrocyte lipid membrane (0.2% spin labels) is 0.747. Therefore, the spectrum should increase 6-fold ($0.747/0.125 = 5.976$) on complete fusion. The ob-

[1] S. Ohnishi and S. Tokutomi, *Biol. Magn. Reson.* 3, 121 (1981).
[2] T. Maeda, A. Asano, K. Ohki, Y. Okada, and S. Ohnishi, *Biochemistry* 14, 3736 (1975).

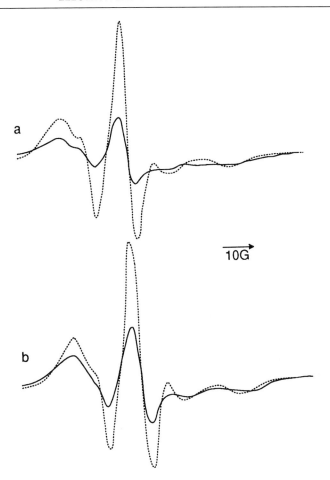

Fig. 2. Change of the ESR spectrum of PC* on envelope fusion. (a) HVJ* and RBC or (b) HVJ, RBC*, and RBC were mixed at 4° and the virus–cell aggregates placed into a capillary. ESR spectra were measured at 22°, after incubation at 37° for 0 min (solid line) or 60 min (dotted line).

served 3-fold increase suggests 50% fusion of virus with RBC. The spectrum is therefore a composite of the larger peak for HVJ*–RBC membranes and the exchange-broadened peak from HVJ*.

Virus-induced cell fusions can also be measured by the same principle. For example, RBC membranes are labeled with PC* vesicles at a high concentration. Then, HVJ, RBC*, and RBC are mixed, and the ESR spectrum of the virus–cell aggregates is measured. The ESR spectrum changes from the exchange-broadened form to the sharp three lines (Fig.

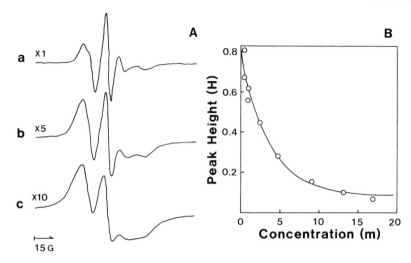

FIG. 3. (A) ESR spectrum of erythrocyte total lipid liposomes containing various concentrations of PC*: 1% (a), 9.2% (b), and 16.9% (c). Each spectrum was recorded at a different gain, as indicated. (B) Normalized central peak height (i.e., central peak height divided by the double-integrated area of the whole spectrum) as a function of the concentration of PC*. The concentration m is given as the mole percent of total phospholipid. The curve is approximated as $H(m) = 0.70 \exp(-0.278m) + 0.081$, for $m < 17$.

2B).[3] The peak height increases 2.4-times in this case, indicating that the PC* in RBC* membranes are mixed with nonlabeled RBC membranes by fusion induced by HVJ.

Hemolysis is generally used as an assay of envelope fusion for viruses that aggregate RBC and cause fusion. The spin label results generally agree with these hemolysis assays. Some viruses do not have hemolytic activity. For example, HVJ grown for 3 days in eggs shows hemolysis, but the virus grown for 1 day in eggs does not cause hemolysis.[4] The spin label results clearly show envelope fusion for the early harvested HVJ as well as the late harvested HVJ.[3]

Mixing of Phospholipids by Cytoplasmic Exchange Proteins

Cells contain phospholipid exchange proteins in the cytoplasm.[5] If these exchange proteins surround membranes, they can exchange phos-

[3] T. Maeda, A. Asano, Y. Okada, and S. Ohnishi, J. Virol. 21, 232 (1977).
[4] M. Homma, K. Shimizu, Y. K. Shimizu, and N. Ishida, Virology 71, 41 (1976).
[5] K. W. A. Wirtz, in "Lipid–Protein Interactions" (P. C. Jost and O. Hayes Griffith, eds.), Vol. 1, p. 151. Wiley, New York, 1982.

pholipids between the membranes. They bind a phospholipid molecule from a vesicle and exchange it with other vesicles. Spin-labeled phospholipids used for the assay of exchange protein indicate a tight binding for PC* bound to PC exchange protein.[6]

In 1983 a PC exchange protein was discovered in erythrocyte hemolysates and was partially isolated.[7] During experiments on the fusion of HVJ* with RBC, HVJ* fuses with RBC and causes hemolysis. The PC exchange protein leaks out of the RBC and mediates the exchange of PC* between the membranes of HVJ* and RBC. The ESR peak height therefore increases, first by envelope fusion and then by exchange of PC*, owing to the exchange protein that has leaked out. When RBC ghosts containing no hemolysate are used, the ESR peak height increases due to envelope fusion only. The ESR peak height increase for HVJ*–RBC fusion is similar to that for HVJ*–ghost fusion in the initial stages, but later it becomes larger due to phospholipid transfer by the PC exchange protein (see below).

Trypsinization of HVJ causes the digestion of its fusion protein F and the loss of hemolytic activity.[8] The ESR peak height for trypsinized HVJ*–RBC does not increase. When HVJ is added to this system (trypsinized HVJ*–HVJ–RBC) the ESR peak height increases slowly, speeding up later.[9] This is due to hemolysis by HVJ, which causes exchange of PC* between the membranes of HVJ*, HVJ, and RBC by the PC exchange protein. Influenza virus does not fuse with cells at neutral pH, but only at acidic pH.[10] When HVJ is added to spin-labeled influenza virus and RBC (influenza virus*–HVJ–RBC), the ESR peak height increases slowly at first at neutral pH, but speeds up later. This is due to hemolysis by HVJ catalyzing transfer of PC* between the membranes.

The ESR peak height increase is caused by envelope fusion in the earlier phase, and later on by the exchange proteins when cells are lysed. If the exchange protein activity is inhibited by some reagents, for example, phosphatidylserine vesicles for PC exchange protein, then only envelope fusion would be examined.

Reduction of Nitroxide Spin Labels

There are some reducing systems in the cytoplasm which cause reduction of the spin-label nitroxide groups, even when they are attached to the

[6] K. Machida and S. Ohnishi, *Biochim. Biophys. Acta* **507**, 156 (1978).
[7] K. Kuroda and S. Ohnishi, *J. Biochem. (Tokyo)* **94**, 1809 (1983).
[8] K. Shimizu and K. Ishida, *Virology* **67**, 427 (1975).
[9] K. Kuroda, T. Maeda, and S. Ohnishi, *Proc. Natl. Acad. Sci. U.S.A.* **77**, 804 (1980).
[10] A. Yoshimura, K. Kuroda, K. Kawasaki, S. Yamashina, T. Maeda, and S. Ohnishi, *J. Virol.* **43**, 284 (1982).

fatty acyl chains and embedded in membranes. The ESR peak height in cells may decrease because of this reduction, the degree of reduction depending on the cells. The peak height increases rapidly but then decreases after 10 min in HVJ*–KB cells.[11] However, the decrease is small in influenza virus*–MDCK cells and essentially nonexistent in erythrocytes.[10] To avoid greater reductions, oxidizing reagents such as $Fe_4 (CN)_6^{4-}$ could be added.

Procedure: Fusion of Virus with Red Blood Cells

Spin Labeling of HVJ with PC*

Spin labeling of virus is carried out by the incubation of the virus with PC* vesicles.[1] PC* vesicles are prepared as follows. PC* in organic solvent is dried on the bottom of a test tube under a stream of N_2 gas, and then under high vacuum to remove the solvent. PC* is dispersed in TBS (Tris-buffered saline: 140 mM NaCl, 5.4 mM KCl, 20 mM Tris-HCl, pH 7.6) by sonication at a concentration of 1 mM. Large multilamellar liposomes are removed by ultracentrifugation (150,000 g for 1 hr). The supernatant (PC* vesicles) is used to label the virus.

HVJ of 10,000 to 20,000 hemagglutinating units (HAU) is suspended in 1 ml of PC* vesicles and incubated for 5 hr at 37°. The virus is washed with a bovine serum albumin solution (10 mg/ml in TBS) to remove fatty acids, especially the spin-labeled ones, that are decomposed from PC*, then washed with TBS to remove albumin. Finally, the virus is suspended in 0.5 to 1 ml TBS and viral aggregates are removed by centrifugation for 15 min at 3000 rpm with a tabletop centrifuge. HVJ* prepared by this procedure contains PC* at a concentration of 10 to 15 mol % of the viral phospholipid.

The biological activity of HVJ is not affected appreciably by the spin-labeling procedure. The hemagglutination activity of HVJ* is essentially the same as that of intact HVJ. The hemolytic activity is lowered only slightly after spin labeling, and HVJ* shows no indication of morphological alterations under electron microscopy.

Assay of Envelope Fusion

Appropriately diluted HVJ* suspension (0.9 ml) is mixed with 0.3 ml of 20% (v/v) erythrocyte ghosts, prepared by hypotonic lysis of human RBC and kept for 15 min for adsorption at 4°. A 10% suspension is used

[11] A. H. Koyama, T. Maeda, S. Toyama, S. Ohnishi, H. Uetake, and S. Toyama, *Biochim. Biophys. Acta* **508**, 130 (1978).

for intact RBC instead of 20% for ghosts. Then 9 ml of TBS is added to the mixture, which is centrifuged for 5 min at 2000 rpm at 4° to remove unadsorbed virus. Finally, the pellet is drawn at 4° into a capillary tube for the ESR measurement. For the X-band cavity, a quartz tube with an outside diameter of 4 mm, an internal diameter of 0.5 mm, and a length of 150 mm is used. Envelope fusion between HVJ and RBC occurs quite efficiently, even at a room temperature higher than 20°. Thus handling of the pellet at room temperature can lead to confusing results.

The amount of virus necessary for this sensitive assay is dependent on the system. A measurable intensity of ESR signal can be obtained at 100 HAU/ml of HVJ* for ghosts, or 20 HAU/ml of HVJ* for RBC.

The ESR spectrum of HVJ*–RBC is measured continuously, or by the quenching method. In the continuous method, the capillary tube containing sample is inserted into the ESR cavity, whose inside temperature is preset at 37°. The central peak heights of the spectrum are measured repeatedly every 30 to 120 sec, or the whole spectrum can be measured continuously. The time course of envelope fusion is obtained from one sample, as shown in Fig. 4. A problem may arise with the time delay for the

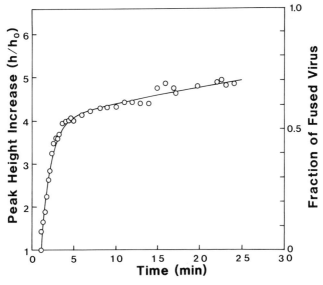

FIG. 4. ESR peak height increase for the HVJ*–ghost system. HVJ* at a final concentration of 101 HAU/ml was mixed with erythrocyte ghosts at 5% (v/v). The central peak height of the ESR spectrum of the virus–cell pellet was measured *continuously* at 37°. The central peak height h divided by the initial peak height h_0 is plotted against time. The curve was drawn by Eq. (8) with the best fit parameter values for α, k_1, k_2, and t_e. On the right-hand ordinate, the fraction of fused virus calculated with Eq. (6) is given.

sample temperature to reach 37° from the initial 4°. Cooled air or N_2 gas can be put into the quartz inset at the cavity to set the gas temperature at 37°. The time delay is about 1 min.

In the quenching method, envelope fusion is started by putting the centrifuge tube containing HVJ*–cell aggregates into a water bath at 37°. After an appropriate period the reaction is stopped by quenching in ice and water. Then the pellet is pipetted into a capillary at 4°, and the spectrum is measured at 15°, where practically no envelope fusion occurs. The lag time of the temperature shift is very short, and it should be possible to follow the initial change in the ESR spectrum more accurately. A disadvantage of the quenching method is the need for more samples to measure the complete fusion kinetics.

Kinetic Analysis of ESR Peak Height Increase Based on Envelope Fusion

When HVJ* fuses with cell membranes, the PC* molecules rapidly diffuse into the membrane and become highly diluted. Lateral diffusion is much more rapid than envelope fusion. Therefore, the PC* molecules of the fused envelope diffuse out in the cell membrane at a low concentration so as not to allow for a spin–spin exchange interaction before the next fusion. The spectrum during the fusion reaction should be the sum of that of HVJ* remaining unfused and that of PC* transferred into the cell membrane by fusion. The central peak height of the composite spectrum is assumed to be the sum of the two components, although the heights for HVJ* and for PC* in erythrocyte membranes do not coincide exactly. A model ESR spectrum is made by the addition of liposomes made of erythrocyte total lipid containing low (0.5%) and high (13.2%) concentration of PC* at various ratios. The results show the approximate proportionality between the central peak height and the concentration of liposomes containing 0.5% PC*.[12]

If envelope fusion follows first-order kinetics, the velocity of envelope fusion is given by

$$dn/dt = k_f(n_0 - n) \qquad (1)$$

and

$$n/n_0 = 1 - \exp(-k_f t) \qquad (2)$$

where n is the number of fused virus particles, n_0 is the number of initially adsorbed virus particles, and k_f is the rate constant or probability of

[12] K. Kuroda, K. Kawasaki, and S. Ohnishi, *Biochemistry* 24, 4624 (1985).

fusion.[12] The ESR peak height (h) can be given as a sum of the height for HVJ* and that for PC* in RBC membranes:

$$h = N[(n_0 - n)H_V + nH_G] \qquad (3)$$

where N is the number of PC* in one particle of HVJ*. H_V and H_G are the normalized peak heights for HVJ* and ghosts, respectively. By dividing both sides by the initial peak height h_0 ($= Nn_0H_V$),

$$h/h_0 = (1 - n/n_0) + (n/n_0)(H_G/H_V) \qquad (4)$$

and by Eq. (2),

$$h/h_0 = f - (f - 1)\exp(-k_f t) \qquad (5)$$

where f ($=H_G/H_V$) is the peak height increase factor (see Fig. 3B). The fraction of fused virus F ($=n/n_0$) can be obtained from h/h_0 by

$$F = (h/h_0 - 1)/(f - 1) \qquad (6)$$

If there are two populations of virus fusing at different rates, then

$$h/h_0 = f - (f - 1)[\alpha \exp(-k_1 t) + (1 - \alpha)\exp(-k_2 t)] \qquad (7)$$

where k_1 and k_2 are the fusion rate constants and α is the fraction of virus with k_1.

In practice, there is a time delay ($\sim$60 sec) in the rise of sample temperature from 4° to 37° in the continuous method. For calibration of these effects, an effective time zero, t_e, may be introduced into the exponent of the kinetic equation. For example, Eq. (7) becomes

$$h/h_0 = f - (f - 1)\{\alpha \exp[-k_1(t - t_e)] + (1 - \alpha)\exp[-k_2(t - t_e)]\} \qquad (8)$$

In the analysis of data by the continuous method, the data before 60 sec are not used.

From the ESR measurement, the central peak height h is plotted as a function of the incubation time. For analysis using Eq. (7) or (8), the value of h_0 has to be estimated. This can be obtained by multiplying the normalized peak height for HVJ* by the double-integrated area of the spectrum (DA) for HVJ*–cell at any time of the fusion reaction

$$h_0 = H_V(DA \text{ for HVJ*–cell}) \qquad (9)$$

Alternatively, h_0 is obtained as an adjustable parameter in the data fitting to the theoretical kinetic equation.

Figure 4 shows the analysis for the HVJ*–ghost system.[12] The experimental data can be fitted very well by Eq. (8) on the basis of two populations of virus, but not by Eq. (5) with a single fusion rate constant. The curve is drawn by Eq. (8) with the best-fit parameter values for α, k_1, k_2, and t_e. The value of h_0 is obtained by the double-integration method [Eq. (9)] and the f value by the ratio of normalized peak heights H_G/H_V, 6.1 (Table I). Similarly, Eq. (8) can be made to fit the data well by taking h_0, α, k_1, and k_2 as the adjustable parameters and using a fixed value for t_e (60 sec) (Table I). The h_0 value for the latter method is close to the value obtained by the double-integration method. In both methods the values for three parameters α, k_1, and k_2 agree. The virus with the faster fusion makes up more than 50%, and the rate constant is 0.86 min^{-1} at 37°.

The data obtained by the quenching method are also analyzed by Eq. (7) or (8) (Table I). The parameter value of $k_1 = 0.83$ min^{-1} is obtained for the HVJ*–ghost system and is consistent with the continuous method.

The kinetic analysis for the HVJ*–RBC system shows the rate constant $k_1 = 0.85$ min^{-1} (Table I). This is again very close to that for the HVJ*–ghost system. On the other hand, the rate constant k_2 is 0.041 min^{-1}, larger than that for the HVJ*–ghost system (0.02 min^{-1}). The larger k_2 value should arise from exchange by the PC* exchange protein.

Release of Tempocholine Preloaded in Virus on Envelope Fusion

Virus–cell interactions can be studied by measuring the release of a small spin label, tempocholine (Fig. 1), previously incorporated into vi-

TABLE I
KINETIC PARAMETERS FOR ENVELOPE FUSION OF
VIRUS WITH ERYTHROCYTES[a]

Source	k_1 (min^{-1})	k_2 (min^{-1})	α	h_0
Ghosts				
a	0.87	0.021	0.53	463
b	0.85	0.019	0.56	450
c	0.83			
RBC				
a	0.87	0.037	0.35	331
b	0.83	0.045	0.36	309

[a] As in Fig. 4, the ESR peak height data were analyzed by Eq. (8) with two different sets of adjustable parameters: (a) α, k_1, k_2, and t_e or (b) h_0, α, k_1, and k_2. h_0 was estimated by the double-integration method [eq. (9)] in (a), and $t_e = 60$ sec was fixed in (b). f was estimated by the ratio of the normalized peak heights H_G/H_V as 6.1. In (c), k_1 was obtained from the quenching data.

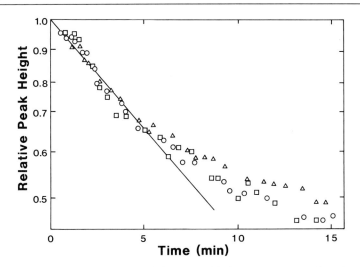

FIG. 5. Assay of tempocholine remaining trapped inside HVJ on incubation with erythrocyte ghosts. HVJ lightly loaded with tempocholine (20 μl, 300–1200 HAU/ml) was mixed with 1 ml of ghosts (3%, v/v), and the ESR spectrum of the pellet was measured continuously at 37° in the presence of 1 mM ascorbate. The concentration of HVJ was 208 (□), 416 (○), and 832 HAU/ml (△).

ruses. The release is closely related to envelope fusion, since the viral contents should be released into the cytoplasm and/or the medium on envelope fusion.[13]

Two methods are used to measure the release of tempocholine. In the first method, the virus is heavily loaded with tempocholine, and the released tempocholine is detected by the increase of the peak height caused by the disappearance of spin–spin exchange interactions. This is similar to the transfer of PC*, as described in the previous section. In the second method, the virus is loaded lightly, and the tempocholine released on fusion is reduced by ascorbate added to the medium. The residual ESR signal corresponds to the amount of unfused viruses. The latter method is especially useful in the system of cells which contain reducing reagents.

Loading Virus Particles with Tempocholine

HVJ (25,000 HAU) is suspended in 0.5 ml of 140 or 1 mM tempocholine solution and incubated at 30° for 4 hr. The higher concentration of tempocholine is used for the heavy loading, whereas the lower concentration is used for the light loading. The virus is then washed three times with TBS and finally suspended in 0.5 ml TBS.

[13] T. Maeda, K. Kuroda, S. Toyama, and S. Ohnishi *Biochemistry* **20**, 5340 (1981).

Assay of Tempocholine Release from Lightly Loaded Virus by Reduction with Ascorbate

The tempocholine-loaded HVJ and erythrocyte ghosts, or RBC, are mixed at 4° in a manner similar to the fusion assay using PC*. After adsorption, the aggregate is spun down, mixed with 1 mM ascorbate, and the ESR peak height measured continuously at 37°. Figure 5 shows the decrease in the peak height over time.[13]

The ascorbate reduces the nitroxide moiety of accessible tempocholine in less than 1 min. However, tempocholine inside the virus particle is protected from the externally added ascorbate. Reduction of the internal tempocholine takes a much longer time, with $t_{1/2}$ of 53 min at 37°. The relative peak height for unfused virus follows the Eq. (10):

$$h/h_0 = \exp(-k_f t) \tag{10}$$

or

$$h/h_0 = \alpha \exp(-k_1 t) + (1 - \alpha) \exp(-k_2 t) \tag{11}$$

for the system with two populations of virus fusing at different rates. The fraction of fused virus, F, can be obtained from h/h_0.

From the data given in Fig. 5, the fusion rate constant k_f is estimated as 0.081 min^{-1}. This is smaller than the value obtained from the analysis using PC* (Table I). There is a possibility for fusion of the outer layers of membranes detected by PC*. The small value in the release assay may be caused by the relatively slow rate of reduction of tempocholine by ascorbate and by some interaction of tempocholine with the viral nucleotides.

[25] Radiation Inactivation Analysis of Virus-Mediated Fusion Reactions

By JOHN LENARD

Introduction

The technique of radiation inactivation analysis offers a unique means by which the function of a protein can be related to its size and structure. The technique has a long and somewhat checkered history,[1] and it has finally emerged only recently as a reliable adjunct to other, more commonly used techniques. Radiation inactivation has been successfully applied to a large number of integral membrane proteins.[2,3] A useful review of radiation inactivation analysis of membrane components, including both theoretical and practical considerations, is provided in a recent volume of this series.[3] This chapter briefly describes the information obtained using radiation inactivation analysis and reviews its application to the specific problem of virus-mediated fusion.

A radiation inactivation experiment is very simple in principle. A series of identical frozen (or lyophilized) samples are subjected to increasing levels of high-energy radiation, and each sample is then assayed for the specific protein and/or functional property of interest. The radiation is administered either as electrons from an accelerator or as γ rays from a ^{60}Co or ^{137}Cs source. These particles are of sufficiently high energy that a single "hit" on a protein molecule can break several covalent bonds, thus obliterating the protein as a discrete entity and totally destroying its activity. The larger the protein, therefore, the more likely it is to be hit (and thereby inactivated) by any given dose of radiation, that is, large proteins are bigger targets than small ones. This model of inactivation by a single hit predicts that a plot of log(protein or activity remaining) versus radiation dose will give a straight line, whose slope is proportional to the "target size" of the protein, as described below. A simple linear relationship is in fact observed in most radiation inactivation experiments.[3]

---

[1] E. C. Pollard, *in* "Target Size Analysis of Membrane Proteins" (J. C. Venter and C. Y. Jung, eds.) (Receptor Biochemistry and Methodology, Vol. 10), p. 1. Alan R. Liss, New York, 1987; D. J. Fluke, *in* "Target Size Analysis of Membrane Proteins" (J. C. Venter and C. Y. Jung, eds.) (Receptor Biochemistry and Methodology, Vol. 10), p. 21. Alan R. Liss, New York, 1987.

[2] J. C. Venter and C. Y. Jung (eds.), "Target Size Analysis of Membrane Proteins" (Receptor Biochemistry and Methodology, Vol. 10). Alan R. Liss, New York, 1987.

[3] E. S. Kempner and S. Fleischer, this series, Vol. 172, p. 410.

METHODS IN ENZYMOLOGY, VOL. 220

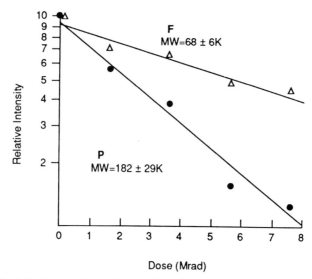

FIG. 1. Radiation-induced loss of Sendai virus P (●) and F (△) proteins from Coomassie blue-stained gels.[4]

Two different "target sizes" are measured by radiation inactivation analysis, depending on the assay used. The first is the structural target size, which is determined from the radiation dependence of destruction of the protein of interest. This is conveniently measured by its disappearance as a discrete band in a gel (Fig. 1).[4] The structural target size is thus independent of any function or activity; it corresponds to the mass of material (containing the protein of interest) that is destroyed by a single radiation "hit." The structural target size may correspond to the monomer molecular weight of the specific protein. Figure 1 shows, for example, that the structural target size for F protein in intact Sendai virus particles corresponds to its monomer molecular weight of 59,000,[5] whereas the structural target size for P protein is several times its monomer molecular weight of 74,000. Each of the other Sendai proteins was found to possess its own characteristic structural target size.[4]

A structural target size that is larger than the monomer molecular weight of the protein (e.g., P protein in Fig. 1) provides evidence that the

[4] S. Gibson, K. Bundo-Morita, A. Portner, and J. Lenard, *Virology* **163**, 226 (1988).
[5] This is the molecular weight of the mature, unglycosylated protein as determined by B. M. Blumberg, C. Giorgi, K. Rose, and D. Kolakofsky, *J. Gen. Virol.* **66**, 317 (1985). The expected monomeric target size for glycoproteins corresponds to the molecular weight of the unglycosylated polypeptide chain [E. S. Kempner, J. H. Miller, and M. J. McCreery, *Anal. Biochem.* **156**, 140 (1986)].

protein exists in physical contact with other proteins in the irradiated sample. The protein-destroying radiation energy may be effectively transmitted from one polypeptide chain to another through disulfide bonds and also through certain noncovalent associations.[6] A more precise interpretation depends on other known properties of the specific protein. In a purified protein preparation, for example, the structural target size may be a simple multiple of the monomer molecular weight, indicating its existence in a multimeric form. The converse, however, is not true: if the structural target size corresponds to a monomer, the protein may or may not be in a complexed or multimeric form. The Sendai virus F protein shown in Fig. 1, for example, has the structural target size of a monomer, but is exists as a tetramer on the surface of the viral particle.[7] It is assumed in such a case that the noncovalent associations that exist between the monomers cannot transmit the radiation energy, so damage is limited to the polypeptide chain that is actually hit. There is no way at present to predict whether radiation energy will be transmitted through any particular set of noncovalent contacts, so the structural target size must be determined by experimentation.

The most important parameter to be obtained from radiation inactivation analysis is the functional target size governing a particular measurable activity. This is determined from the slope of the graph of log(activity remaining) versus radiation dose. The empirical factor used to convert slope to molecular weight is the same as that employed for the structural target size, so the functional and the structural target sizes can be directly compared. A functional target size can be determined for any activity that can be measured in the frozen, irradiated, and then thawed samples; hence, functional target sizes can be measured in crude membrane preparations, whole virus particles, and intact cells, as well as in purified protein preparations.[3]

It should be clear from the above that the functional target size cannot be smaller than the structural target size, which is the relevant mass destroyed by a single radiation hit. In effect, the structural target size limits the resolution of the radiation inactivation technique. If the structural target size of a particular enzyme is determined to be a dimer, for example, and the functional target size is also a dimer, then the minimum structure required for enzyme activity is no larger than a dimer; it could be either a monomer or a dimer. If, on the other hand, the functional target size is larger than the structural target size (and if the protein analyzed is truly

[6] H. T. Haigler, D. J. Woodbury, and E. S. Kempner, *Proc. Natl. Acad. Sci. U.S.A.* **82**, 5327 (1985).
[7] O. Sechoy, J. R. Philppot, and A. Bienvenue, *J. Biol. Chem.* **262**, 11519 (1987).

responsible for the measured function), then the functional target size represents the minimum size of the structure required to carry out the measured activity.

Application of Radiation Inactivation Analysis to Fusion Reaction: General Considerations

Radiation inactivation analysis has been successfully used to study many membrane enzymes and transport proteins.[2,3] Discrete, reasonable target sizes were determined for integral membrane proteins that were irradiated while present in native membranes. This established the generalization that radiation energy is not efficiently transferred through the bilayer by lipid molecules. Transport functions have been successfully analyzed after irradiation of membrane vesicles,[8,9] providing evidence that the permeability barrier properties of the bilayer can be preserved after exposure to radiation doses that destroy transport proteins.[8]

These generalizations provide the basis for the study of protein-mediated membrane fusion reactions using radiation inactivation analysis. Many such fusion processes occur in living cells; specific fusion of targeted vesicles is central to endo- and exocytosis, membrane and lysosomal protein processing, secretion, etc. With the development of cell-free systems to study many of these vesicular transport processes,[10-12] the proteins that catalyze specific fusion reactions are being identified and characterized, and should shortly be amenable to study in functional form.

At present, however, only virus-mediated fusion is sufficiently well-characterized that the mechanism of action of specific fusion proteins can be studied. The enveloped viruses all possess a lipid bilayer membrane, acquired from the host cell during virus assembly.[13] The bilayer contains only a few (usually only one or two) types of integral glycoproteins, one of which functions as a fusion protein during the process of viral entry into

---

[8] B. R. Stevens, A. Fernandez, B. Hirayama, E. M. Wright, and E. S. Kempner, *Proc. Natl. Acad. Sci. U.S.A.* **87**, 1456 (1990).

[9] B. K. Chamberlain, C. J. Berenski, C. Y. Jung, and S. Fleischer, *J. Biol. Chem.* **258**, 11997 (1983); M. Takahashi, P. Malathi, H. Preiser, and C. Y. Jung, *J. Biol. Chem.* **260**, 10551 (1985); see also Ref. 2, chapters by Jung (p. 137), Hah (p. 173), and Chamberlain *et al.* (p. 181).

[10] J. E. Rothman (ed.), this series, Vol. 221, in preparation.

[11] R. Diaz, L. S. Mayorga, M. I. Colombo, J. M. Lenhard and P. D. Stahl, this series, Vol. 221 [16].

[12] S. Pind, H. Davidson, R. Schwaninger, C. J. M. Beckers, H. Plutner, S. L. Schmid, and W. E. Balch, this series, Vol. 221 [17].

[13] J. Lenard and R. W. Compans, *Biochim. Biophys. Acta* **344**, 51 (1974); J. Lenard, *Annu. Rev. Biophys. Bioeng.* **7**, 139 (1978).

cells. The best characterized of these is the influenza hemagglutinin (HA) protein, for which the X-ray crystal structure is known. Other well-characterized fusion proteins of enveloped viruses include the F protein of Sendai virus, the G protein of vesicular stomatitis virus, and the glycoproteins of Sindbis and Semliki Forest virus. These proteins can catalyze the fusion of virus membranes or reconstituted lipid vesicles with cell or erythrocyte membranes or with target liposomes free of protein; no additional proteins are needed to carry out the fusion reaction.[14-16]

Because the fusion proteins of several different enveloped viruses have been identified unambiguously, and suitable assays for fusion have been developed, it has been possible to apply radiation inactivation analysis to the study of virus-mediated fusion. Results of these studies are described below.

Radiation Inactivation Analysis of Virus-Mediated Fusion

Radiation inactivation analysis has proved to be uniquely suitable for addressing the question of the stoichiometry of virus-mediated fusion: How many fusion protein molecules participate in a fusion event? If several, does a single radiation hit inactivate a discrete functional "fusion unit," or is there a more complex radiation dose dependence?

As an example, radiation inactivation profiles for some Sendai virus fusion functions are shown in Fig. 2.[17] Three fusion-related reactions are assayed: (1) fusion with erythrocytes, using as an assay the fluorescence dequenching of octadecylrhodamine B incorporated into the virus membrane (owing to dilution of the probe into the target membrane)[18], (2) erythrocyte hemolysis (i.e. virus-induced leakage of hemoglobin); and (3) fusion with cardiolipin liposomes, using as an assay the decrease in resonance energy transfer (i.e., dilution) of fluorescent lipids incorporated into the target liposomes.[17]

A unique radiation inactivation profile is found for each of the activities measured. (1) The functional target size for fusion with erythrocyte membranes is 60 kDa, identical with the structural target size of F protein (Fig. 1). This indicates that a single monomer of F protein (not the tetramer that constitutes the F protein "spike") constitutes the "functional unit" for this fusion reaction. A similar result is obtained for influenza virus, namely, a single HA monomer constituted the functional target for

[14] J. White, M. Kielian, and A. Helenius, *Q. Rev. Biophys.* **16**, 151 (1983).
[15] J. M. White, *Annu. Rev. Physiol.* **52**, 675 (1990).
[16] D. Hoekstra, *J. Bioenerg. Biomembr.* **22**, 121 (1990).
[17] K. Bundo-Morita, S. Gibson, and J. Lenard, *Biochemistry* **26**, 6223 (1987).
[18] D. Hoekstra and K. Klappe, this volume [20].

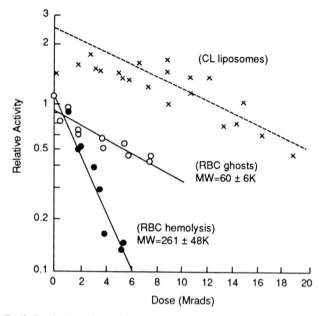

Fig. 2. Radiation-induced loss of fusion and hemolytic activities from Sendai virus: O, fusion with erythrocyte ghosts; ×, fusion with pure cardiolipin liposomes at pH 7.0; ●, hemolysis.[14]

fusion with erythrocytes.[17] (2) The functional target size for Sendai virus-induced hemolysis was 261 kDa, probably corresponding to the tetrameric F protein "spike." This indicates that hemolysis is governed by a discrete protein complex that is different from, and larger than, that governing fusion. Hemolysis therefore does not arise as a passive consequence of insertion (by fusion) into the erythrocyte membrane of a leaky virus membrane, as was previously believed. (3) The inactivation profile for fusion with cardiolipin liposomes is not linear, suggesting that the simple model of single-hit destruction of a functional unit is not applicable to this process.

Because this fusion reaction is also mediated by F protein,[4] and because the structural target size of F protein is known and does fit the single-hit model (Fig. 1), this curve may be interpreted in terms of a "multiple target–single hit" model.[19] According to this model, the functional unit consists of about 3 molecules of 52 ± 9 kDa (i.e., 3 monomers of F protein), since this is the only Sendai protein possessing this structural target size. The enhancement of activity seen at low radiation doses is not

[19] R. Oliver and B. J. Shepstone, *Phys. Med. Biol.* **9**, 167 (1964).

explained by this interpretation.[20] Alternatively, the increased activity at low radiation doses could arise as a result of destruction of a high molecular mass inhibitor.[21] By subtracting the experimental points from the extrapolated line, the size of this putative inhibitor can be determined to be approximately 162 kDa.[14] The existence of such an inhibitor does not seem compatible, however, with other known facts about the Sendai virus fusion system. This discussion points up the difficulties that are encountered in attempting to interpret nonlinear radiation inactivation curves.

Similar experiments have been done with influenza virus[17,22] and with vesicular stomatitis virus.[23] Similar but not identical functional target sizes to those in Fig. 2 are found for influenza virus: fusion with cardiolipin liposomes[22] and erythrocyte ghosts[17] both have functional units corresponding to a single monomer of HA protein, whereas hemolysis has a nonlinear radiation inactivation profile, indicating a complex functional unit composed of 3–7 structures of about 350 kDa each.[22] Vesicular stomatitis virus, on the other hand, has quite different properties. Fusion with either erythrocytes or cardiolipin liposomes has a substantial radiation-independent component, perhaps indicating that the "hit" proteins retain some activity. After correcting for the radiation-insensitive component, both fusion activities and hemolysis show similar, very large functional target sizes of approximately 900 kDa, suggesting that fusion of this virus might proceed by a quite different mechanism from that of Sendai or influenza viruses.[23]

Both the strengths and the limitations of radiation inactivation analysis are illustrated by the results described above. The strength of the technique is indicated by the fact that the results of these few experiments call into question two long-standing assumptions, namely, that viral fusion is mediated by a number of fusion proteins acting in concert and that hemolysis or leakage (fusion-induced hole formation) is a passive consequence of fusion with a leaky virus. The major limitation of the technique is that it provides only two relevant parameters: the structural and functional target sizes. Additional limitations are that nonlinear radiation inactivation profiles (e.g., top curve in Fig. 2) do not allow unambiguous interpretation and

---

[20] A plateau, but no elevation, in activity is seen if fusion is carried out at pH 5.0, or if the cardiolipin vesicles contain 10 mol % (w/w) of gangliosides. The values calculated for the fusion unit using the "multiple target–single hit" model remain essentially the same, however.[14]

[21] J. T. Harmon, C. R. Kahn, E. S. Kempner, and W. Schlegel, *J. Biol. Chem.* **255**, 3412 (1980); R. L. Kincaid, E. Kempner, V. C. Manganiello, J. C. Osborne, and M. Vaughan, *J. Biol. Chem.* **256**, 11351 (1981).

[22] S. Gibson, C. Y. Jung, M. Takahashi, and J. Lenard, *Biochemistry* **25**, 6264 91986).

[23] K. Bundo-Morita, S. Gibson, and J. Lenard, *Virology* **163**, 622 (1988).

that external information about the system is required if any useful interpretations are to be obtained.

A more specific limitation of the experiments is that fusion assays, including those discussed here, are somewhat ambiguous as regards their relationship with "fusion." Fusion is generally considered to be complete after the contents of fusing vesicles have mixed. Although contents mixing assays for fusion have been developed,[24] their use in radiation inactivation analysis has not yet been reported. Lipid dilution techniques have been widely and successfully used,[18,25] but it has been suggested that lipid exchange (leading to dilution) might actually be a property of some kind of activated prefusion state, and thus might not measure the completion of fusion under all circumstances.[15] Interpretation of radiation inactivation studies of fusion are thus limited by continuing uncertainty about what is meant by "fusion," and what the standard assays actually measure. Nevertheless, the technique of radiation inactivation analysis has already contributed significantly to our limited understanding of the mechanism of virus-mediated fusion, and it holds promise of providing further insights into this and other well-defined fusion processes.

[24] N. Düzgüneş and J. Wilschut, this volume [1].
[25] A. Puri, M. J. Clague, C. Schoch, and R. Blumenthal, this volume [21].

[26] Fluorescence Photobleaching Recovery to Probe Virus–Cell Fusion and Virus-Mediated Cell Fusion

By Yoav I. Henis

Introduction

Membrane fusion events are involved in a variety of physiological and pathological processes. In spite of the importance of fusion, many aspects of the fusion mechanism are still poorly understood.[1,1a] Fusion of viral envelopes with the cellular plasma membrane or with endocytic vacuoles after endocytosis of the virions constitutes a crucial step in the penetration of enveloped viruses into cells and in viral infectivity.[1-3] Fusion of the virions with the cellular plasma membrane is followed by virus–mediated

[1] J. White, M. Kielian, and A. Helenius, Q. Rev. Biopys. 16, 151 (1983).
[1a] N. Düzgüneş, Subcell. Biochem. 11, 195 (1985).
[2] P. W. Choppin and A. Scheid, Rev. Infect. Dis. 2, 40 (1980).
[3] M. Marsh, Biochem. J. 218, 1 (1984).

cell fusion, leading to the formation of large syncytia. A thorough investigation of the mechanisms underlying these processes requires development of methods capable of direct and quantitative measurement of the fusion between viral particles and cells, as well as measurement of the motion of fusion-promoting viral envelope proteins in the target cell membrane.

In this chapter, the application of fluorescence photobleaching recovery (FPR)[4-6] to measure the lateral motion of fluorescence-labeled viral envelope components in the target cell membrane is described. Measurement of the degree of lateral mobility in the cell membrane of a fluorescent lipid probe initially incorporated into the viral envelope enables a direct and quantitative determination of the extent of virus–cell fusion. This method can distinguish unambiguously viral particles fused with the plasma membrane from those internalized by the cells, and it allows studies on the distribution of the fusion level within the cell population and on specific cell regions. On the other hand, measurements of the lateral mobility of viral envelope proteins in the target cell membrane provide information on the role of the motion of these proteins in the induction of cell–cell fusion. Both types of measurements are described, with emphasis on the system of Sendai virus fusion with human erythrocytes.

Fluorescence Photobleaching Recovery

The measurements described are based on the FPR technique; a general description of the instrumentation and physical principles of this method has been given.[4-6] We have used the single-spot laser irradiation version of FPR, which is both simple and capable of detecting variations between different regions on the surface of a single cell. The specific instrumental setup employed in our laboratory has been described.[7]

FPR has been used to measure the lateral diffusion of viral envelope components in the cell membrane, following the diffusion of either fluorescent lipid probes initially introduced into the viral envelope or viral envelope proteins labeled with fluorescent Fab' antibody fragments. The labeling procedures are detailed in the following sections. In the FPR experiment, the fluorescence in a small region on the cell surface is excited by an argon ion laser beam focused on the cell through a fluorescence microscope (we employed a beam focused to a Gaussian radius of 0.93

[4] N. O. Petersen and E. L. Elson, this series, Vol. 130, p. 454.
[5] N. O. Petersen, S. Felder, and E. L. Elson, in "Handbook of Experimental Immunology" (D. M. Weir, L. A. Herzenberg, C. C. Blackwell, and L. A. Herzenberg, eds.), p. 24.1. Blackwell Scientific, Edinburgh, 1986.
[6] T. M. Jovin and W. L. C. Vaz, this series, Vol. 172, p. 471.
[7] Y. I. Henis and O. Gutman, Biochim. Biophys. Acta 762, 281 (1983).

μm); the beam is attenuated to an intensity that does not induce significant bleaching of the fluorophores during the measurement (the monitoring beam). Typical intensities are 0.1 μW at 488 nm for N-(7-nitrobenz-2-oxa-1,3-diazol-4-yl)phosphatidylethanolamine (N-NBD-PE) fluorescence and 1 μW at 529.5 or 515 nm for tetramethylrhodamine (TMR) or octadecylrhodamine B (R_{18}) fluorescence. After measuring the baseline fluorescence intensity, 50–70% of the fluorophores in the illuminated region are irreversibly bleached by a brief pulse (usually up to tens of milliseconds) of a high-intensity laser beam (the bleach beam; about 5000-fold stronger than the monitoring beam). The fluorescence in the bleached region recovers, owing to lateral motion of unbleached fluorescent-tagged components into the irradiated region, and is measured by the monitoring beam. The lateral diffusion coefficient (D) and the mobile fraction (R_f) of the labeled component are extracted from the rate and extent of fluorescence recovery by fitting the data to the equation for two-dimensional diffusion using non-linear regression analysis.[5]

Fusion of Reconstituted Sendai Virus Envelopes with Cells

Fusion of viral envelopes containing fluorescent lipid analogs with the cell membrane is accompanied by dispersal of the fluorescent lipids in the target membrane. Prior to fusion, when the viral envelopes are merely adsorbed to the cells, the lateral motion of the fluorescent lipids incorporated into them is limited to the viral membrane; after fusion, these lipids are free to diffuse over the cellular plasma membrane. The FPR experiments employ a laser beam focused on the cell surface to a size considerably larger than that of a viral particle, and the lateral diffusion has to be on the same distance scale (microns) to be detected. Thus, fluorescent lipids originally incorporated into the viral envelope should be immobile in FPR measurements prior to fusion, then become laterally mobile in the cell membrane following viral envelope–cell fusion. This provides the basis for the measurement of virus–cell fusion by FPR.[8]

The first requirement for fusion measurement by FPR is the incorporation of a slow-exchanging fluorescent lipid analog into the viral envelope. N-NBD-PE labeled at the head group (Avanti, Alabaster, AL) is a good choice, since it does not exchange readily between liposomes and cell membranes.[9] The procedure followed for Sendai virus, which can be adapted to other enveloped viruses, is based on the method for preparation

[8] B. Aroeti and Y. I. Henis, *Biochemistry* **25**, 4588 (1986).
[9] D. K. Struck and R. E. Pagano, *J. Biol. Chem.* **255**, 5404 (1980).

of reconstituted Sendai virus envelopes (RSVE) described by Vainstein *et al.*[10]

A pellet of Sendai virions (Z strain; 10 mg protein) is dissolved in 0.5 ml of 4% (w/w) Triton X-100 in 10 mM NaCl, 50 mM N-[tris(hydroxymethyl)methyl]glycine (Tricine), pH 7.4. After centrifugation (100,000 g, 1 hr, 4°), the supernatant (containing viral lipids and about 2.5 mg viral glycoproteins) is added to a thin layer of 20 μg N-NBD-PE formed by drying an ethanolic solution of the fluorescent lipid under a stream of nitrogen. The mixture, containing 0.6–0.8 mol % N-NBD-PE relative to viral lipids, is shaken vigorously for 5 min. The detergent is removed by direct addition of SM2 BioBeads (Bio-Rad, Richmond, CA) in two steps, adding 140 mg beads and shaking vigorously 1.5 hr at room temperature each time. The solution is taken up from the beads by a syringe, and the fluorescent RSVE are collected by centrifugation (100,000 g, 1 hr, 4°). They are suspended in 160 mM NaCl, 20 mM Tricine, pH 7.4 (solution A), and stored at −70°. The low N-NBD-PE/lipid ratio is required to avoid self-quenching of its fluorescence, which occurs at higher fluorophore densities in the viral envelope and complicates the interpretation of the FPR fusion measurements (see below). N-NBD-PE-bearing RSVE prepared as described above exhibit less than 10% quenching of N-NBD-PE fluorescence.[8]

Binding of the fluorescent RSVE to cells is carried out in the cold, to avoid fusion during this step. Binding to human erythrocytes is performed in solution A; 2% (v/v) red cells are incubated 15–30 min with 400 hemagglutinating units (HAU)/ml RSVE (30,000–40,000 HAU/mg envelope proteins). Free RSVE are removed by washing twice with cold solution A in a tabletop centrifuge.[8] Fusion is achieved by a 30-min incubation at 37°. An analogous procedure is applied to other cell types in suspension, such as splenic lymphocytes, except that solution A is replaced by the appropriate medium used for growing the cells, buffered to pH 7.2–7.4 by N-(2-hydroxyethyl)piperazine-N'-2-ethanesulfonic acid (HEPES). The medium should be devoid of serum, which in many cases inhibits virus binding; however, 0.2% bovine serum albumin (BSA) may be included to reduce nonspecific binding. Tissue culture cells are grown on glass coverslips; after washing twice with cold HEPES-buffered medium (without serum), 100 μl cold medium containing 1000 HAU of RSVE is layered on the coverslip and the RSVE are allowed to bind to the cells for 1 hr. The monolayers are washed twice with the same cold medium, then incubated at 37° (30 min) to induce fusion.

[10] A. Vainstein, M. Hershkovitz, S. Israel, S. Rabin, and A. Loyter, *Biochim. Biophys. Acta* **773**, 181 (1984).

It should be noted that the procedure outlined above was designed for viruses that fuse with cells at neutral pH, such as paramyxoviruses; the fusogenic activity of many other enveloped viruses requires a brief exposure to a low pH buffer (the exact value of which depends on the specific virus) for its activation.[1] In such cases, an additional step of exposure (1–3 min, 37°) to a low pH medium (usually after virus binding to the cells) is required prior to the 37° incubation in the normal pH medium in order to achieve fusion.[1,11]

The cells incubated with the N-NBD-PE-containing RSVE (either before or after fusion) are taken for FPR experiments to determine the extent of viral envelope–cell fusion. In the case of tissue culture cells, the experiments are performed on the coverslips carrying the cells, wet-mounted in the appropriate incubation medium. Suspended cells are first attached to glass coverslips precoated with polylysine [10 min incubation of the coverslip with 5 μg/ml poly (L-lysine) in 20 mM phosphate buffer, pH 8]. Alternatively, a drop of the cell suspension can be mounted in a "sandwich" between a microscope slide and a coverslip; however, care has to be taken to replace the sample after each 10–15 min, to avoid drying. The FPR experiments can be performed at room temperature, since virus–cell fusion is very slow around 20–22°. Thus, viral lipid mobilization in the cell membrane due to fusion during the FPR measurement is negligible. The FPR experiments employ a laser beam considerably larger than the size of the viral particle. The N-NBD-PE originally incorporated into the RSVE is therefore expected to be immobile before fusion, when its motion is limited to the viral envelope, and to become laterally mobile in the plasma membrane after fusion (as the N-NBD-PE molecules are incorporated by fusion into the plasma membrane).

These expectations are met, as demonstrated in Fig. 1 for the fusion of RSVE with human erythrocytes. The fraction of cell-associated RSVE fused with the plasma membrane in the region illuminated by the laser beam is given by the ratio R_f/R_f^o, where R_f is the mobile fraction measured for the RSVE-incorporated N-NBD-PE on the cell surface and R_f^o is the mobile fraction of N-NBD-PE incorporated directly into the cell membrane and determined in a separate FPR experiment. R_f^o, which is in the range of 0.8–0.9 for most cell types around room temperature, represents the highest R_f value to be obtained if all the RSVE-incorporated N-NBD-PE is inserted into the cellular plasma membrane. Direct incorporation of N-NBD-PE into the plasma membrane (for the determination of R_f^o) is achieved by incubating the cells in serum-free buffered medium (without BSA, which would bind the fluorescent lipid) with a 1:100 dilution of

[11] J. White, K. Matlin, and A. Helenius, *J. Cell Biol.* **89**, 674 (1981).

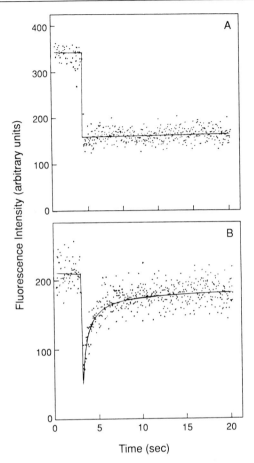

FIG. 1. Representative FPR measurement of the fusion of RSVE containing N-NBD-PE with human erythrocytes. The points are the actual fluorescence intensities (photons/40 msec dwell time). Solid lines are the best-fit fluorescence recovery curves obtained by fitting the experimental data to a lateral diffusion process with a single diffusion coefficient using nonlinear regression.[5] (A) Erythrocytes with bound (unfused) RSVE containing N-NBD-PE. Incubation was in the cold, and the FPR measurement was at 22°. No mobility of N-NBD-PE in the cell membrane could be detected ($D \leq 5 \times 10^{-12}$ cm^2/sec). (B) Erythrocytes fused with RSVE containing N-NBD-PE (4° incubation, followed by 30 min at 37°). The FPR measurement was at 22°. The specific curve shown yielded $D = 5.0 \times 10^{-9}$ cm^2/sec, and $R_f = 0.84$. The R_f^o value for N-NBD-PE incorporated directly into the membrane of human erythrocytes is 0.87; thus, $R_f/R_f^o = 0$ for bound RSVE (A) and 0.97 after fusion (B), suggesting that almost all of the viral envelopes in the measurement spot have fused with the cell membrane in the latter case.

2 mg/ml N-NBD-PE in ethanol (15 min, 22°). Excess dye is removed by washing twice with the same medium supplemented with 0.2% BSA.

It should be noted that R_f/R_f^0 measures the fraction of fused RSVE out of the population of cell-associated (adsorbed, fused, or internalized) viral particles. Because the laser beam is focused on the cell surface, the contribution of fluorescence from free viral envelopes is negligible. This is contrary to the situation encountered in spectroscopic measurements of viral envelope–cell fusion (e.g., by fluorescence dequenching), which are performed on suspensions containing cells and viral particles and include a contribution from viral particles not associated with the cells. Thus, when comparing results from such methods with those obtained by FPR, care has to be taken to wash away free viral particles prior to the spectroscopic measurement. This has to be done not only after the binding step, but also after the 37° incubation, since at least in the case of Sendai virus some of the viral envelopes dissociate during the 37° incubation owing to the viral neuraminidase activity.[12]

The determination of viral envelope–cell fusion by FPR has several advantages. Because FPR measurements are performed on single cells and on relatively small cell surface regions, they enable both mapping of the surfaces of single cells regarding fusion with viral envelopes and determination of the distribution of the fusion level within populations of cells. This feature is absent in fusion methods that employ large cell masses and yield only average values. Moreover, the FPR measurements can distinguish fused from internalized viral envelopes, since fluorescent probes in the latter are immobile in FPR experiments.[8] This is not the case with fusion measurements by fluorescence dequenching or energy transfer, which can be complicated by signals arising from the degradation of internalized viral particles[13] or from fusion between internalized viral envelopes and endosomal membrane.[1,14]

Fusion of Native Virions with Cells

The use of N-NBD-PE to determine viral envelope–cell fusion is limited to reconstituted viral envelopes, since it does not incorporate readily into the membrane of native virions. This problem can be eliminated by the use of another fluorescent probe, R_{18} (Molecular Probes, Eugene, OR), which has been shown to incorporate directly into the membrane of native virions.[14] We outline below the procedure to measure fusion between R_{18}-labeled virions and cells by FPR for the system of Sendai virions and

[12] K. J. Micklem, A. Nyaruwe, and C. A. Pasternak, *Mol. Cell. Biochem.* **66,** 163 (1985).
[13] N. Chejanovsky and A. Loyter, *J. Biol. Chem.* **260,** 7911 (1985).
[14] D. Hoekstra, T. de Boer, K. Klappe, and J. Wilschut, *Biochemistry* **23,** 5675 (1984).

human erythrocytes; the same procedure is applicable to other virus–cell systems.

Virions are labeled with R_{18} essentially as described by Hoekstra et al.,[14,15] except that a lower R_{18} concentration is used. In a typical labeling procedure, Sendai virions (0.9 mg protein/ml in solution A; 13,000 HAU/mg) are incubated with a 1:100 dilution of a solution of 0.4 μM R_{18} in ethanol (60 min, 22°). Excess dye is removed by passing the mixture (0.5 ml) through a 5-ml column of Sephadex G-75 equilibrated with the same buffer.[16] These conditions result in virions containing 0.05–0.07 mol % R_{18}, as determined by measuring R_{18} fluorescence after solubilizing the labeled virions with Triton X-100 (1%, w/w), using R_{18} solutions of known concentrations (in the presence of 1% detergent) for calibration. The concentration of the viral lipids is calculated assuming 400 nmol viral lipid per milligram viral protein.[17] The use of a low molar ratio of R_{18} is imperative since R_{18} fluorescence is quenched in the envelope of Sendai virions even at low concentrations, most likely owing to preferential insertion of R_{18} into specific domains in the viral membrane.[16] This does not occur in the erythrocyte membrane, since direct incorporation of up to 1 mol % R_{18} into the plasma membrane of human erythrocytes (performed as described above for N-NBD-PE, using varying concentrations of R_{18}) does not lead to detectable fluorescence quenching.

The above labeling conditions of the virions result in 50% quenching of R_{18} fluorescence in the viral membrane. It is desired to keep the fluorescence quenching in the viral envelopes as low as possible, since this leads to a lower contribution to the fluorescence signal of R_{18} in unfused bound virions relative to R_{18} in fused virions. This in turn causes an overestimation of the fusion level determined by R_f/R_f^o, and a correction which takes into account the lower quantum yield of R_{18} in unfused virions has to be introduced. It is therefore desired to determine in advance the degree of R_{18} fluorescence quenching in the viral membrane, and to choose conditions that minimize the required correction. It is also advisable to ensure that no significant quenching takes place in the target cell membrane at the probe concentrations employed. The latter phenomenon (quenching in the target cell membrane) is much less likely, since the R_{18} is highly diluted in the target cell membrane following fusion relative to its concentration in the viral membrane.

The correction of the FPR fusion quantification for the quenching of R_{18} fluorescence in the viral membrane is as follows.[16] Let ΔF be the amount of fluorescence bleached at the initiation of an FPR measurement.

[15] D. Hoekstra, K. Klappe, T. de Boer, and J. Wilschut, *Biochemistry* **24**, 4739 (1985).
[16] B. Aroeti and Y. I. Henis, *Exp. Cell Res.* **170**, 322 (1987).
[17] A. Loyter and D. J. Volsky, *Cell Surf. Rev.* **8**, 215 (1982).

The contribution to ΔF is in part by fused virions (ΔF_{fus}) and in part by adsorbed unfused viral particles (ΔF_{unf}). The fraction of ΔF arising from bleaching of R_{18} originating in fused virions is $\Delta F_{fus}/\Delta F$. Of this, only a fraction R_f^o is laterally mobile in the cell membrane (R_f^o is the highest R_f value to be obtained if all the virus-incorporated probe has fused with the cell membrane). Because the R_{18} in the membrane of bound virions is immobile in the FPR experiment, only fused virions contribute to the measured R_f value, given by

$$R_f = (\Delta F_{fus}/\Delta F)\, R_f^o \tag{1}$$

Because $\Delta F = \Delta F_{fus} + \Delta F_{unf}$, Eq. (1) can be rearranged to

$$(\Delta F_{unf}/\Delta F) = 1 - (R_f/R_f^o) \tag{2}$$

The sensitivity to bleaching of R_{18} in fused and unfused virions is similar within the experimental accuracy, as verified for R_{18}-labeled Sendai virions by comparing the amount of bleaching in FPR experiments performed on bound versus fused virions. Thus, ΔF_{fus} and ΔF_{unf} are proportional to the densities of fused and unfused virions in the laser-illuminated region on the cell surface. Owing to the quenching of R_{18} fluorescence in the viral envelope, the fluorescence quantum yield for fused virions is higher by a factor of N. Therefore, the proportionality constants relating ΔF_{fus} and ΔF_{unf} with the density of the virions differ by the same factor N, and ΔF_{unf} should be multiplied by N when ΔF_{fus} and ΔF_{unf} are converted to cell-surface densities of fused and unfused virions:

$$FV = \frac{\text{fused virions}}{\text{fused virions} + \text{unfused virions}} = \frac{\Delta F_{fus}}{\Delta F_{fus} + N\,\Delta F_{unf}} \tag{3}$$

where FV is the fraction of cell-associated virions fused with the cells. Dividing the numerator and denominator by ΔF, introducing Eqs. (1) and (2), and rearranging, one obtains

$$FV = 1/[1 + N(R_f^o/R_f - 1)] \tag{4}$$

Because N is directly derived from the degree of fluorescence quenching in the viral membrane (in R_{18}-labeled Sendai virions, 50% quenching means an N value of 2), Eq. (4) enables the calculation of the fraction of fused virions, correcting for the increase in R_f owing to the relief of fluorescence quenching in the fusion process. When $N = 1$, Eq. (4) reduces to $FV = R_f/R_f^o$, the simpler relation employed in the case of RSVE containing low levels of N-NBD-PE.

The adsorption of the R_{18}-labeled virions to the cells in the cold and the ensuing induction of fusion at 37° are performed exactly as described above for RSVE and human erythrocytes; however, experience with Sendai virions shows that it is advisable to eliminate large viral aggregates prior to incubation with cells by a mild sonication (three bursts of 30 sec each with Model W-10 sonicator, Heat Systems Ultrasonics, Plainview, NY) followed by filtration through 0.45 μm Millipore (Bedford, MA) filters. Otherwise, the presence of large aggregates stuck on the cells may occasionally interfere with the measurement. The FPR experiments are performed at room temperature either before (to obtain a prefusion measurement) or after the fusion-promoting incubation at 37°, as described for RSVE. Typical FPR curves of such experiments are depicted in Fig. 2.

Measurement of Lateral Mobility of Viral Envelope Proteins in Target Cell Membrane

Virus-mediated membrane fusion is induced by specific viral glycoproteins.[1] Unlike the situation with the viral envelope lipids, the envelope proteins do not necessarily diffuse freely in the target cell membrane following envelope-cell fusion. Their mobility in the target cell membrane may play a crucial role in the induction of cell–cell fusion, as demonstrated for Sendai virus-mediated fusion of human erythrocytes.[18-22] The ability to measure directly the lateral motion of specific viral glycoproteins in the target cell membrane and to compare it with the viral fusogenic activity is therefore important. Such measurements can be achieved by FPR, as is outlined for the system of Sendai virus and human erythrocytes.

Incubation of the cells with Sendai virions (or RSVE) is performed exactly as described in the previous sections. After virus adsorption and agglutination in the cold, or after the induction of fusion at 37°, the cells are washed to remove unattached viral particles, then allowed to attach to polylysine-coated glass coverslips as detailed above (this step is not required in the case of tissue culture cells grown directly on coverslips). In the case of human erythrocytes, which are lysed by the viral hemolytic activity during fusion, it may be desired to generate ghosts from the unfused preparation prior to attachment to the coverslips, in order to work on similar systems (ghosts) both before and after fusion. In such cases, we have employed the hypotonic lysis and resealing procedure described by Steck and Kant[23]; however, aside from the somewhat higher background fluorescence contributed by hemoglobin, we did not find significant differences

[18] Y. I. Henis and T. M. Jenkins, *FEBS Lett.* **151**, 134 (1983).
[19] Y. I. Henis, O. Gutman, and A. Loyter, *Exp. Cell Res.* **160**, 514 (1985).
[20] Y. I. Henis and O. Gutman, *Biochemistry* **26**, 812 (1987).
[21] B. Aroeti and Y. I. Henis, *Biochemistry* **27**, 5654 (1988).
[22] Y. I. Henis, Y. Herman-Barhom, B. Aroeti, and O. Gutman, *J. Biol. Chem.* **264**, 17119 (1989).

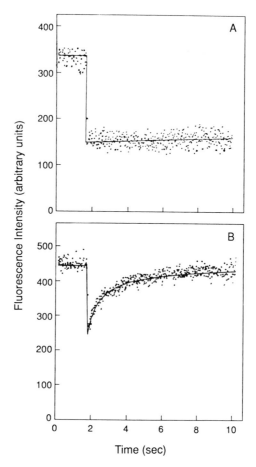

FIG. 2. Representative FPR measurement of the fusion of R_{18}-labeled Sendai virions with human erythrocytes. R_{18}-labeled Sendai virions were prepared and incubated with human erythrocytes as described in the text. FPR measurements were performed at 22°. Points are the measured fluorescence intensities (photons counted/20 msec dwell time). Solid lines are nonlinear regression best-fit fluorescence recovery curves for a lateral diffusion process with a single diffusion coefficient.[5] (A) Erythrocytes with adsorbed (4° incubation) R_{18}-labeled virions. $R_f = 0$, and $D \leq 5 \times 10^{-12}$ cm²/sec. (B) Erythrocytes fused with R_{18}-labeled Sendai virions (incubation at 4° followed by 30 min at 37°). The specific curve shown yielded $R_f = 0.93$ and $D = 3.7 \times 10^{-9}$ cm²/sec. The R_f^o value at 22° for R_{18} incorporated directly into the membrane of human erythrocytes is 0.94. Thus, no fusion is detected in (A), whereas the R_f/R_f^o value in (B) (0.93/0.94) yields through Eq. (4) a value of 0.98 for the fraction of fused virions in the illuminated region on the cell surface.

TABLE I

LATERAL MOBILITY OF SENDAI VIRUS GLYCOPROTEINS ON HUMAN ERYTHROCYTES[a]

Envelope protein labeled	Treatment	Measurement temperature	R_f	D (cm²/sec × 10¹⁰)
F	4°	22°	0.03 ± 0.04	≤0.05
F	4° + 37°	22°	0.60 ± 0.05	3.4 ± 0.2
F	4° + 37°	37°	0.61 ± 0.04	6.4 ± 0.5
HN	4°	22°	0.04 ± 0.04	≤0.05
HN	4° + 37°	22°	0.59 ± 0.05	3.2 ± 0.3
HN	4° + 37°	37°	0.60 ± 0.03	6.2 ± 0.6

[a] Incubation of cells with virions and labeling with anti-F or anti-HN TMR–Fab' were carried out as described in the text. A 4° treatment stands for incubation of cells with virions only in the cold (adsorbed, unfused virions). A treatment of 4° + 37° represents cells after fusion (additional 30 min at 37°). The results are means ± S.E.M. of 50–60 measurements in each case. The measurement temperature is the temperature during the FPR experiment. As can be seen, the measurement temperature affects the diffusion coefficient D, but not the R_f value measured after the completion of fusion. Data are taken from Henis et al.[19] and from Aroeti and Henis.[21]

between the lysed and unlysed preparations regarding the lateral mobility of Sendai virus glycoproteins.

The specific viral glycoprotein whose mobility is to be measured is labeled by incubating the cells on the coverslip with tetramethylrhodamine (TMR)-tagged monovalent Fab' antibody fragments directed against it. We have employed 100 μg/ml TMR–Fab' directed against the fusion protein (F) or the hemagglutinin–neuraminidase protein (HN) of Sendai virus, performing the labeling for 30 min at 22° (at which temperature fusion is negligible for several hours) in solution A containing 0.2% (w/v) BSA. The presence of BSA eliminates adsorption of the TMR–Fab' to the polylysine-coated coverslip. If other cell types are employed, solution A is replaced by the appropriate HEPES-buffered medium supplemented with BSA. It should be noted that only monovalent antibody fragments should be used, since polyvalent immunoglobulin G (IgG) species cross-link the viral glycoproteins and lead to their immobilization.[22,24] We have routinely prepared Fab'₂ fragments from rabbit or goat IgG by pepsin digestion,[25] processing them to monovalent Fab' by 2-mercaptoethanol followed by alkylation with iodoacetamide.[26] Labeling of the Fab' with TMR isothio-

[23] T. L. Steck and J. A. Kant, this series, Vol. 31, p. 172.

[24] Z. Katzir, O. Gutman, and Y. I. Henis, *Biochemistry* **28**, 6400 (1989).

[25] G. M. Edelman and J. J. Marchalonis, *Methods Immunol. Immunochem.* **1**, 405 (1967).

[26] J. Braun, R. I. Shaafi, and E. R. Unanue, *J. Cell Biol.* **82**, 755 (1979).

cyanate (Molecular Probes) is performed according to standard proce-dures.[27] After incubation with the fluorescent Fab', the cells are washed twice with the BSA-containing incubation medium, and the labeled cells are then used for the FPR experiments. Typical results obtained for the lateral mobility of Sendai virus glycoproteins, F and HN, in the mem-branes of human erythrocytes are depicted in Table I.[19,21] As in the case of virus–cell fusion, the distribution of the dynamic parameters characteriz-ing the viral glycoproteins over the cell population and over different regions of the same cell can be established.

[27] P. Brandtzaeg, *Scand. J. Immunol.* **2,** 273 (1973).

[27] Electron Microscopy of Virus–Liposome Fusion

By Koert N. J. Burger, Lesley J. Calder, Peter M. Frederik, and Arie J. Verkleij

Introduction

Whereas most biochemical and biophysical techniques offer only aver-age information on the composition and structure of a sample, informa-tion on local structure and composition can be obtained by using morpho-logical techniques. In membrane fusion research, morphological techniques may provide the ultimate proof of membrane fusion by directly showing a change in size of the membrane structures participating in the fusion process. By taking advantage of the resolution of the electron mi-croscope, information can be obtained on the structure and composition of a sample at the actual fusion site. When combined with fast-freezing and (immuno)cytochemistry, electron microscopy (EM) additionally offers the unique possibility of localizing a fusogen in relation to the fusion process, both in time and space.

Clearly, data from electron microscopy play a dominant role in the way in which we think membrane fusion occurs.[1-3] It should be realized, however, that ultrastructure as visualized by electron microscopy is corre-lated with, but certainly not identical to, the native ultrastructure of the

[1] G. E. Palade and R. R. Bruns, *J. Cell Biol.* **37,** 633 (1968).
[2] D. E. Chandler, *Curr. Top. Membr. Transp.* **32,** 169 (1988).
[3] K. N. J. Burger and A. J. Verkleij, *Experientia* **46,** 631 (1990).

sample. In fact, membrane fusion events have proved to be extremely susceptible to many of the pretreatments that are required to stabilize the intrinsically unstable structure of a sample prior to electron microscopic analysis.[3-5] Because of this, detailed knowledge of sample processing is an absolute requirement for the correct interpretation of electron microscopic data, especially in the field of membrane fusion research.

The aim of this chapter is to review the electron microscopic methods that can be used to study influenza virus–model membrane fusion, with special emphasis on sample processing and on the advantages and disadvantages of each technique. Although focused on influenza virus–model membrane interaction, the technical considerations should apply to enveloped virus–membrane fusion in general.

Influenza Virus–Membrane Fusion

Influenza virus (an orthomyxovirus) enters the cell by receptor-mediated endocytosis.[6] The low pH of the endosomal compartment instigates an irreversible conformational change of the influenza virus hemagglutinin glycoprotein (HA), leading to fusion of viral and endosomal membranes and release of the viral nucleocapsid into the cytoplasm.[7-10] Although detailed information on the low pH-induced conformational change of the HA spike protein is accumulating, the mechanism by which HA induces membrane fusion after its conformational change remains unsolved.

To shed some light on the fusion mechanism in general, and on possible intermediates of the actual joining of the two lipid bilayers in particular, we have characterized the low pH-induced fusion of influenza virus with pure lipid model membranes. Four independent morphological techniques were used: negative staining, fast-freeze freeze–fracturing, fast-freeze freeze–substitution low-temperature plastic embedding, and cryo-electron microscopy (cryo-EM).

[4] H. Plattner, *Cell Biol. Int. Rep.* **5,** 435 (1981).

[5] G. Knoll, A. J. Verkleij, and H. Plattner, *in* "Cryotechniques in Biological Electron Microscopy" (R. A. Steinbrecht and K. Zierold, eds.), p. 258. Springer-Verlag, Berlin, 1987.

[6] K. S. Matlin, H. Reggio, A. Helenius, and K. Simons, *J. Cell Biol.* **91,** 601 (1981).

[7] J. M. White and I. A. Wilson, *J. Cell Biol.* **105,** 2887 (1987).

[8] D. C. Wiley and J. J. Skehel, *Annu. Rev. Biochem.* **56,** 365 (1987).

[9] S. A. Wharton, W. Weis, J. J. Skehel, and D. C. Wiley, *in* "The Influenza Virus" (R. Krug, ed.), p. 153. Plenum, New York, 1989.

[10] M. J. Gething, J. Henneberry, and J. Sambrook, *Curr. Top. Membr. Transp.* **32,** 337 (1988).

Sample Preparation

Influenza B/USSR and influenza A (X-31) are grown in embryonated chicken eggs, harvested, and purified by isopycnic centrifugation on sucrose and potassium tartrate density gradients according to standard procedures.[11,12] Virus stocks (10–20 mg protein/ml) are stored in phosphate-buffered saline (PBS, pH 7.4) or TN (50 mM Tris-HCl, 100 mM NaCl, pH 7.4) at 4°, and used within a few days after the final purification. Alternatively 0.02% NaN$_3$ can be added to stabilize the virus stocks for several months (at 4°). Because impurities in the virus suspension are easily mistaken for products of virus–liposome interaction, the purity of the virus suspension should always be checked carefully, for example, by freeze–fracture electron microscopy (required concentration of virus stock ~ 10 mg protein/ml).[12]

Large unilamellar vesicles are prepared in PBS or TN by *n*-octylglucoside dialysis,[13] or by high-pressure extrusion of freeze–thawed multilamellar vesicles through two stacked polycarbonate filters[14] (pore size 100–400 nm; Nuclepore Corp., Pleasanton, CA). Small unilamellar vesicles are prepared by probe sonication as described.[15] Lipids may be obtained from Sigma Chemical Co (St. Louis, MO).

Virus (0.4 mg protein; concentration of virus stock 10–20 mg/ml) and liposomes (0.15 μmol phospholipid phosphorus; concentration of liposome suspension 5–10 mM phospholipid phosphorus) are mixed at pH 7.4. To prevent ganglioside breakdown and release of liposome-bound virus, the viral neuraminidase may be inhibited by adding 2-deoxy-2,3-dehydro-*N*-acetylneuraminic acid (Boehringer Mannheim GmbH, Mannheim, Germany). The inhibitor is added from a 12 mg/ml solution in PBS, adjusted to pH 7.4, to give a final concentration of 10 mM.[12,16] Virus–liposome fusion is initiated by mixing equal amounts of the virus–liposome suspension and a low pH buffer at 37° (50 mM succinic acid–NaOH, 100 mM NaCl, pH 4.8 and 5.2 for influenza A and B, respectively). The pH of the buffer is chosen so as to obtain a final pH after mixing close to the pH optimum of the fusion activity of influenza virus, namely, pH 5.0 and 5.5 for influenza A and B, respectively. Alternatively, a suboptimal pH and/or temperature may be used to slow down virus activation and

[11] J. J. Skehel and G. C. Schild, *Virology* **44**, 396 (1971).
[12] K. N. J. Burger, G. Knoll, and A. J. Verkleij, *Biochim. Biophys. Acta* **939**, 89 (1988).
[13] G. van Meer, J. Davoust, and K. Simons, *Biochemistry* **24**, 3593 (1985).
[14] L. D. Mayer, M. J. Hope, and P. R. Cullis, *Biochim. Biophys. Acta* **858**, 161 (1986).
[15] S. A. Wharton, J. J. Skehel, and D. C. Wiley, *Virology* **149**, 27 (1986).
[16] G. van Meer and K. Simons, *J. Cell Biol.* **97**, 1365 (1983).

membrane fusion.[16a] Samples are taken and immediately fast-frozen, before and at various time points after the pH drop, and subsequently processed for freeze–fracture, freeze–substitution, or cryoelectron microscopy.

For the negative stain technique more dilute virus and liposome suspensions are used, namely, 0.05–0.1 mg protein/ml and 20–50 nM phospholipid phosphorus, respectively. The exact amounts should be determined empirically. The pH is lowered and fusion induced by adding the appropriate amount of 0.1 M citric acid. At various time points after the pH drop the sample is reneutralized using 1 M Tris-HCl, pH 8.0, added in quantities up to 10% of the original volume of the sample. Once the sample is reneutralized, fusion events will cease, and samples can be conveniently processed for negative staining.

Conventional Techniques Using Electron Microscopy

Biological samples can only be analyzed by electron microscopy after stabilization of their structure. In conventional EM techniques the structure of a sample is stabilized by chemical means. The chemical pretreatments used in conventional thin sectioning (chemical fixation, dehydration, and plastic embedding), in conventional freeze–fracturing (chemical fixation and glycerol impregnation), and in negative staining (embedding in an electron-dense matrix and air-drying) may have great impact on the ultrastructure of the sample. As a consequence, the value of conventional EM techniques in membrane fusion research is rather limited.[17-19] Fortunately, most of the artifacts induced by the chemical pretreatments can be avoided using one of the alternative techniques developed more recently. In the modern "pure cryo" EM techniques, chemical pretreatments are omitted, and samples are processed by a combination of fast-freezing with freeze–fracture, with freeze–substitution, or with cryo-EM.

Nevertheless, one conventional technique, namely, negative staining, has proved to be extremely valuable especially in virus research. This is in spite of potential preparation artifacts and the fact that cryo-EM is a more reliable alternative. The practical procedures involved in negative staining are quick and relatively simple; reliable information may be obtained in particular on the structure and structural changes of viral spike proteins.

[16a] T. Stegmann, J. M. White, and A. Helenius, *EMBO J.* **9**, 4231 (1990).
[17] H. Plattner and H. P. Zingsheim, *Subcell. Biochem.* **9**, 1 (1983).
[18] B. P. M. Menco, *J. Electr. Microsc. Techn.* **4**, 177 (1987).
[19] J. C. Gilkey and L. A. Staehelin, *J. Electr. Microsc. Techn.* **3**, 177 (1986).

Negative Staining

Negative staining involves treating a specimen with an electron-dense salt, the negative stain. Because the negative stain will partially penetrate into the hydrophilic parts of the specimen, an image is obtained in which the unpenetrated regions of the specimen appear as electron-translucent structures against an electron-dense background.

Prior to negative staining, a virus–liposome mixture is incubated at low pH for the time required, and subsequently reneutralized. Since fusion events do not continue after neutralization, the fusion process can be studied over defined time periods.

Before negative staining, the sample is adsorbed to a supporting film. Several methods exist for doing this; the method favored in our laboratory is that described by Valentine *et al.*[20] A carbon film is deposited on a piece of freshly cleaved mica. The viral particles and liposomes are adsorbed to the carbon film by dipping the mica, film uppermost, into the sample. The film is then stripped from the mica by floating off onto a bath of suitable buffer such as PBS. A 400-mesh copper grid (TAAB Laboratories, Reading, UK) covered with a thin layer of adhesive is placed on top of the film, and a piece of adsorbent paper is used to pick up the grid with its now adhering film. Before the grid dries it is floated, film side down, on a stain bath for about 30 sec and then allowed to air-dry. It is now ready for examination in the electron microscope.

There are a number of different negative stains that can be used, and often it is a good idea to try two or three stains with a new sample. We routinely use sodium silicotungstate (SST, Agar Scientific Ltd., Stansted, Essex, UK) made to a concentration of 1% (w/v) with double-distilled water. This stain is preferable as it has a natural pH of 7; most other stains have to be corrected to neutrality, which can diminish their staining properties. In addition, SST produces a good, even spread with fine granularity. Phosphotungstic acid (PTA) and ammonium molybdate have the same fine granularity but give less contrast. Uranyl salts such as uranyl acetate tend to have a very low pH creating a serious problem for pH-sensitive samples; in general granularity is coarse, but contrast is high.

When examining the grid in the microscope and taking micrographs, it is important to preserve, and then image, as much fine detail as possible. We routinely use the low-dose, accurate defocus method of imaging described by Wrigley *et al.*[21] Essentially the procedure is to scan the grid at low magnification with a desaturated filament to reduce the specimen-damaging electron dose until an area of interest is found. The beam is then

[20] R. C. Valentine, B. M. Shapiro, and E. R. Stadtman, *Biochemistry* **7,** 2143 (1968).
[21] N. G. Wrigley, E. B. Brown, and R. K. Chillingworth, *J. Microsc.* **130,** 225 (1983).

deflected to a neighboring area so that the image can be focused at high magnification and fine alignment and astigmatism corrections made, while the area of interest is not irradiated. This area is then exposed to the electron beam only during the photographic exposure time. Thus, specimen damage by the beam is limited, and the microscope is aligned and focused sufficiently to image the preserved detail. Many modern microscopes have a low-dose facility as a standard fixture, and most others require only minor modifications.

Negative staining may influence the ultrastructure of a sample as visualized by electron microscopy. With respect to the analysis of influenza virus–liposome fusion, the most serious artifacts may occur in the final step of the negative stain procedure. During air-drying the specimen is exposed to high concentrations of a heavy metal salt and progressively dehydrated. Both steps could, in principle, affect the fusion process and influence the ultrastructure of the sample.

Modern Pure Cryo Electron Microscopy Techniques

Fast-freezing stabilizes the structure of biological specimens without any chemical pretreatments. Fast-freezing as the basis of modern "pure cryo" EM techniques avoids the artifacts induced by the chemical pretreatments during conventional sample processing,[18,19] and furthermore it offers a time resolution better than 1 msec.[5,22] It should be noted, however, that even if special fast-freezing devices are used, offering freezing rates in excess of 10,000°/sec, only a superficial layer of the sample (between 3 and 10 μm thick) can be frozen properly, that is, without large ice crystals being formed.

Freeze–Fracture

During freeze–fracture, either membrane structures are cross-fractured or the fracture plane runs through the middle of the membrane and the membrane interior is revealed.[23] Particles on the fracture faces of a membrane may indicate the presence of integral membrane proteins, protein–lipid complexes, or pure lipid structures.[24,25] With respect to the latter, it is worth mentioning that the presence of particles on the fracture faces of the membrane in pure lipid systems under fusion conditions has resulted in an

[22] J. E. Heuser, T. S. Reese, M. J. Dennis, Y. Jan, L. Jan, and L. Evans, *J. Cell Biol.* **81**, 275 (1979).

[23] P. Pinto da Silva and D. Branton, *J. Cell Biol.* **45**, 598 (1970).

[24] A. J. Verkleij and P. H. J. Th. Ververgaert, *Biochim. Biophys. Acta* **515**, 303 (1978).

[25] A. J. Verkleij, *Biochim. Biophys. Acta* **779**, 43 (1984).

interpretation of these lipidic particles as being fusion intermediates[25]; however, their interpretation is not unequivocal.[3]

Prior to freeze–fracture, virus–liposome samples (1 μl) are rapidly transferred to a sandwich composed of a plastic disk, a 15 μm thick copper spacer, and a copper cover.[26] The plastic disks, 3 mm in diameter, are punched out of Thermanox (Lux Scientific Corp., Naperville, IL). The spacers are made by punching a rectangular hole out of the center of 200-mesh copper grids; hat-shaped copper covers are punched out of 50 μm thick copper strips (CU 000495/3, Goodfellow, Cambridge, UK). Before punching, Thermanox and copper strips are roughened with sandpaper. After punching, plastic disks and spacers are cleaned with acetone and air-dried; copper covers are treated with dilute chromic/sulfuric acid [a 1/10 dilution of 10% (w/w) H_2SO_4, 25% (w/w) CrO_3], distilled water, and acetone, then air-dried. Commercially available sample holders (100 μm thick copper covers with and without a central depression, 12054-T and 12056-T, Balzers Union AG, Liechtenstein) may also be used, although in general the quality of freezing is slightly lower.

The sandwich samples are fast-frozen by plunging them into liquid propane ($-190°$), using a spring-operated freezing device (KF80, Reichert Jung, Vienna, Austria).[27] Alternatively, the samples may be fast-frozen by using a one-sided propane jet directed onto the copper cover of the sandwich.[26] Sandwiches can then be stored in liquid nitrogen. For freeze–fracture, the sandwiches are transferred to a Balzers BAF300 freeze-etch machine, fractured at $-105°$ by removing the plastic disk and spacer with the cold knife, and replicated following established procedures.[24] Replicas are cleaned by carefully floating the copper cover on chromic/sulfuric acid,[28] made at least 3 days in advance. After dissolution of the copper cover, the chromic/sulfuric acid is gradually replaced by distilled water, and the replicas are picked up with 300-mesh copper grids. Before use the grids are cleaned and made hydrophilic with dilute chromic/sulfuric acid, washed with distilled water, treated with acetone, and air-dried. The replicas were examined with a Philips EM420 at an acceleration voltage of 80 kV.

Two major problems may occur using this fast-freeze, freeze–fracture approach. First, the quality of freezing may be insufficient, yielding ice crystals larger than the resolution of the platinum–carbon replica (~ 3 nm); this problem has been discussed in detail elsewhere.[19,27] A second

[26] P. Pscheid, C. Schudt, and H. Plattner, J. Microsc. **121**, 149 (1981).

[27] H. Sitte, L. Edelmann, and K. Neumann, in "Cryotechniques in Biological Electron Microscopy" (R. A. Steinbrecht and K. Zierold, eds.), p. 87. Springer-Verlag, Berlin, 1987.

[28] M. J. Costello, R. Fetter, and M. Höchli, J. Microsc. **125** (1982).

problem relates to the use of thin sandwiches during fast-freezing. Interaction of the sample, especially with the copper cover, often results in a fracture plane at, or very near to, the copper interface. It should be realized that components of the sample, for example, lipid vesicles, are often enriched at the copper interface and may even collapse as a result of the interaction with the copper cover. Therefore, in the final EM analysis only areas of the replica in which the fracture plane has run through the ice should be considered to be reliable.

Freeze–Substitution Low-Temperature Plastic Embedding

The need for chemical fixation limits the time resolution of the classic plastic embedding/thin-sectioning method (in the range of seconds, at best). In addition, dehydration of the sample in organic solvents at room temperature results in an almost complete extraction of (membrane) lipids.[29] Both disadvantages can be overcome by taking advantage of the time resolution of fast-freezing (millisecond range) and by performing sample dehydration and plastic embedding at low temperature. Phospholipids are not extracted using this technique, and after thin-sectioning a reliable cross-sectional view of membrane systems may be obtained.[29,30] It should also be noted that during freeze–substitution only bulk water is removed; the retention of bound water limits protein denaturation and in many cases ensures an excellent preservation of antigenicity.[31]

Freeze–substitution and low-temperature plastic embedding are performed essentially according to Müller et al.[32] and Humbel et al.[33] using a commercially available freeze–substitution device (Auto CS, Reichert Jung). Alternatively, specimen temperature may be controlled using a Balzers spray freeze unit equipped with a brass block (with holes for insertion of Eppendorf vials). Sandwich samples are assembled and fast-frozen as described for freeze–fracture. The sandwiches are split under liquid nitrogen ("freeze–fracture"), removing the copper cover. Because only the plastic part of the sandwich is freeze-substituted, material adher-

[29] C. Weibull, A. Christiansson, and E. Carlemalm, J. Microsc. 129, 201 (1983).

[30] A. J. Verkleij, B. Humbel, D. Studer, and M. Müller, Biochim. Biophys. Acta 812, 591 (1985).

[31] R. A. Steinbrecht and M. Müller, in "Cryotechniques in Biological Electron Microscopy" (R. A. Steinbrecht and K. Zierold, eds.), p. 149. Springer-Verlag, Berlin, 1987.

[32] M. Müller, T. Marti, and S. Kriz, in "Electron Microscopy, Proceedings of the Seventh European Congress on Electron Microscopy" (P. Brederoo and W. DePriester, eds.), Vol. 2, p. 720. Leiden, The Netherlands 1980.

[33] B. Humbel and M. Müller, in "Proceedings of the Eighth European Congress on Electron Microscopy" (A. Csanády, P. Röhlich, and P. Szabó, eds.), Vol. 3, p. 1789. Budapest, 1984.

ing to the copper cover is lost. This loss may be reduced by precoating the copper cover with a lipid film,[34] by dipping clean copper covers in 10 mg lecithin/ml chloroform, blotting against filter paper, and allowing them to air-dry. As a consequence the fracture plane will run through the lipid film close to the copper cover, and the entire specimen is recovered together with the plastic disk.

The plastic disks together with the spacer and adhering parts of the sample are transferred to the substitution medium at $-80°$. The substitution medium consists of 1% OsO_4 (Drijfhout & Zoon, Amsterdam), 0.5% uranyl acetate (8473 Merck, Darmstadt, Germany), and 3% glutaraldehyde (20100, LADD Research Ind. Inc., Burlington, VT) in anhydrous methanol. Anhydrous methanol is obtained by treating absolute methanol with molecular sieve (3 Å pore size, 5704, Merck). If methanol is used as the substitution medium, the addition of both OsO_4 and uranyl acetate is recommended in order to ensure minimal extraction of phospholipids.[29,30]

It is important to note that the chemical fixation applied during freeze–substitution differs significantly from the chemical prefixation used in the classic plastic embedding/thin sectioning method. The mode of action of the chemical fixatives appears to be strongly temperature-dependent, and, in contrast to the classic thin section method, the use of chemical fixatives during freeze–substitution does not seem to affect membrane ultrastructure.[30]

After a period of 10 hr at $-80°$, the temperature is slowly raised to $-45°$ ($4°$/hr), and the specimens are washed with methanol and gradually infiltrated with Lowicryl HM20 (B8010 13032, Balzers Union AG; HM20/methanol, 1 : 1, 2 : 1, 100% HM20 at $-45°$ for 1–2 hr each). After an additional change of 100% Lowicryl and overnight incubation, the specimens are transferred to an embedding mold filled with Lowicryl HM20. Subsequently, UV polymerization is initiated ($-45°$, 24 hr). The specimens are brought to room temperature and allowed to harden under UV illumination for at least another 24 hr. The plastic disk is carefully removed using a razor blade, and thin sectioning is performed according to standard procedures. Ultrathin sections are stained for 10 min in 2% potassium permanganate in distilled water. Ultrathin sections were examined with a Philips EM420 at 80 kV.

A serious problem encountered during processing of virus–liposome samples is the washing out of material during freeze–substitution. As yet a satisfactory solution to this problem has not been found. However, if a ratio of virus particles to liposomes is chosen at which some liposome aggregation occurs, washout is kept to a minimum.

---

[34] G. Knoll, K. N. J. Burger, R. Bron, G. van Meer, and A. J. Verkleij, *J. Cell Biol.* **107**, 2511 (1988).

Cryoelectron Microscopy

In cryoelectron microscopy a sample is not fractured and replicated, or dehydrated and embedded in plastic. Rather, the sample is fast-frozen and visualized at low temperature in its original (hydrated) state.[35,36]

Carbon-coated grids, holey carbon grids, or bare grids may be used as the specimen support. The use of bare grids excludes adsorption artifacts, and in our hands more reproducible results are obtained than with holey carbon grids. A 700-mesh copper grid (thin bar, honeycomb pattern, G2760C, Agar, Scientific Ltd.) is simply dipped into, and withdrawn from, a virus–liposome suspension. Excess fluid is removed by blotting against filter paper, and the specimen is immediately (within 1 sec) fast-frozen by plunging it into liquid ethane cooled to its melting point with liquid nitrogen (using a gravity-powered guillotine[37]). The specimen grids are stored in liquid nitrogen.

For cryo-EM the specimen grid is mounted on a cryoholder with double shielding (Philips PW6599/00 or Gatan 626) and subsequently transferred to the electron microscope. To prevent contamination of the specimen and to reduce thermal drift of the cryoholder, the shields covering the specimen are kept closed for at least 30 min.[38,39] The specimens are examined at a specimen temperature of $-170°$; a Philips CM12 electron microscope was used at 100 kV. Imaging of small vitrified objects almost completely depends on phase-contrast, and large defocus values are essential, especially if large details are to be visualized. Therefore a defocus value of $2-6$ μm is used to resolve structural details $4-6$ nm in size optimally.[35,36]

In analogy to negative stain specimens, frozen hydrated specimens, especially those containing high concentrations of organic material, are very sensitive to electron beam damage. Ideally, images are obtained using a minimal electron dose. It should be noted that the cryo-EM images presented in this chapter were not made under strict low-dose conditions (in contrast to the negative stain images).

For the correct interpretation of cryo-EM data obtained by the bare-grid method, it is essential to consider the early steps of sample preparation in more detail. During withdrawal of a bare grid from a suspension containing surface-active components (lipids, denatured proteins, etc.) a thin film spontaneously forms spanning the holes between the grid bars. This

[35] J. Dubochet, M. Adrian, J. J. Chang, J. C. Homo, J. Lepault, A. W. McDowall, and P. Schultz, *Q. Rev. Biophys.* **21**, 129 (1988).

[36] J. Lepault and J. Dubochet, this series Vol. 127, p. 719.

[37] J. Dubochet, J. Lepault, R. Freeman, J. A. Berriman, and J. C. Homo, *J. Microsc.* **128**, 219 (1982).

[38] P. M. Frederik and W. M. Busing, *J. Microsc.* **144**, 215 (1986).

[39] J. Trinick, J. Cooper, J. Seymour, and E. H. Egelmann, *J. Microsc.* **141**, 349 (1986).

thin film consists of a thin aqueous layer enclosed by two monolayers of surface active molecules, one at each air–water interface. As soon as the grid is withdrawn from the suspension the film starts to thin. Thinning is initially driven by gravity and capillary forces. At a film thickness below 100 nm, London and van der Waals attractive forces along with electrostatic and hydration repulsive forces between the surface layers dominate, and thinning will proceed until these counteracting forces are at balance.[40,41] The resulting thin film is very thin near the center of the grid hole (nominally some 20 nm thick) and increases in thickness toward the grid bars (>200 nm). Thin films are easily fast-frozen by plunging them into a suitable coolant. In fact, ice crystals do not form at all, and the specimen may be considered to be vitrified.[37]

Thin film formation may influence the structure of a sample as observed by cryo-EM.[40] During thin film formation part of the sample may disintegrate to form the monomolecular surface layers at the air–water interface;[41a] in the case of a virus–liposome suspension thin film formation will probably involve collapse of part of the liposomes. During draining of the thin film the water content drops significantly (nominally from 99 to 30% water in the thinnest parts of the film), which might affect the structure of a sample.[41] Draining of a thin film will in many cases influence the composition of a sample as observed by cryo-EM. Only parts of the thin film having a thickness less than 200 nm can be imaged at high resolution by cryo-EM. These thinner parts of the film are relatively enriched in small structures since large structures or aggregates are squeezed out from these areas during draining of the thin film. This explains why predominantly loose virus particles and single liposomes are visualized in virus–liposome samples, and only occasionally are (small!) virus–liposome aggregates seen.

Summary of Experimental Results

The binding and fusion of influenza virus with pure lipid model membranes can be followed quite conveniently with all four morphological techniques described in this chapter. The morphological data agree with most biochemical studies on influenza virus–membrane fusion in showing

---

[40] P. M. Frederik, M. C. A. Stuart, P. H. H. Bomans, and W. M. Busing, *J. Microsc.* **153**, 81 (1989).

[41] P. M. Frederik, M. C. A. Stuart, A. H. G. J. Schrijvers, and P. H. H. Bomans, *Scanning Microsc.* **3**, 277 (1989).

[41a] P. M. Frederik, K. N. J. Burger, M. C. A. Stuart, and A. J. Verkleij, *Biochim. Biophys. Acta* **1062**, 133 (1991).

no obvious fusion activity of influenza virus at neutral pH[7-10,42,43]; neither continuity of viral and liposomal membranes nor diffusion of viral spike proteins into the liposomal membrane are observed. Interestingly, the freeze–fracture data indicate that the binding of influenza virus to liposomes at neutral pH results in local protrusions of the convex fracture face of the liposomal membrane, sometimes bearing an irregularly shaped central particle of 9 to 14 nm in size (Fig. 1b) (for details, see ref. 12). Because fusion at neutral pH can be virtually excluded, this central particle is interpreted as being a prefusion structure; a local point contact between viral and liposomal membranes may exist in spite of the high density of spike glycoproteins in the intact virion.

Lowering the pH results in a fast and efficient fusion of viral and liposomal membranes, and large fusion products are formed (Fig. 2). Continuity of the viral and liposomal membranes is obvious from the results of negative staining (Fig. 2a), freeze–fracture (Fig. 2b), freeze–substitution (Fig. 2c), and cryo-EM (Fig. 2d). Both freeze–fracture and cryo-EM indicate that membrane continuity is accompanied by the formation of a large aqueous channel, fairly constant in size (~ 50 nm in diameter). Smaller aqueous channels are rarely observed, suggesting that the initially formed aqueous connection enlarges very rapidly. Viral spike proteins diffuse into the liposomal membrane as shown by the results of both negative staining and cryo-EM. Lateral diffusion of viral spike proteins is also supported by freeze–fracture data showing the presence of viral intramembrane particles on the concave fracture face of the liposomal membrane; these intramembrane particles probably represent the hemagglutinin or neuraminidase spike proteins, or both. Freeze–substitution shows that as a result of virus–membrane fusion the viral nucleocapsid may be released into the liposomal interior.

The principal aim of the morphological characterization of influenza virus–liposome fusion was to find intermediates of the actual joining of viral and liposomal membranes. These intermediates were not found.[12] In contrast, during hemagglutinin-mediated fusion of liposomes with influenza virus-infected cells[34] and during fusion of influenza virus with erythrocyte ghosts,[44] possible fusion intermediates were observed. The fact that fusion intermediates were not observed during influenza virus–liposome fusion could be due to an extremely short lifetime of these intermediates in the virus–liposome system. The detection of fusion intermediates during

[42] T. Stegmann, D. Hoekstra, G. Scherphof, and J. Wilschut, *Biochemistry* **24**, 3107 (1985).
[43] A. M. Haywood, and B. P. Boyer, *Proc. Natl. Acad. Sci. U.S.A.* **82**, 4611 (1985).
[44] K. N. J. Burger, G. Knoll, P. M. Frederik, and A. J. Verkleij, NATO ASI Series, **H40**, 185 (1990).

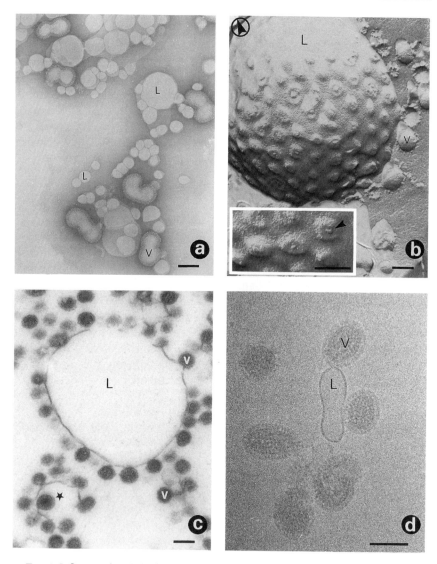

FIG. 1. Influenza virus (V)–liposome (L) interaction at neutral pH. (a) Binding of influenza A (X-31 strain) to predominantly small unilamellar vesicles of palmitoyloleoylphosphatidylcholine and cholesterol (molar ratio 0.7) with 1% (w/w) mixed bovine brain gangliosides (5 min, 37°) as visualized by negative staining using 1% SST. (b–d) Binding of influenza B (USSR strain) to large unilamellar vesicles composed of egg phosphatidylcholine, egg phosphatidylethanolamine, cholesterol, and the ganglioside G_{D1a} in a molar ratio of $1:1:2:0.2$. Samples were incubated for 30 min on ice in the presence of a neuraminidase inhibitor and subsequently fast-frozen and processed for freeze–fracture (b), freeze–substitution (c), or

liposome–cell and virus–ghost fusion might then be related to an increased lifetime of the fusion intermediates, for example, as the result of steric constraints put on liposome–cell and virus–ghost fusion by membrane–cytoskeleton interactions.[44]

Comparison of Techniques

The choice of technique depends mainly on the information that is required and on the specific advantages and disadvantages of each technique (Table I). Using negative staining or cryo-EM a sample is visualized in its projection, and information may be obtained in particular on the structure and structural changes of the viral spike proteins.[45] Freeze–fracture, on the other hand, exposes the membrane interior and is ultimately suited for the identification of intermediates of the actual coalescence of viral and liposomal membranes.[25,34,44] A reliable cross-sectional view of a membrane system may be obtained after freeze–substitution and low-temperature plastic embedding[12,30]; internal structures (e.g., the nucleocapsid) are visualized, and, in combination with (immuno)cytochemical techniques, macromolecules can be localized both in time and space.

Although the information obtained by using modern "pure cryo" EM techniques is in general more reliable, a conventional EM technique like negative staining may still prove valuable. The negative staining technique is quick, relatively simple, and can be applied to very dilute samples. Negative staining can be used to monitor the low pH-induced conformational change of the hemagglutinin spike proteins.[45] In addition, negative staining easily demonstrates whether virus–liposome fusion has or has not occurred. However, potential drying artifacts and artifacts due to interac-

[45] R. W. H. Ruigrok, N. G. Wrigley, L. J. Calder, S. Cusack, S. A. Wharton, E. B. Brown, and J. J. Skehel, *EMBO J.* **5**, 41 (1986).

cryo-EM (d). At neutral pH influenza virus binds to but does not fuse with pure lipid model membranes. Negative staining (a) and cryo-EM (d) optimally resolve the viral spike proteins. Freeze–fracture (c) resolves the membrane interior; binding of influenza virus leads to protrusions on the convex fracture face of the liposomal membrane, occasionally bearing a central particle [inset in (b), marked by arrowhead]. Freeze–substitution (c) offers a cross-sectional view, and occasionally a membrane-surrounded virion is observed inside a liposome [most likely as the result of a cross section through an almost entirely engulfed virion; marked by the star in (c)]. Direction of the Pt/C shadow is indicated by the encircled arrowhead in (b). Bars, 0.1 μm. [(b) and (c) were taken from Burger *et al.* (1988)[12] and reproduced with permission.]

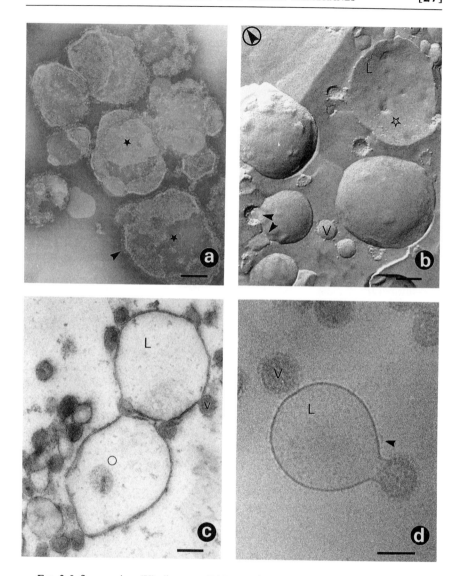

FIG. 2. Influenza virus (V)–liposome (L) interaction at low pH. (a) Fusion of influenza A (X-31 strain) with small unilamellar vesicles (for composition see legend to Fig. 1) after incubation at pH 5.0 (5 min, 37°), as visualized by negative staining using 1% SST. (b–d) Fusion of influenza B (USSR strain) with large unilamellar vesicles (for composition see legend to Fig. 1); after incubation at pH 5.4 (2 min, 37°) samples were fast-frozen and processed for freeze–fracture (b), freeze–substitution (c), and cryo-EM (d). Lowering the pH initiates influenza virus–liposome fusion, and large fusion products ae formed [e.g., in (a), marked by stars]. The results of both negative staining and cryo-EM indicate the presence of

tion of the negative stain with the sample do not allow final conclusions to be drawn on the actual fusion mechanism using this technique. In the final analysis of virus–liposome fusion, and of membrane fusion in general, the results of conventional EM techniques should always be verified by using "pure cryo" EM techniques.

Unquestionably all EM techniques, including the modern "pure cryo" EM techniques, suffer from preparation artifacts. The possibility of being misled by a preparation artifact will be reduced by using more than one technique. In addition it should be noted that the data obtained by the four techniques discussed here supplement each other. It is only by using several morphological techniques that a complete picture of virus–liposome fusion may be obtained.

Conclusions and Prospects

Morphological techniques offer the unique possibility of obtaining information on the structure and composition of a sample at the actual fusion site. A local point contact between viral and liposomal membrane was observed, as were local point fusion intermediates during HA-mediated liposome–cell fusion and virus–erythrocyte ghost fusion. Thus, the morphological data obtained with three model systems used to study influenza HA-mediated membrane fusion support the idea that HA-mediated membrane fusion involves local point adhesion followed by local point fusion.

The local point fusion concept fits in well with an important role of nonbilayer preferring lipids in facilitating a local bilayer to nonbilayer lipid structure transition and membrane fusion.[3,25] Furthermore, the local lipid structure transition may be directly initiated by a (fusion) protein. With respect to the latter, it is important to note that the local point fusion mechanism implies that only a few HA spike proteins are likely to be directly involved in the induction of membrane fusion. Cryo-EM has the potential to resolve viral spike proteins together with viral and target

---

viral spike proteins in the liposomal membrane [marked by arrowheads in (a) and (d)]. In addition, viral intramembrane particles appear on the concave fracture face of the liposomal membrane [marked by star in (b)]. Continuity between viral and liposomal membranes is illustrated most convincingly by freeze–fracture [marked by arrowheads in (b)]. Freeze–substitution shows that after virus–membrane fusion the nucleocapsid may be released into the liposomal interior [marked by circle in (c)]. Direction of the Pt/C shadow is indicated by the encircled arrowhead in (b). Bars, 0.1 μm. [(b) and (c) were taken from Burger *et al.* (1988)[12] and reproduced with permission.]

TABLE I

ELECTRON MICROSCOPY TECHNIQUES USED IN MORPHOLOGICAL CHARACTERIZATION OF INFLUENZA VIRUS–LIPOSOME FUSION[a]

Technique	Information	Advantages	Disadvantages
Negative stain	Structures excluding negative stain seen in projection	Quick, relatively simple Low sample concentration required Macromolecules visualized at high resolution	Selective adhesion to support film Interaction with support film (shape change) Drying artifacts (dehydration, high salt) Artifacts due to negative stain interaction Low time resolution Radiation sensitive
Fast-freezing		Artifacts of chemical pretreatments avoided High time resolution (msec range)	Elaborate Only superficially well-frozen
Freeze–fracture	Extended view of membrane interior	Membrane ultrastructure visualized Replica inert	Interaction with sample holder High sample concentration required
Freeze–substitution	Cross-sectional view of membrane systems	No phospholipid extraction Optimal possibilities for cytochemistry	Occasional washout of material High sample concentration required
Cryo-EM	Sample in original composition, seen in projection	Hydrated sample in original composition Macromolecules visualized at high resolution	Thin-film formation: disintegration (minor), dehydration, selectivity: exclusion of structures > 200 nm High sample concentration required Radiation sensitive

[a] Techniques differ in information that is obtained about sample ultrastructure and have specific advantages and disadvantages.

membrane at relatively high resolution.[46,47] In the near future cryo-EM may play an important role in the further unraveling of the mechanism of HA-mediated membrane fusion by monitoring the conformational change of viral spike proteins at the actual fusion site.

[46] F. P. Booy, R. W. H. Ruigrok, and E. F. J. van Bruggen, *J. Mol. Biol.* **184,** 667 (1985).
[47] P. M. Frederik, M. C. A. Stuart, and A. J. Verkleij, *Biochim. Biophys. Acta* **979,** 275 (1989).

[28] Kinetics and Extent of Fusion of Viruses with Target Membranes

By SHLOMO NIR

Introduction

Procedures for analyzing fusion experiments are described in this chapter. The results of fusion studies between influenza and Sendai viruses and liposomes, erythrocyte ghosts, and suspension cells are presented.

The mass action model for membrane fusion[1-3] views the overall fusion reaction as the sequence of a second-order process of liposome–liposome, virus–liposome, or virus–cell adhesion or aggregation, followed by a first-order fusion reaction. The analysis of the final extent of fluorescence intensity can yield the percentage of virions capable of fusing with certain target membranes at a given pH. Analysis of the kinetics of fusion enables us to elucidate to what extent the action of viral glycoproteins goes beyond the promotion of contact between apposed membranes, as well as details of virus inactivation.

Final Extents of Fusion: Virus–Cell Fusion

The analysis of the final extents of fusion can give answers to two important questions: (1) what percentage of virions are capable of fusing with the given cells as a function of pH and temperature, and (2) what number of virions can fuse per single cell. In this chapter we introduce equations that provide answers to these questions based on the final extents of fluorescence intensity increase. We employ an assay based on mixing of

[1] S. Nir, J. Bentz, and J. Wilschut, *Biochemistry* **19,** 6030 (1980).
[2] J. Bentz, S. Nir, and J. Wilschut, *Colloids Surf.* **6,** 333 (1983).
[3] S. Nir, J. Bentz, J. Wilschut, and N. Düzgüneş, *Prog. Surf. Sci.* **13,** 1 (1983).

membranes, namely, the octadecylrhodamine B chloride (R_{18}) assay.[4,5] It is assumed that the virus is initially labeled.

Let I be the fractional increase in the final fluorescence intensity of R_{18} molecules arising from their dilution as a result of fusion. I is given by

$$I = 1 - X \tag{1}$$

in which X is the relative surface concentration of R_{18} molecules, provided that their initial concentration is sufficiently low. Let us denote the surface areas of a single virus particle and a cell by S_v and S_e, respectively, and let M be the average number of virus particles per cell. If all virus particles have fused, then[6]

$$I = 1 - MS_v/(MS_v + S_e) = S_e/(S_e + MS_v) = 1/(1 + MS_v/S_e) \tag{2}$$

If N is the number of virions that can fuse per single cell, and $p > 1$ is the number of virus particles per cell divided by N, then Eq. (2) is modified to

$$I = 1/[p(1 + NS_v/S_e)] \tag{3}$$

Another modification of Eq. (2) is obtained if the fraction of virus particles capable of fusing is q,

$$I = q/(1 + qMS_v/S_e) \tag{4}$$

If S_v/S_e is known, then Eq. (2) does not involve any unknown parameter and can be used to generate predicted I values. In the case of Sendai virus fusing with erythrocyte ghosts at 37° and pH 7.4, the use of Eq. (4) indicates that $q > 0.86$, that is, practically all virus particles are fusion active. The use of Eq. (3) yields $N = 100-200$, that is, about 100 Sendai virus particles fuse[6,7] per single cell (depending on the estimated number of particles), in contrast to about 1500 that can bind per cell.[7,8] The use of Eq. (3) to study the fusion of Sendai virus with erythrocyte ghosts yielded good simulations and predictions for many ratios of virions to cells. The predictions[6] were reasonably good for different batches of virus and ghosts, despite large differences in the rates of fusion for the different batches. Furthermore, in a solution of 4% polyethylene glycol (PEG), the rate of fusion was significantly enhanced without affecting the number of virions fusing per single cell.[9]

[4] D. Hoekstra, T. De Boer, K. Klappe, and J. Wilschut, *Biochemistry* **23**, 5675 (1984).
[5] D. Hoekstra, K. Klappe, T. De Boer, and J. Wilschut, *Biochemistry* **24**, 4739 (1985).
[6] S. Nir, K. Klappe, and D. Hoekstra, *Biochemistry* **25**, 2155 (1986).
[7] D. Hoekstra and K. Klappe, *J. Virol.* **58**, 87 (1986).
[8] D. Wolf, I. Kahan, S. Nir, and A. Loyter, *Exp. Cell Res.* **130**, 361 (1980).
[9] D. Hoekstra, K. Klappe, H. Hoff, and S. Nir, *J. Biol. Chem.* **264**, 6786 (1989).

The application of Eq. (2) to the case of influenza virus fusing with erythrocyte ghosts at 37° and pH 5 yielded good simulations and predictions for I values for virus/cell numbers of up to 400.[9a] The surface area of influenza virus is about two-thirds smaller than that of the Sendai virus, the respective radii being 50 and 75 nm.

The analysis of final extents of fusion of influenza virus (at pH 5) and Sendai virus (above pH 7.5) with nonendocytosing suspension cells indicated that all virus particles could fuse, provided that the number of virions per cell was not large (e.g., $\sim 50-100$).[9b] However, it was difficult to estimate the number of virions that can fuse with a single cell, because of the significant degree of cell lysis that occurred after long incubation times.

Final Extents of Fusion: Virus–Liposome Fusion

Fusion of influenza or Sendai viruses with liposomes resulted in fusion products consisting of a single virus and several liposomes. This conclusion was first reached by comparing the final extents of fluorescence increase with calculated values obtained by the application of several models for which analytical solutions were found [see Ref. 10 and Eq. (5) below]. The experimental results included liposome/virus ratios that varied by two orders of magnitude.

Another demonstration of this phenomenon has been achieved by adding liposomes or virions (influenza[10] and Sendai[11]) to a system of virions and liposomes after long incubation times. The addition of blank liposomes resulted in an increase in fluorescence intensity which was close to the final level predicted for the final virus/liposome ratio. This process was repeated several times and indicated that the added liposomes fused with the fusion products.

A direct consequence of the fact that virus–liposome fusion products consist of a single virus and several liposomes is that a certain fraction of the virus population will not fuse, unless the liposome/virus ratio is large. The fraction of fully active virions which fuse depends on the number of liposomes and virions according to[10]

$$V_f/V_0 = 1 - \exp(-L_0/V_0) \tag{5}$$

in which L_0 and V_0 are initial molar concentrations of liposomes and virions, and V_f is the concentration of fused virions. Thus, for a 1/1

[9a] S. Nir, unpublished, 1985, using data of T. Stegmann and J. Wilschut.

[9b] N. Düzgüneş, M. C. Pedroso de Lima, L. Stamatatos, D. Flasher, D. Alford, D. S. Friend, and S. Nir, *J. Gen. Virol.* **73**, 27 (1992).

[10] S. Nir, T. Stegmann, and J. Wilschut, *Biochemistry* **25**, 257 (1986).

[11] K. Klappe, J. Wilschut, S. Nir, and D. Hoekstra, *Biochemistry* **25**, 8262 (1986).

population the fraction of fused virions is $[1 - \exp(-1)]$, whereas for $L_0/V_0 = 4$ most of the virions fuse. This result also implies that for given liposomal and viral lipid concentrations an increase in virus size will result in a smaller value of V_0, and hence in a smaller percentage of unfused virions. This effect is illustrated in Table I.

The reason why virions cannot fuse with the fusion products remains unclear, but it may be due to mutual interference of the viral glycoproteins, as determined by their surface density in the fusion products. Thus, fusion products consisting of a single virus particle and a large number of liposomes may fuse with additional virions. In fact, many virions can fuse with a single erythrocyte ghost whose surface area is three orders of magnitude larger than that of a virion. In this context it is of interest that the binding of the hemagglutinin (HA) glycoproteins from influenza virus to liposomes is limited. Calculations indicate that the limit[12] on the binding of HA corresponds to ratios of less than 1/8 virus/liposome in the fusion products. Thus, it appears that results of binding of HA to liposomes support the suggestion that fusion of a virus with the fusion product is an unfavorable event because of the limit on the number of HA molecules in the membrane.

Equation (5) was derived by assuming that all virions are capable of fusing with the given liposomes at the given pH. Equation (5), which does not involve any parameters, gave good predictions for the final extents of fluorescence increase for the fusion of influenza virus with cardiolipin (CL)[10,13] and phosphatidylserine (PS)[13] liposomes, and for the fusion of Sendai virus[14] with CL liposomes at pH 5 for ratios of V_0/L_0 varying by about two orders of magnitude, irrespective of whether the virus or the liposomes were labeled. In many other cases, however, the fraction of fused virions was smaller than unity even in the presence of a large excess of liposomes.[10,11,13-15] This can be expressed by stating that a fraction of the virions will remain unfused even with a large excess of liposomes. Equations (6)–(8) were derived by assuming that all liposomes can fuse, unless they are irreversibly bound to inactive virus particles. If the binding of liposomes to virions is fully reversible, then the liposomes bound to inactive virions will eventually be released and will fuse with the active virions or with the fusion products. In this case the fraction of fused virions, V_f/V_0, is[14]

$$V_f/V_0 = (1 - \alpha)(1 - \exp\{-L_0/[V_0(1 - \alpha)]\}) \qquad (6)$$

[12] R. W. Doms, A. Helenius, and J. White, *J. Biol. Chem.* **260**, 2973 (1985).
[13] T. Stegmann, S. Nir, and J. Wilschut, *Biochemistry* **28**, 1698 (1989).
[14] S. Nir, K. Klappe, and D. Hoekstra, *Biochemistry* **25**, 8261 (1986).
[15] S. Amselem, Y. Barenholz, A. Loyter, S. Nir, and D. Lichtenberg, *Biochim. Biophys. Acta* **860**, 301 (1986).

TABLE I

CALCULATED FINAL LEVELS OF FLUORESCENCE INTENSITY OWING TO FUSION OF
R_{18}-LABELED VIRUS PARTICLES WITH LIPOSOMES

System	Active virus (%)	Ratio of liposomal to viral lipid		
		16:1	4:1	1:1
Homogeneous mixture, $R_v = R_L$	100	94.1	80	50
$R_v = R_L$	100	94.1	78.8	38.7
$R_v = 75$ nm, $R_L = 50$ nm	100	94.1	80	47.2
Reversible binding, $R_v = R_L$	50	48.5	44.4	33.1
Irreversible binding, $R_v = R_L$	50	47.1	39.4	19.4
Reversible binding, $R_v = R_L$	25	24.6	23.5	19.7
Irreversible binding, $R_v = R_L$	25	23.5	19.7	9.7

In the case of irreversible binding, the fraction of fused lipsomes, L_f/L_0, is

$$L_f/L_0 = 1 - \alpha \qquad (7)$$

and the fraction of fused virions is

$$V_f/V_0 = (1 - \alpha)[1 - \exp(-L_0/V_0)] \qquad (8)$$

Table I demonstrates that the effect of irreversible binding of liposomes to inactive virions becomes more pronounced at smaller ratios of liposomal to viral lipids. In the case of Sendai virus fusing with PS and CL/dioleoylphosphatidylcholine (DOPC) liposomes at pH 5 and 7.4, or with CL liposomes at pH 7.4, the final extents of fluorescence intensity are explained by assuming an essentially irreversible binding of liposomes to inactive virions.[14] Another demonstration of this pattern was achieved by means of a long incubation of Sendai virus with small labeled PS vesicles at pH 7.4. Sudden acidification to pH 4 resulted in an increase in fluorescence intensity owing to the dilution of the probe, indicating that the system initially included certain unfused liposomes, which were not free as deduced from the fact that addition of virions did not result in an increase in fluorescence.[15]

Fusion of Sendai virus with PS and PS/dioleoylphosphatidylethanolamine (DOPE)/cholesterol liposomes[15a] showed that bound, unfused virions are capable of fusion after their release and separation (on a sucrose gradient), followed by subsequent incubation with added liposomes. In these cases, the unfused virions are not considered fusion-inactive particles, but rather are particles bound to liposomes via "inactive" sites.

[15a] S. Nir, K. Klappe, H. Hoff, and D. Hoekstra, unpublished.

Kinetics of Fusion: Virus–Cell Fusion

The goal of kinetic studies is to find the dependence of the fusion rate constant, f (sec^{-1}), and adhesion rate constant, C (M^{-1} sec^{-1}), on the virus–cell combination, pH, and temperature. Deadhesion processes described by the rate constant D (sec^{-1}) must be taken into account in most situations. In principle, the number of parameters (i.e., rate constants) may be larger, depending on the number of virions that are associated with a cell and on the programs based on numerical solution of the differential equations allow for such variations. Because it is preferable to employ few parameters, we describe the equations having only three parameters, and we emphasize the variations in the experimental procedure which are most suited for the purpose of analysis. We also ignore cell–cell interactions.

The molar concentration of an aggregation–fusion product consisting of I virus particles adhering to a cell, and J virus particles which have fused with it, is denoted by $AF(I,J)$. The molar concentration of virus particles is denoted by V. Initially, at time zero, $V(t) = V_0$. Mass conservation of virus particles is expressed by

$$V_0 = V + \sum_{I+J=1}^{N} AF(I,J)(I+J) \tag{9}$$

in which N is the largest number of virus particles that can be associated with a single cell. About 1500 (N) Sendai virus particles can bind to one erythrocyte ghost, but only about 100 (N_f) particles can fuse with it. In studying the kinetics of fusion it is preferable to employ virus/cell ratios below the N_F value. The molar concentration of free cells is denoted by $G_0 = AF(0,0)$. Initially, $G(t) = G_0$. Mass conservation for the ghost particles gives

$$G_0 = \sum_{I+J=0}^{N} AF(I,J) \tag{10}$$

The kinetics of adhesion, de-adhesion, and fusion is described[6] by the Eqs. (11) and (12):

$$\frac{d}{dt} AF(I,J) = CVAF(I-1,J)(N+1-I-J)/N$$
$$- CVAF(I,J)(N-I-J)/N + DAF(I+1,J)(I+1)$$
$$+ fAF(I+1,J-1)(I+1) - (f+D)AF(I,J)I \tag{11}$$

$$\frac{d}{dt} V = CV \sum^{N-1} AF(I,J)(N-I-J)/N + D \sum_{I+J=1}^{N} AF(I,J) I \tag{12}$$

For input the program accepts the concentrations G_0, V_0, the rate constants, and the effective ratio $S_e/S_v = D_0$ or dilution factor. The program calculates the kinetics of fluorescence increase owing to probe dilution, assuming that the virus is initially labeled. The program yields, for any time t, the fluorescence intensity I and the distribution of adhesion–fusion products $AF(I,J)$, as well as the virus concentration, V.

The value of $I(t)$ is well-approximated by the following expression[16]:

$$I(t) = \frac{G_0 N_F T}{V_0[1 + N_F T/D_0 + (D_0 + N_F T)^{-1})]} \, \mathscr{F}_1 \, (\hat{K}, \hat{\tau}) \tag{13}$$

where

$$T(\hat{\tau}) = \frac{\exp\{\alpha\hat{\tau}\} - 1}{(1 + \alpha)\exp\{\alpha\hat{\tau}\} - 1} \tag{14}$$

$$\mathscr{F}_1 = 1 + \frac{\exp\{-\hat{K}\hat{\tau}\}[\alpha \exp\{\alpha\hat{\tau}\})^2 - (\hat{K} - \alpha) (\exp\{\alpha\hat{\tau}\} - 1)^2] - \alpha^2 \exp\{\alpha\hat{\tau}\}}{(\hat{K} - \alpha)(\exp\{\alpha\hat{\tau}\} - 1)[(1 + \alpha) \exp\{\alpha\hat{\tau}\} - 1]} \tag{15}$$

and

$$\begin{aligned}
\hat{\tau} &= \hat{C}V_0 t \\
\hat{K} &= \hat{f}/(\hat{C}V_0) \\
\hat{C} &= C'/(1 + D/F) = C/[N_F(1 + D/f)] \\
\hat{f} &= f(1 + D/f); \quad \alpha = (N_F G_0/V_0) - 1
\end{aligned} \tag{16}$$

If deaggregation processes can be ignored ($D \ll f$; $D \ll CV_0$) and if there is little delay in the fusion of adhered virus particles, then adhesion is the rate-limiting step, and a simple analytical solution is obtained[6]:

$$V(t) = V_0 \exp(-CG_0 t) \tag{17}$$

Then the fraction of fused virus particles is

$$V_f/V_0 = 1 - \exp(-CG_0 t) \tag{18}$$

When $CG_0 t \ll t$,

$$[V_0 - V(t)]/V_0 \approx CG_0 t \tag{19}$$

Equation (19) implies a linear increase in the fraction of fused virus particles with time and with the initial concentration of cells. If fusion results in an infinite dilution of a certain fluorescent label (e.g., R_{18} molecules), then the expected increase in fluorescence intensity due to fusion should be proportional to the time and to the initial concentration of cells.

[16] J. Bentz, S. Nir, and D. G. Covell, *Biophys. J.* **54**, 449 (1988).

In simple cases where deadhesion can be ignored, the determination of the parameters C and f is a simple task. First C is determined from I values at sufficiently long times, where fusion of adhered virions is practically complete. The function $\mathscr{F}_1$ in Eq. (13) is close to unity, and Eq. (13) or Eq. (18) provides the value of C. In principle, one concentration of virions and cells is sufficient, but the use of two concentrations (or more) of cells enables a test, or a prediction. The earlier time points ($ft \ll 1$) provide the value of f, provided that the values of C and N_F are known.

In general, the data described by Eqs. (13)–(16) are insufficient for the determination of C and f values. Inspection of Eqs. (13)–(16) indicates that I values are expressed in terms of $\hat{C}$ and $\hat{f}$. This means that the analysis can only yield $\hat{C}$ and $\hat{f}$ values. To determine C and f values, it is necessary to find D values from data on dissociation of virions from cells. This is done by pre-incubating Sendai virus particles with ghosts on ice,[6,9] under conditions where binding occurs without fusion, and then fusion is initiated by diluting the suspension into a warm buffer. It is found that in certain cases about one-third of the bound virions dissociate from the cells within several minutes, during which fusion also occurs.[7] The determination of D values enables the determination of C and f values from I values by employing Eq. (13), or a numerical solution. Furthermore, the procedure of analysis employed a program which calculates the distribution of adhesion products $AF(I,0)$ during the initial period of preincubation in the cold. The second part of the program considers the dilution into a warm buffer, at which stage fusion is initiated. In this procedure Eqs. (13)–(16) cannot be applied, and the program is based on numerical solution of Eqs. (11) and (12), employing a Taylor expansion. Another complication is encountered in certain cases owing to a time lag in fusion activity of the precooled particles.[9] However, an approximate estimate of D values combined with a knowledge of $\hat{f}$ and $\hat{v}$ alues can yield C and f values with respective relative errors of 20–50% and 50–100%.

The procedure described above is suitable for the analysis of fusion kinetics between Sendai virus and cells. However, it may not be suitable for studies employing influenza virus, because of a reduction in fusion activity during incubation at low pH (e.g., pH 5).[17–23] In this case a very simple

[17] J. White, A. Helenius, and J. Kartenbeck, *EMBO J.* **1**, 217 (1982).
[18] S. B. Sato, K. Kawasaki, and S. I. Ohnishi, *Proc. Natl. Acad. Sci. U.S.A.* **80**, 3153 (1983).
[19] P. R. Junankar and R. J. Cherry, *Biochim. Biophys. Acta* **854**, 198 (1986).
[20] T. Stegmann, D. Hoekstra, G. Scherphof, and J. Wilschut, *J. Biol. Chem.* **261**, 10966 (1986).
[21] T. Stegmann, F. P. Booy, and J. Wilschut, *J. Biol. Chem.* **262**, 17744 (1987).
[22] S. Ohnishi, *in* "Membrane Fusion in Fertilization, Cellular Transport, and Viral Infection" (N. Düzgüneş and F. Bronner, eds.), p. 257. Academic Press, New York, 1988.
[23] T. Stegmann, S. Nir, and J. Wilschut, *Biochemistry* **28**, 1698 (1989).

procedure can be applied. First, the (labeled) virus is incubated with the cells at neutral pH at a given temperature for several minutes, and the amount of virus bound to the cells is determined. No fusion with the external membrane of the cell occurs during the preincubation stage. No increase in fluorescence intensity of R_{18}-labeled influenza virus occurs during the preincubation stage when nonendocytosing cells are employed. On sudden reduction in the pH of the suspension, there is an increase in fluorescence intensity (see Fig. 1), the rate and extent of which depend on the virus–cell combination (e.g., different strains), pH, temperature, and time of preincubation at netural pH.

The analysis can focus on short time periods (e.g., 5–30 sec) following the reduction in pH. During these times the fraction of virus bound does

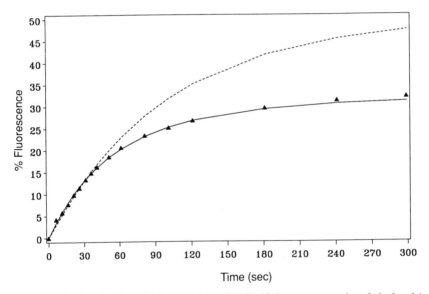

FIG. 1. Kinetics of fusion of influenza virus with HL-60 (human promyelocytic leukemia) suspension cells. Influenza virus of strain PR8 and labeled with R_{18} (5 μg viral protein/ml) was incubated at pH 7.5 and 37° with 2×10^7 HL-60 cells/ml. After 10 min the pH was lowered to pH 5 and fusion was initiated, as monitored by fluorescence increase. The value of %F is defined by %$F = 100[I(t) - I_0]/[I(\infty) - I_0]$, in which I_0 is the fluorescence at the beginning of preincubation and $I(\infty)$ is the fluorescence at infinite dilution achieved by the addition of excess detergent. A small increase in fluorescence intensity occurs owing to an increase in the light scattering of cells on reduction of pH. A small decrease in $I(\infty)$ values occurs owing to a dilution effect. No increase in fluorescence intensity occurred during 10 min (or even 20 min) of preincubation of virus and cells at neutral pH. Experimental data points are shown by triangles. The dashed curve is calculated by using Eq. (20) with $f = 0.01$ $\sec^{-1}$. The solid curve is calculated by using Eq. (24) with $f = 0.012$ $\sec^{-1}$ and $\gamma = 0.012$ $\sec^{-1}$.

not change significantly. Hence the increase in fluorescence intensity may reflect a first-order process given by Eq. (20):

$$I(t) = [1 - \exp(-ft)]B \qquad (20)$$

in which $I(t)$ is the normalized fluorescence intensity increase and B is the fraction of virus bound, which is obtained separately from the binding experiments (see Fig. 1). For very short times, virus inactivation at low pH is not significant. However, inactivation of influenza virus at low pH (where viral fusion activity is extensive) is seen even at a t of 15 sec, when the target membranes are erythrocyte ghosts or liposomes having several compositions. Visual inspection of Fig. 1 cannot detect such inactivation because it is expected that the slope of the function $I(t)$ should decrease with time according to Eq. (20). However, an attempt to simulate the curve of $I(t)$ in Fig. 1 with Eq. (20) indicates that the calculated values overestimate the experimental values at times above 30 sec.

In a previous work (employing numerical solutions) we accounted for the low pH inactivation of influenza virus–liposome fusion by either one of two functional forms, both giving a reduction in the fusion rate constant.[24]

$$f(t) = f(0)[\exp(-\gamma t)] \qquad (21)$$

or[23]

$$f(t) = f(0)/(1 + \gamma t)^2 \qquad (22)$$

Both expressions can yield analytical solutions of the first-order equation

$$dF/dt = (1 - F)f(t) \qquad (23)$$

in which F is the fraction of prebound virus that has fused. The solution of Eq. (23) for $f(t)$ according to Eq. (21) is

$$I(t) = (1 - \exp\{f[\exp(-\gamma t) - 1]/\gamma\})B \qquad (24)$$

The data in Fig. 1 are simulated by the use of $f = 0.012$ sec^{-1} and $\gamma = 0.012$ sec^{-1}. The value of $B = 0.5$ is taken from binding measurements. It should be emphasized that Eqs. (21) and (22), and hence also the solution. Eq. (24), are approximate forms suitable for a certain time range, which in the case of Fig. 1 is fully covered. A residual fusion activity is retained even after long incubation (20 min) of influenza virus at low pH.[23,24] The use of Eq. (21) with $\gamma = 0.01$ sec^{-1} would yield zero values for f

[24] S. Nir, T. Stegmann, D. Hoekstra, and J. Wilschut, in "Molecular Mechanisms of Membrane Fusion" (S. Ohki, D. Doyle, T. D. Flanagan, S. W. Hui, and E. Mayhew, eds.), p. 451. Plenum, New York, 1988.

at $t = 20$ min or even at earlier times. In addition, virus binding to the cells proceeds after lowering the pH, and eventually B will reach unity. Thus the analysis should focus on times of up to a few minutes. It is of interest to note that in fusion studies[24] of influenza virus with liposomes, γ values vary slightly between 0.04 and 0.05 sec^{-1} (except for CL where $\gamma = 0$–0.003 sec^{-1}), whereas the result for HL-60 cells is $\gamma = 0.012$ sec^{-1}. The value of γ reflects processes related to changes in the HA molecules in the membrane of influenza virus. These changes affect the fusion rate constant in a similar fashion for several liposomes and erythrocyte ghosts, but a smaller degree of inactivation occurs with HL-60 cells in suspension. The value of $f = 0.012$ sec^{-1} is about an order of magnitude below that obtained for influenza virus–liposome fusion, but it is clear that a certain enhancement in the fraction of phosphatidylcholine in the liposomes used can yield significantly lower f values.

Kinetics of Fusion: Virus–Liposome Fusion

The equations describing the kinetics of virus–liposome fusion were derived earlier than those of virus–cell fusion [Eqs. (11) and (12)] but are more complex, owing to the fact that a liposome can adhere to and fuse with a fusion product and aggregation–fusion products can further aggregate. Although the programs were originally written for general cases, we have only employed a version in which the fusion products consist of one virus and several liposomes.

The following notations[10] are used. The molar concentration of an aggregate of I liposomes and J virus particles is denoted by $A(I,J)$; with this notation $L = A(1,0)$ and $V = A(0,1)$. A fusion product consisting of I liposomes and J virus particles is denoted by $F(I,J)$, whereas $AF(I_1,J_1,I_2,J_2)$ denotes the concentration of a composite particle consisting of I_1 unfused liposomes, J_1 unfused virus particles, and I_2 liposomes fused with J_2 virus particles. The quantity $FF(I_1,J_1,I_2,J_2)$ denotes the concentration of an aggregation product of $F(I_1,J_1)$ and $F(I_2,J_2)$, that is, an aggregate of two fusion products consisting of I_1 liposomes and J_1 virus particles and I_2 liposomes and J_2 virus particles, respectively.

Here we illustrate a few of the initially occurring reactions:

$$L + V \rightleftharpoons A(1,1) \tag{25}$$

$$A(1,1) \rightarrow F(1,1) \tag{26}$$

$$A(1,1) + L \rightleftharpoons A(2,1) \tag{27}$$

$$A(1,1) + V \rightleftharpoons A(1,2) \tag{28}$$

$$A(2,1) \rightarrow AF(1,0,1,1) \rightarrow F(2,1) \tag{29}$$

In practice we use only three parameters $C = C_{11} = C_{ij}$, $f = f_{11} = f_{ij}$, and $D = D_{11} = D_{ij}$ as in virus–cell fusion. The differential equations describing the processes are given elsewhere.[10]

The number of nonlinear differential equations describing the overall fusion reaction with aggregation–fusion products consisting of up to 8 particles amounts to several hundred. The method of numerical solutions[10,14,23,24] is the same as for virus–cell fusion. Here we present an analytical solution[10] which satisfies several restrictions: (1) all virions are fusion active; (2) aggregation is rate-limiting; (3) deaggregation is ignored; and (4) the fusion products consist of a single virus and several liposomes. In this case

$$L(t) = L_0 \exp(-CV_0 t) \tag{30}$$

and

$$V(t) = V_0 \exp[-(L_0/V_0)][1 - \exp(-CV_0 t)] \tag{31}$$

where $V_f = V_0 - V$ and $L_f = L_0 - L$. Note that Eq. (5) is obtained from Eq. (31) at $t = \infty$.

The programs allow for a certain fraction, α, of the virus to be fusion inactive. Inactive virions participate in aggregation. Another version[23,24] of the program explicitly takes into account virus inactivation (e.g., at low pH), which in principle may amount to a time-dependent reduction in both C and f. In most cases studied, virus–liposome binding was unaffected by preincubation of influenza virus at low pH.

We have considered two forms of viral inactivation. The first [Eq. (21)] shows a reduction in f owing to a first-order process which could amount to a folding and binding of segments of the HA molecules to the viral membrane, or could merely represent any conformational change which can evolve in time before the attachment (and probably penetration) of the HA molecules into the target membranes. The second possibility expressed by Eq. (22) amounts to intramembrane clustering[19] involving the HA molecules at low pH and above freezing temperatures. As pointed out, we have applied Eq. (21) or Eq. (22) for simulating the first 2 min of the process of fusion and inactivation. At later times a residual fusion activity can be retained. In fact, long inactivation[23] (for 20 min) results in unaltered final extents of fusion or a reduction by up to 50%. The simulation yielded similar values of γ [see Eqs. (21) and (22)], with or without preincubation of the virus at low pH, yielding lower values of $f(0)$ with preincubated virus. The analysis of kinetics of fusion requires as an input the fraction of virions capable of fusion, hence, the determination of final extents of fusion for each target membrane and pH.

The study of the kinetics of virus–liposome fusion has provided[10,11,13-15] useful information on the process of virus inactivation and on the effect of target membrane composition on the separate stages of aggregation and fusion. A few conclusions are mentioned below for the case of influenza virus fusing with liposomes at pH 5.[23] The addition of cholesterol at the expense of neutral phospholipids does not yield a dramatic effect, resulting in a 2-fold increase in the rate constant of fusion. The incorporation in liposomes of the ganglioside G_{D1a}, which is a receptor for the virus, causes an enhancement in the overall rate of fusion, which is due to a 3-fold enhancement in the rate constant of its adhesion. However, G_{D1a} does not affect the percentage of particles capable of fusing or the rate constant of the fusion per se.

Acknowledgments

This work was supported by grants from the National Institutes of Health (AI 25534, N. Düzgüneş; GM 31506, J. Bentz) and the United States–Israel Binational Science Foundation (BSF), Jerusalem, Israel (Grant 86-00010, S.N. and N.D.). The expert typing of Mrs. Sue Salomon is gratefully acknowledged.

Author Index

Numbers in parentheses are footnote reference numbers and indicate that an author's work is referred to although the name is not cited in the text.

Subject Index

J

JKR theory, 133

K

Kinetic studies
  of Ca^{2+}-induced lipid-phase separation, 29
  of cardiolipin/dioleoylphosphatidylcholine large unilamellar liposomes, 13–14
  of enveloped virus–cell fusion, 277–287
    analysis of, 286–287
  of erythrocyte–virus fusion, ESR peak height increase in, based on envelope fusion, 337–340
  of fusion, mass action model, 22
  of HL-60 (human promyelocytic leukemia) suspension cell fusion with influenza virus, 387
  of influenza virus–cell fusion, 386–389
  of influenza virus–liposome fusion, 390–391
  of Sendai virus–cell fusion, 383–389
  of vesicular stomatitis virus–cell fusion, analysis of, 286–287
  of vesicular stomatitis virus fusion
    with erythrocyte ghosts, 284–285
    with phospholipid vesicles, 274–275
    with Vero cells, 281–282
  of virus–cell fusion, 286–287, 384–389
  of virus fusion, 286–287
    lipid mixing assay for, 274
    octadecylrhodamine B dequenching assay, 283–285
    R$_{18}$ dequenching assay for, 283–285
  of virus–liposome fusion, 389–391
  of virus–membrane interaction, 264–265, 286–287
    electron spin resonance probes as markers for, 264–265

L

Langmuir–Blodgett deposition, of insoluble surfactants or lipids at known surface coverage, 136
Large unilamellar vesicles
  phosphatidylserine, fusion of, monitored by terbium/dipicolinic acid assay, 10–11
  phosphatidylserine/dipalmitoylphosphatidylcholine, fusion of, monitored by terbium/dipicolinic acid assay, 10–11
  preparation of, 364
    for fusion assays, 6–7, 12–13
  in solution, hemifusion, 141
Laser interferometry, in measurement of distance between membranes, 105
Light microscopy, of virus binding, endocytosis, and fusion, 259
Lipid bilayers. See also Bilayer membranes; Phospholipid bilayers
  adsorbed, fusion, stages of, 138, 140–141
  fusion, observation of, 134–135
  lamellar-to-hexagonal phase transitions, examination of, micromanipulation system for, 129
  preparation of, 135–136
  spontaneous curvature, examination of, micromanipulation system for, 129
  supported, interbilayer force measurements with, 130–142
    advantages and disadvantages of, 142–143
  thinning
    with dilution of lipid monomers in aqueous solution below CMC, 135–136, 139–140
    fusion-inducing effects of, 135–136, 139–142
Lipid bilayer–vesicle fusion, studies of, 79–80
Lipid mixing
  intermembrane, 32
  between labeled lipid (or lipid/protein) vesicles and membranes, calibration of, 39–40
  during liposome fusion, assays of. See Lipid mixing assay
  during membrane fusion, 16–18, 30
    fluorescence lifetime measurement in, 42–49
  during phospholipid vesicle fusion, monitoring, 3. See also Lipid mixing assay
Lipid mixing assay
  fluorescence lifetime measurements in, 42–49
  intermembrane
    advantages of, 33
    aggregation in, 40–41

ISBN 0-12-182121-8

9 780121 821210

90018